Projects, Government, and Public Policy

Many governments have effectively organized public project implementation systems in their jurisdictions. At the same time, many other countries remain at a less advanced level of public project management. Globally, there is a need for project management knowledge to be transferred between governments. However, no systematic review of these practices has been developed to date. *Projects, Government, and Public Policy* was written to fulfill this need and presents a review of project management practices in countries with developed project-based capabilities. This book uses its own rigorous model to present this review systematically. This book's practical purpose is to give a structured overview of government-level project management practices. This knowledge can be used in the work of governments to improve the management of public projects and the implementation of public policies.

Many professionals working in public institutions understand project management concepts differently than project management professionals. Therefore, this book begins with a chapter that describes the differences between the conceptual basis of public administration and project management. The body of this book has five parts. Part I is mainly intended for those involved in government and public administration who want to acquire or increase knowledge about project management. Part II provides an overview of the basic concepts from the theory of public administration, public policies, and development management. Part III describes what makes public projects unique and the success factors specific to projects of this sector. Knowledge about effective government project management practices is covered in Part IV. The concluding Part V begins with a general overview of the maturity model concept. Its main part covers the description of a maturity model showing ways to systematically improve the implementation of public projects.

This book is written for governments and government administrators, including the most influential decision-makers, who craft policies to guide a country's development as well as how to implement projects. This book is also intended for supporters and enthusiasts of project management in government and public administration by providing them with a description of the solutions used by project management in public administration. This book is intended, too, for all project management practitioners working for public projects: project managers, team members, sponsors, and middle-level executives of project-delivering private companies. By knowing public administration concepts, they can manage their projects better and use a common language with their clients.

Projects, Government, and Public Policy

Stanisław Gasik

CRC Press
Taylor & Francis Group
Boca Raton London New York

CRC Press is an imprint of the
Taylor & Francis Group, an **informa** business

AN AUERBACH BOOK

First edition published 2023
by CRC Press
6000 Broken Sound Parkway NW, Suite 300, Boca Raton, FL 33487-2742

and by CRC Press
4 Park Square, Milton Park, Abingdon, Oxon, OX14 4RN

CRC Press is an imprint of Taylor & Francis Group, LLC

Library of Congress Cataloging-in-Publication Data
Names: Gasik, Stanislaw, author.
Title: Projects, government, and public policy / Stanislaw Gasik.
Description: 1 Edition. | Boca Raton, FL : Taylor and Francis, 2023. | Includes bibliographical references and index.
Identifiers: LCCN 2022018618 (print) | LCCN 2022018619 (ebook) | ISBN 9781032343419 (hardback) | ISBN 9781032232683 (paperback) | ISBN 9781003321606 (ebook)
Subjects: LCSH: Project management. | Economic development.
Classification: LCC HD69.P75 G388 2023 (print) | LCC HD69.P75 (ebook) | DDC 658.4/04--dc23/eng/20220815
LC record available at https://lccn.loc.gov/2022018618
LC ebook record available at https://lccn.loc.gov/2022018619

ISBN: 978-1-032-34341-9 (hbk)
ISBN: 978-1-032-23268-3 (pbk)
ISBN: 978-1-003-32160-6 (ebk)

DOI: 10.1201/9781003321606

Typeset in Garamond
by MPS Limited, Dehradun

Contents

Preface: Why Do We Need This Book?

The Importance of Public Projects

The largest investor in any economy is the public sector (Crawford and Helm, 2009; Grandia, 2018). Governments employ 18% of all employees and make up 34% of the global GDP (Allas, 2018). Since the global GDP in 2021 was around US$ 70 trillion (IMF, 2021), the world's estimated spending in 2021 amounted to about US$ 23 trillion. Projects have a significant, though difficult to estimate, share in this. Turner et al. (2010) estimated the share of projects in GDP at 20–40%. 60% of the EU budget is spent and managed by various forms of projects and programs (Hodgson et al., 2019), and in the United Kingdom, projects consume about 30% of the GDP (Schuster, 2015). So, what is a project?

> A project is a discrete initiative undertaken by an organization to meet a business need (NSW DFSI, 2011).[1]

Many types of public projects can be used to implement changes and reforms in the public sector (Sjöblom, 2006). Public projects can also be used to implement statutory tasks of Public Administration (Grundy, 1998), delivering goods and services for citizens such as, for example, organizing cultural or sporting events, building monuments, fighting epidemics, or distributing one-off material help to citizens in need. Projects are becoming so popular that they are treated as "business as usual" in public organizations (Fred, 2015). To implement them, public sector employees, management specialists, and politicians are mobilizing in many ways. Large investments, such as building roads, airports, mines, and many other investments that drive economic growth in many countries, constitute projects, and spending on them will continue in the coming years (United Nations, 2019). The management of projects deals with implementing one-off, usually unique works (or their collections), in contrast to operational management, which aims to ensure an organization's routine, continuous operation.

Another component of project management is the implementation of project programs,[2] i.e., groups of related projects with a common objective, for example, building a highway network or implementing a new type of weapon in the army, which may require the reconstruction of many military facilities. Organizations must make decisions about the set of projects they implement. Such decisions define the organization's project portfolio. In order for projects, programs, and portfolios to work together effectively for the organization's benefit, it must undertake activities

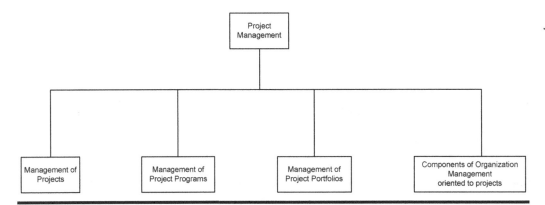

Figure 0.1 Structure of Project Management.

enabling it, for example, establishing organizational units supporting projects and their sets mentioned above. Together, these activities make up **project management** (Figure 0.1).

> Project management is a name that covers the management of projects, management of project programs, management of project portfolios, and the activities that support all of this in an organization.

The quality of governments largely depends on the quality of their projects and programs (Abuya, 2016). All of the phenomena involved in introducing project management into organizations are collectively called **projectification** (Midler, 1995). The public sector has been intensively projectified for some time – this is one of the essential methods of structural development in the public sector (Godenhjelm et al., 2015). The level of projectification in the public sector in developed countries lies between 14% and 33% (Schoper et al., 2018). Hence, the success of projects is one of the main issues for governments and societies (Alzahrani and Emsley, 2013).

Public management is the use of management techniques (often invented in the private sector) that increase the value of public services for money (Bovaird and Löffler, 2009). Public Administration, as a scientific field, is the study of what Public Administration does (Hill and Hupe, 2014). In addition, according to Frederickson et al. (2012), public management consists of formal and informal processes guiding human interactions toward public organizational goals. One of the main issues in the contemporary management of public organizations is the mobilization and effective management of resources and opportunities to implement the changes necessary to achieve strategic objectives and goals and meet society's expectations (Patapas and Smalskys, 2014). This is just another definition of the concept of a project. Hence, studying projects must be an integral part of Public Administration science.

Public Policies as Projects

From another point of view, governments' goals are achieved by implementing public policies. No single definition of public policy would be accepted by all (or even most of) the practitioners and researchers of this subject (Subchapter 11.1). The broadest definition says that this is all that government does (or does not) (Dye, 2013; Colebatch, 1998). Cochran et al. (2010) add intentions to the public policy concept. However, according to Cochran and Malone (2010), public policy also means implementation decisions. A policy contains enacted law and subsequent actions implementing these acts (Anderson, 2015). According to Jenkins (1978), public policies are political decisions about achieving specific goals. Policies can be substantial regarding creating, purchasing, and distributing goods, or procedural, defining ways of working and approaches to problems (Anderson, 2015; Howlett, 2005). The outcomes of policy-making almost always take the form of structures; they create administrative capacity at the management level and implementation of services (Lynn and Robichau, 2013). Developing a product in a limited time is an essential feature of projects. Blomkamp (2018) identifies policies with a project. Fischer (2015) considers projects to be a unit of governmental action. Shiferaw and Klakegg (2012) and Kuokkanen (2014) view projects as a tool for implementing public policies. Implementation is the transformation of policy into outcomes (Sætren and Hupe, 2018). The use of projects in public policy creates hope for obtaining rational, controllable products (Jałocha and Prawelska-Skrzypek, 2017; Moutinho and Rabechini, 2020).

> All of these statements show that projects are the main components in the implementation of public policies.

Therefore, projects – their problems, benefits achieved through them, and their implementation methods – should be one of the main subjects of interest for researchers dealing with public policies, particularly with their implementation and those dealing with Public Administration. But it appears that interest in this is negligible, fragmented and that researchers from these areas even present the wrong idea of what the projects and project management are. It is interesting that, in the book that lays the foundation of the (policy) implementation sciences, Pressman and Wildavsky (1973) point to typical project management problems: coordination, too many decision points, too many creating, purchasing, and distributing decision-makers, too many intermediaries in policy implementation, and too lengthy a negotiation duration. According to them, problems can be made by "organizational machinery," which may be reframed into project management processes in the contemporary language. But in later public policy research, researchers turned away from project management issues. For example, in the index of terms of one of the most important contemporary books on public policy implementation by Hill and Hupe (2014), the word "project" does not appear. Yet Pressman and Wildavsky were actively researching when the theory and practice of project management were much less developed. Thus, as project management knowledge became more and more developed, it could be expected that public policy researchers would have used this knowledge in their research. However, this did not happen. New Public Management as a school of Public Administration introduced project management to the arsenal of Public Administration tools, but this was not accompanied by an increased interest in project management by public policy and Public Administration researchers.

> According to Howlett (2005, 2020), new instruments for implementing public policies and attaining policy implementation capabilities by governments are needed. Projects should be such instruments.

The Role of Project Management

The only way to implement changes is to implement projects and, above all, to manage them. So far, no more effective way of implementing changes than project management has been invented. Modern project management began in the US public sector with military projects – the construction of the atomic bomb during World War II (Gosling, 2010) or the construction of ballistic missiles, for instance. In the 1950s, the American Navy introduced modern project management methodologies with the Polaris project. In the 1960s, the US administration began treating project management as one of its most important management tools (Lenfle and Loch, 2010; Morris, 1994). The best use of public resources requires, among other things, project management skills which are recognized as one of the main areas of management (Van Der Merwe, 2002). Project management was the most important management tool introduced into public policies by EU programs (Jałocha, 2019). Pells (2018) believes that the growing awareness of project management among (some) governments is the fifth revolution in project management. The owner of public projects is either the state or other public entity acting under the authority of the government or is the society as a whole (Van de Walle and Scott, 2009), so governments can and should influence their investments and other changes by establishing institutions and implementing institutional reforms (Adelman, 1999). As the responsible entity, the government sets rules and structures to ensure that taxpayers' money is well spent (Klakegg, 2010). Therefore, governments initiate programs to improve their project management (UK OPSR, 2003), and some enact relevant laws (USA Congress, 2015). Implementing any change, including improving the state's capabilities, is itself a project or project program. Hence, we may call project management meta-capability, i.e., a capability that enables the development of other capabilities and the state as a whole.

> Effective project management adds immense value to governments and societies under their jurisdiction.

Problems with Public Projects

Failure of public projects is a problem for entire societies (Klakegg, 2010). Strategies fail not when they are analyzed and designed but due to the failure of projects implementing them (Van Der Merwe, 2002). There is plenty of examples and extensive literature describing the failures of public projects: in 2010, only 39% of World Bank projects were successful (Chauvet et al., 2010), and according to Kuipers et al. (2014), 70% of changes in public organizations fail. In turn, according to the McKinsey (2018) report, the level of failures in government transformation projects reached up to 80%. In 2014, according to Ika and Saint-Macary (2014), the success rate of projects in Africa was only about 50%. Still, problems with projects occur not only in developing countries but also in developed European countries (Reichard, 2003; Stentoft et al., 2015). An interesting phenomenon is the belief of almost every government that it manages projects the worst of all governments (Flyvbjerg, 2018). There is a lack of resources in developing countries; administrative systems and procedures are poorly organized (Yanwen, 2012). Public projects are managed in developed countries much worse than private ones (UK Lord Browne of Madingley, 2013).

As a primary development tool, projects attract a lot of public interest. Because many projects fail, people think many unnecessary projects are being implemented. Several projects provide wrong or only partial solutions to existing problems, sometimes even creating more problems than solutions (Shiferaw and Klakegg, 2012). The percentage of problems and failures in the public sector seems even more significant due to the illusion that results from noticing and exaggerating publicly implemented projects (Williams, 2019). The public discusses many failures in implementing projects, including public ones, because it's easier to talk about project failures than operational activity failures. After a project is completed, it can be unambiguously stated whether the scope, schedule, and budget were implemented according to its plan. But there is no specific point in time in operational activities (apart from those designated by the calendar) that would require such verifications to be carried out; operational activities are, in fact, determined by events, not by a calendar. The role of projects in shaping opinion about governments is vital.

Project failures result from, among other things, the low-quality institutions involved in their implementation (Baum and Tolbert, 1985; Bowden, 1986). Poor front-end management processes, lack of relevance, and sustainability issues are the main problems of public projects (Klakegg, 2009). Bad management is one of the factors hindering the implementation of development projects (Edmunds, 1984). According to Othman (2014), Mackhaphonh and Jia (2017), Kohsaka (2006), and many others, one of the main problems in developing countries is the lack of ability to manage public projects. Most development projects are late and exceed costs (Ahsan and Gunawan, 2008). Shiferaw and Klakegg (2012) claim that common project problems in developing countries are lack of justification for the need of the project, unclear definition of the project objectives, lack of identification and comparison of alternative solutions, or non-involvement of relevant stakeholders at the beginning of the project. For mega-projects which are most often implemented by governments, there is an iron law: "Over budget, over schedule, over and over again" (Flyvbjerg, 2014).

> Therefore, the capabilities of implementing public projects must be improved.

Benefits of Project Management for Public Policies

If all governments improved their productivity to the level of the best countries, then US$ 3.5 trillion could be saved without adversely affecting the effects (Allas, 2018). Since projects account for one-third of public sector activity globally (Schoper et al., 2018), the potential saving is over US$ 1 trillion. International development projects change people's lives worldwide (Zeitoun, 2002). With better project implementation, citizens and communities would simply be happier with their governments. Teorell et al. (2017) determined that project implementation is a component of government quality. There are also less measurable benefits of efficient project management. For example, higher procedural quality promotes support for democracy and reduces the likelihood of cognitive dissonance between ideal and experienced democracy (Boräng et al., 2017). Organizations focused on specific capabilities achieve an excellent efficiency level (Brown et al., 2015), and project management is one such capability. Therefore, it is worth considering possible ways to improve the implementation of projects in the public sector so that they are implemented as effectively as possible.

All countries need regularly improving project management systems (Bayiley and Teklu, 2016). Many publications confirm the impact of proper management on project success. Most generally, it can be stated that the projectized approach is a flexible, useful management tool in the public sector, regardless of the economic system, type of government, or stage of development

(Baum and Tolbert, 1985). A well-organized hierarchical "Weberian" administration, the foundation of institutions of which implementation and efficient project management are a part (although in Weber's time, there were no effective methods of project management), can contribute to the development of a country (Evans and Rauch, 1999). In a 2018 study by McKinsey, operational and project management skills were found in over 50% of successful projects and less than 40% of unsuccessful ones. According to Medeiros et al. (2017), government institutions' implementation of project management practices contributed to improving project effects. Project management is one of the main factors influencing the success of development projects (Diallo and Thuillier, 2004). Du and Yin (2010) indicate two main routes for improving the efficiency of public projects: improvement of management and improvement of governance. Successful changes in organizations can only be achieved through adequately implemented project management processes (Van Der Merwe, 2002). Also, Edmunds (1984) and Prasad et al. (2013) claim that problems with projects can be solved by introducing proper management methods. In particular, the essential factors for the success of development projects are monitoring, coordination, design, training, and institutional environment (Ika et al., 2012). The management, law, regulations, and capabilities of the implementing agency influence the success of development projects (Ika and Donelly, 2017). Good project organization and the project's environment affect development success (Conyers and Kaul, 1990). Economic development cannot be achieved without a proper policy being implemented through projects (Obradović, 2019). An increase in the efficiency of projects is correlated with an increase in the rate of economic development (Denizer et al., 2013).

Therefore, not only projects themselves but also their effective and efficient management should be introduced to public organizations. There are many specific project management techniques, processes, and knowledge areas that could be useful in implementing public policies. For instance, there are many places where the SMART (stating that requirements should be Specific, Measurable, Achievable, Reasonable, and Time-related; see Section 3.2.1) model for requirements specification, one of the fundamental techniques in a project management, can be used. This includes, for example, a clear and unambiguous statement of the agency's missions and policies (Andrews et al., 2017). Policies should also have priorities – for that purpose, the MoSCoW model (Must, Should, Could, and Will Not priorities) (see Section 3.2.2; Clegg and Barker, 1994) may be applied. The traceability matrix showing the relationship between different sets of items (Section 3.2.3) may be used, for instance, for assuring consistency between components of policy definition such as aims, objectives, targets, and implementation tools (Howlett, 2009) or for controlling the fulfillment of Tinbergen's law which states that for every policy objective there must be at least one instrument that implements it (Tinbergen, 1952). An agency's mission must refer to policy goals (Masci, 2007). Here the traceability matrix would be the right tool for assuring coherency between the mission and goals. When creating policies consisting of many instruments, the issue is to match them and check whether they are non-contradictory, supportive, or non-redundant (Howlett, 2014). Here, the activities, processes, and techniques of project program management that have the objective, among others, of coordinating component projects would be helpful (PMI, 2017b). The requirement to consider the weight of goals and the (weighted) value of individual instruments to achieve these goals when identifying individual mixes of instruments (Mees et al., 2014) implies the use of portfolio analysis techniques (PMI, 2017c). Treisman (2002) believes that governments can be wrong in two ways: they can do things they shouldn't do or fail to do something they should do. This also makes an implicit reference to project portfolio management.

These are just some examples of allying project management tools with public policy implementation. The use of projects, specific techniques, and project management methods could be beneficial for implementing public policies. More on this subject you may find in Part I.

The exemplary application of some project management techniques and processes shows that project management should be widely used in Public Administration. Unfortunately, this does not happen. According to Peters (2010), the traditional (Weberian) model of Public Administration is currently not the right approach. It is hard to agree with this, especially when considering policies that need solid, well-defined project management procedures. Hierarchical authority is more important than was previously recognized by scientists (Lynn and Robichau, 2013; Howlett, 2020). British interest in implementation theory declined significantly in the second half of the 1990s (Hill and Hupe, 2014). Lindquist (2006) also complains that implementation literature has lost its momentum. It is difficult not to relate these facts to a lack of interest in the most important policy implementation tool – project management. Analyzing public policies and Public Administration without considering project management can be compared to analyzing cars' performances without interest in their engines. Further examples of discrepancies between Public Administration and project management knowledge can be found in Chapter 1 of this book.

Different Types of Public Policies

In some public policies, projects play a fundamental role – perhaps the best examples are transport and natural resources policies. The main element of the former is a road, railroad, and airport construction project, while of the latter is a project of constructing mines extracting natural resources. On the other hand, for instance, in fiscal, health, and educational policies, continuous tax collection processes, patient treatment, or school education, respectively, play a major role. Their effects are observed after implementing these processes for a longer time. It seems natural then that the methods of planning, implementing, and exploiting these two types of policies should be different. The contemporary Public Administration primarily deals with policies based upon operations. Because of this, knowledge of public policies developed in this environment is almost useless for implementing many policies where project are their main tools. This is an essential issue for Public Administration.

Countries undergo three development phases (Rolland and Roness, 2009; Rose, 1976; Premfors, 1999): ensuring security and territorial integrity, infrastructure development, and providing welfare. In the first two phases, projects and policies based upon them are essential. States must first designate their territory, organize the coercive apparatus (phase one), then build roads, mines, waterworks, and all the rest of the infrastructure (phase two). The policies of the first two phases are therefore based upon project implementation. Only in the third phase, to ensure welfare, first and foremost, public policies are needed where continuous operations play the most important role, e.g., participation in cultural life, improving the level of education, and improving the health services. The contemporary Public Administration and Political Science have only just been created in highly developed welfare countries and likely, therefore, does not sufficiently acknowledge the role of policies based upon project implementation, which are characteristic of the countries at earlier development phases.

> To address this issue, this book is devoted to the role of projects in the implementation of public policies.

Governments and Project Management

Public sector projects are performed in every country. Projects of all sizes are implemented, from a school trip or road repair to the organization of elections and the implementation of IT systems to support the work of the Public Administration to the construction of major airports or flights to Moon and planets; these are understood as project programs. Many countries have developed their approaches, organizational structures, and project management processes. Some practices not necessarily created initially for the needs of the public sector, such as PMBOK® Guide[3] (PMI, 2017a), Prince2®[4] (UK OGC, 2017), Logical Framework Approach (Practical Concepts Incorporated, 1979), or UK OGC Gateway®[5] Process (UK OGC, 2007) have spread between countries. However, the Public Administration academic community does not adequately support these activities.

An organization's approach to project management is not the simple sum of projects' implementations. This was understood a long time ago on the level of "ordinary" organizations. The Project Management Institute published the first document containing the Organizational Project Management Maturity Model (OPM3®[6]) describing the preparation of organizations for project management in 2003 (PMI, 2003; Schlichter et al., 2003). As a part of the work on OPM3®, the concept of organizational project management (OPM) was defined, but the relevant standard was only published in 2018 (PMI, 2018a).[7] For an organization to implement projects effectively, it must, for example, prepare project management professionals, define the rules of project selection, or manage the knowledge needed to implement projects. These are activities carried out at the level of the entire organization and not in individual projects.

The state is a specific organization. The government does not implement projects directly. Usually, for the implementation of large projects, separate organizations are created. Meanwhile, relatively small projects are implemented by governmental organizations: ministries, departments, or other government agencies. Consequently, as the organization that directs the functioning of the entire state apparatus, the government acts differently than other types of organizations. The government may, for example, define the procedures for project implementation applicable in the public sector; it may establish organizations supporting and controlling project implementation, or it may determine ways of involving stakeholders in public projects. Such activities are carried out, depending on the country's governance system, by enacting laws or resolutions by the legislature or by issuing executive orders by the executive branch. This area can be treated as a derivative of public management and organizational project management. Until now, the area of government involvement in implementing projects has been modestly addressed by research. Significant achievements in this area have been achieved mainly by Norwegian researchers and their foreign partners (Klakegg et al., 2008; Williams et al., 2010; Shiferaw and Klakegg, 2012; Shiferaw, 2013; Volden and Samset, 2017).

> This book tries to identify and systematize government-level practices related to project implementation.

Goals of this Book

Practical Goals

Many governments have developed and implemented project management processes and methods. At the same time, many other countries remain at a less advanced level of public project management. So, there is a need for project management knowledge to be transferred between governments. However, no systematic review of these practices has been developed to date. For such knowledge to be systematically presented, two actions are necessary. First, you should review these practices. The basis for writing this book was, among others, review of project management methods in many countries. Second, there should be a model to present this knowledge systematically. I have also created such models.

> The practical purpose of writing this book was to provide a structured overview of project management methods. This knowledge can be used in the work of governments to improve the management of public projects and the implementation of public policies.

The practices used to manage projects in the public sector are often not known and not applied by governments in need of this knowledge. Also, knowledge about Public Administration and public policies is often foreign to the project management communities. As a result of this lack of knowledge, public sector management is much less effective and efficient than it might be.

> Another goal of this book is to combine knowledge from these two areas and thus enable better implementation of public policies by better project implementation in the public sector.

Theoretical Goals

There are three types of entities and three levels of participation in scientific involvement in the social sciences (Shapira, 2011; Ostrom, 2010): conceptual frameworks, models, and theories. The conceptual framework contains the essential concepts from a given area, thus defining the area of interest and the scope of plausible research. Defining a conceptual framework starts research in a given (sub)discipline of study. Models describe the relationships between the concepts that make up the conceptual framework. Within a single conceptual framework, many models can and usually do arise. Theories explain why these relationships occur. There may be many theories explaining how it works for any particular model.

Government-level project management is a well-defined area of expertise. Therefore, it can be expected that a conceptual framework for it will also be well defined. This conceptual framework will make it possible to build general models, which will be the basis for building theories regarding government-level project management. In particular, the Comparative Governmental Project Management could be developed.

Based on these models, it will be possible to create further models within this conceptual framework and build theories explaining the actions of governments in the field of project management, which will contribute to the improvement of project implementation in the government-directed public sector. In this way, science can try to catch up with public project implementation practice. Because implementing projects is one of the main components of implementing public policies, it will indirectly contribute to clarifying and improving many issues vexing Public Administration.

> Identifying the government level of project management by creating a conceptual framework and proposing some models for this area of research is one of the goals of this book.

Recipients of this Book

There are several groups of addressees in this book. The first and most obvious group is members of governments and government administrations, including the most influential decision-makers, who make decisions about a country's development, including how to implement projects.

A particular addressee group is policy analysts and policy-makers. It is worth distinguishing this group, although they are usually government employees. These audiences will learn how to use projects and their management methods in defining and implementing policy instruments and public policies as a whole.

Organizational changes are often met with resistance, but in the case of projectification, there is an opposite tendency – bottom-up support (Fred and Hall, 2017). This book is also intended for supporters and enthusiasts of project management in government and Public Administration by providing them with a description of the solutions used by project management in Public Administration. Ultimately, this book is intended for all public sector employees involved in project implementation.

This book is intended, too, for all project management practitioners working for public projects – project managers, team members, sponsors, and middle-level executives of project-delivering private companies. By knowing Public Administration concepts, they will be able to manage their projects better and use a common language with their clients.

The other target group is supranational organizations like the World Bank or the United Nations Office for Project Services. These would be able to support governments in the systemic improvement of their project management capabilities. They would provide them with a fishing rod (preparing the government and administration for effective project implementation) instead of a fish (implementing single projects and project programs).

Structure of this Book

People in the public sector often understand project management concepts differently than project management professionals. These discrepancies hinder or prevent the effective use of the capabilities created by projects and proper project research by the Public Administration and Political Science communities. Therefore, the first chapter, "Discrepancies between Public Administration and Project Management," describes the differences between the "conceptual basis" of the Public Administration (PA) (and to some extent Political Science) community and the Project Management (PM) community.

After this initial chapter, this book has five main parts. The first part, entitled "Project Management for Public Administration" (Chapters 2–6), is mainly intended for people involved in government and Public Administration who want to acquire or increase knowledge about project management. The essential concepts of project management, supported by examples of their application in Public Administration, are presented there.

The second part, entitled "Public Administration for Project Managers" (Chapters 7–14), provides an overview of the basic concepts of Public Administration, public policies, and development management. This way, project management staff working for public governments will have an opportunity to (better) understand their clients. And the Public Administration staff will be able to operationalize project management knowledge better.

Once readers have an idea of projects and Public Administration, this knowledge can be used to present the specifics of public projects. Part III "Public Projects Are Specific" (Chapters 15–18) is devoted to this subject. It presents an approach to differences and describes the specificity of public projects and the success factors specific to projects of this sector. This also includes in Chapter 16 a specific, different than for other types of organizations, model of differences between public projects and projects from other sectors.

Knowledge about effective government project management practices is very much needed by many governments, but has not yet been gathered and systematized. This is what Part IV Practices of Public Project Management is about. It is mainly based on solutions implemented by government organizations in several dozen countries and is also supported by some literature review. These practices are divided into governance (Chapter 19), delivery (Chapter 20, the biggest in this book), support (Chapter 21, one of the smallest as not too many governments actively support their projects), and development (Chapter 22). Together they were placed in and constitute a Governmental Project Implementation System (GPIS) (Chapter 23).

But some (or perhaps most of) governments are not optimally effective in implementing their public projects. So, let's create a tool to improve this system. In recent years, in practice and management literature, maturity models have become increasingly important as they approach the issue of improving the operation of the entire organizations as a whole or their separate areas. The first chapter (24) of Part V "The Way Ahead. How to Improve It?" generally describes the concept of maturity models and their application in the public sector. In the last chapter of this part, I describe a model showing the consecutive stages of more and more effective stages of GPIS development (Chapter 25).

I believe that this book will contribute to the improvement of the implementation of public projects, and thus the development of entire countries and the improvement of the level of well-being of the inhabitants.

Notes

1. For a comprehensive review of project definitions, see http://www.maxwideman.com/pmglossary/PMG_P12.htm#Project
2. In project management literature, usually when writing about programs, the word "project" is not added before them. I add this term when it is necessary due to a different role of programs in the area of project management and Public Administration (Section 11.4.1).
3. PMBOK® is a trademark registered to Project Management Institute.
4. Prince2® is a trademark registered to Axelos Limited.
5. OGC Gateway® is a trademark registered to UK Cabinet Office.
6. OPM3® is a trademark registered to Project Management Institute.
7. A rare case where a maturity model was defined earlier than the baseline model of the domain concerned.

Acknowledgments

I want to thank everyone who helped me gain the knowledge described in this book.

First, those who are no longer with us: unforgettable Russell Archibald and Ginger Levin, who strongly encouraged, supported, and helped me in my research in various ways; David Smith from Cowra, NSW, Australia, without whom I would not have performed any research in his country. I regret so much that they will not enjoy this book.

I want to thank Winnie Liem from Toronto, Canada and Marta Gaino from Salvador, Bahia, Brazil who helped me organize the research in their countries. Without help from Prof. Jan Fazlagić, Poznań University of Economics and Business, Poland, I would not be able to prepare this book.

I would also like to thank everyone who took the time to tell me how projects are managed in their countries.

This book uses some results of the research project UMO-2012/07/D/HS4/01752 financed by the National Science Center of Poland, performed when the author worked at Vistula University, Warsaw, Poland.

Author

Stanisław Gasik is a practitioner with more than 30 years of experience in project management. He holds the PMP® certificate issued by the Project Management Institute. He obtained his PhD from the University of Warsaw for developing a holistic model of project knowledge management. He was a significant contributor to PMI project management standards (PMBOK® Guide; Standard for Program Management). He was also an expert for project management in Government Accountability Office, an institution of the US Congress.

Chapter 1

Discrepancies between Public Administration and Project Management

1.1 Introduction

The staff of each Public Administration implements projects. And project management specialists operate in the public sector. These activities should be supported by knowledge about the common part of these activities, i.e., public project management. Hence, we may expect that the engaged parties – Public Administration and project management community – have the same understanding of project and project management concepts. But people in the Public Administration have a different understanding of project management concepts, even the basic ones, than project management professionals. These discrepancies hinder or prevent the effective use of the capabilities created by project specialists and researchers by the Public Administration community. Improving the ways of public project management may start from indicating these conceptual differences and then, based on their perception, defining managerial methods which would be in the same way understood by both these parties. This chapter describes a systematic attempt to organize the concepts related to differences in understanding project management between public-sector staff and researchers and project management communities.

Several "schools" of different understandings of these concepts not conducive to developing integrated knowledge about managing public projects may be identified.

1.2 Anti-management

The first to be mentioned is the "school" of anti-management which explicitly negates the importance of project implementation processes. Kettl (2010) wants governments to focus on the performance and results generated by the contracts, not on contract implementation processes. This approach is contradictory, for example, with the US government's policy for more than

DOI: 10.1201/9781003321606-1

30 years, suggesting with CMMI®[1] awarding project contracts only to companies that are mature in project management (Paulk et al., 1993; SEI, 2010). Disregarding projects and their management is justified by possible depoliticizing public policies and less democratic accountability (Hodgson et al., 2019). It is difficult to agree that efficient project implementation would be contrary to democracy.

Fortunately, representatives of this school are few.

1.3 Ignoring

Probably the numerically dominant "school" is ignoring project management. For instance, in the volumes of the *Journal of Public Policy* from 2010 to 2018, only two (Oudot, 2010; Zhang, 2015) out of 143 articles try to draw on project management knowledge. Searching for "project management" in all editions of the *International Journal of Public Management* gives only 15 results. This is a very low number for an estimated number of around 300 articles. Jreisat (2010) can be mentioned at the level of individual works, for whom only operational management, not project management, is relevant for public services. In the Multiple Governance Framework, Hill and Hupe (2014) include three levels of policy implementation: system, organization, and individual, but they do not consider the level of projects. Ramesh and Howlett (2017), who use this framework, also do not consider project management as a necessary capability to implement policies, and therefore are unable to indicate, for example, project risk management or project quality management as capabilities needed for the efficient implementation of policies. Asquer and Mele (2018) negate the importance of managers in the area of policy-making – while the project management community has long known that the skills of managers (PMI, 2017d) and the ability of the organization as a whole to implement their projects are necessary for the efficient implementation of projects (PMI, 2003, 2018a).

Rand (2014) does not include project management in proposals to measure governments' policy implementation capabilities. In the book on public sector efficiency, Bouckaert and Halligan (2008) use each of the terms "project management" and "program management" in the body of their book only once. Grandia (2018) and Di Pierro and Piga (2017) do not indicate the issue of management as the one worth dealing with in the area of public purchasing while, in fact, every acquisition is a project and must be effectively managed. Howlett (2009) gives examples of problems with implementing public policies but does not notice any problems arising from their project nature. Koning (2016) analyzes changes in public organizations without referring to projects – which are the most important tool for implementing changes. Bredgaard et al. (2003) do not indicate mismanagement (of projects) as the cause of public policy failures. In their research on successful implementation, Struyk (2007) and Giacchino and Kakabadse (2003) do not include management of projects as an independent variable. Efficient management, in particular of projects, is not recognized as a value in the public sector (Jørgensen and Bozeman, 2007). Approaches to policy analysis lack techniques associated with the projectized approach (Stewart et al., 2007; Dunn, 1981, 2004; Parsons, 1995; Dye, 2001). Specialists in improving the functioning of Public Administration do not consider project management as a component of policy implementation capability (Andrews et al., 2016).

The number of articles in Public Administration and public policy literature that explicitly or implicitly ignore project management's importance and methods is truly immeasurable.

1.4 Basic Discrepancies

Another "school" is misunderstanding the project and project management concepts. Here we can distinguish two "streams": practical and theoretical.

1.4.1 Practical Stream

The practical stream includes works in Public Administration and public policies that would not be considered projects in the project management community. The Policy Agendas Project (PAP; www.policyagendas.org) does not meet the definition of the project used in the area of managerial knowledge (PMI, 2017a) because it is not time-limited and the product whose production causes its end is not defined (groups of data are added year after year). Similar considerations apply, for example, to the Climate Guide Project (https://climatedataguide.ucar.edu/climate-data/gpcp-monthly-global-precipitation-climatology-project), Government Performance Project (https://www.govexec.com/magazine/2000/03/about-the-government-performance-project/6316/), or Institutions and Elections Project (IAEP; https://havardhegre.net/iaep/). This is not just a matter of naming. Considering continuous activities that do not have a time perspective as projects make applying techniques developed in the project management community impossible. For example, since the activity has no time perspective, it is impossible to calculate the critical path. You can't build its limited budget, too. Hence, project-specific techniques, like the Earned Value Analysis (Subchapter 3.5), for controlling progress based on a finite schedule and budget cannot be applied there.

The Center of Excellence in Project Management of the European Commission developed Project Management Methodology PM^2. As defined in this document, a project must create a "unique product or service (output) within certain constraints such as time, cost, and quality" (European Commission CoEPM, 2018, p. 5). Moreover, this document explicitly states that project is "not operations, not a work activity, not a program, etc." (p. 18). The European Union awards entities in the Member States grants, i.e., "funding for projects" (European Commission, 2017, p. 8; European Commission, n.d.). None of these documents refers either directly to the concept of PM^2 project definition nor does it refer to the project characteristics defined by this document – e.g., uniqueness or non-recognition of operational activities as projects. There are also no such requirements in the official grant application forms. In the official list of grants received by Poland from the EU (www.mapadotacji.gov.pl), you may find grants for projects conforming to PM^2 definition of project, but it should be treated as a coincidence and not as deliberate policy. Completed EU financed "projects" include, for example, staff costs of EU programs (program is not a project according to PM^2), HR support, support for socially endangered people, street-working, processes of career counseling and job placement, or providing environmental forms of assistance – they are operations which are also not projects according to PM^2. The plethora of EU-financed "projects" has nothing to do with the uniqueness required by PM^2. Such an understanding of the concept of "project" contributes to the illusion that projects are ubiquitous in EU countries. In fact, ubiquitous are grants that do not pay attention to the managerial form of their implementation. This shows that even in one organization – the European Union – there are substantial discrepancies between administration and the project management community in understanding the concept of a project.

A side effect of non-standard usage of the "project" concept prevents the progress of the project management culture among stakeholders of these activities.

1.4.2 Theoretical Stream

Foundational discrepancies between Public Administration and Project Management are also present in research and scientific works. That is what may be called the "theoretical stream" of discrepancies. On the Public Administration side, it is often believed that all activities carried out anywhere are projects (a phenomenon called "hyper-projectification"). Meinert and Whyte (2014) consider permanent organizations to be projects. Abrahamsson and Agevall (2010) also believe that every activity is a project, and they also do not take into account aggregated forms of a project-based management approach. On the other hand, Biggs and Smith (2003) do not consider projects to be organizations. According to Edelenbos and Klijn (2009), decisions are made in projects without consulting stakeholders (at the Project Management side, stakeholder management is one of the ten most important project management areas; PMI, 2017a). They generally contrast projects with processes, while on the project management side, it is considered that every project is a process. They write a statement that does not cover reality about the "military mode" of making decisions in projects ("Decide, Announce, Defend"). An interesting case is a paper by Wilson (2011). He explicitly says that his research program of comparing public management between countries consists of projects – but projects are not the subject of his study. O'Toole (2004), in one paragraph (p. 309), calls public policies "program" and "project" simultaneously and interchangeably – and these for the project management community are obviously different concepts.

1.5 Parallel Works

Still another "school" of discrepancies results from the non-penetration of the works of these two communities and parallel work on similar concepts. I will discuss two such cases here.

1.5.1 Maturity Models

A maturity model is a structured set of elements describing the characteristics of effective processes at various levels of development of entities implementing them (Pullen, 2007; SEI, 2010; UK OGC, 2010; see Chapter 24). There is abundant literature on maturity models (cf. Wendler, 2012), a sub-area of project management (but maturity models have become more and more popular in other areas). The mainstream researchers in Public Administration revolve around this concept of the maturity model. Still, they do not explicitly use it. The authors who build maturity models without naming them expressly are, for example, Andrews et al. (2017) – for state capacities, from the level of negative ability to the ideal level. Also, Bouckaert and Halligan (2008) create a model of efficiency management maturity in the public sector, but they do not explicitly use this approach. Andrews (2010) complains that reforms fail because all good practices are implemented simultaneously. Governments do not know which best practices to implement. And yet, just maturity models answer the questions about the order in which best practices should be implemented. It should be noted that recently maturity models are beginning to enter Public Administration (see examples in Subchapter 24.4). Still, it is happening mainly in the common part of the discussed areas, i.e., in areas related to implementing public projects.

1.5.2 A Non-traditional Approach to Implementation

The second area of knowledge in which these two communities parallelly develop their expertise and knowledge is non-classical, different from waterfall (Royce, 1970) approaches to implementation.

A waterfall (project) implementation model is one in which consecutive phases (initiation, planning, implementation, etc.) are executed strictly sequentially. The next phase may start only after the previous is formally completed (Section 2.4.1). The most popular non-traditional approach to project implementation (Section 2.4.3) is currently the agile one. Small packages of work are defined, planned, and implemented in the course of the project (Beck et al., 2001). It should be noted that a similar approach was proposed by Berman (1978) just in the area of Public Administration. He distinguished two approaches to policy implementation: programmed (in project management called "waterfall") and adaptive, allowing the adaptation of original plans to emerging events and decisions (more than 20 years later in the area of software project management called "agile"). But Berman's achievements probably have been forgotten by the Public Administration community. In modern times when Jensen et al. (2017) write in *Public Policy and Administration* journal about ambiguously defined policies, they do not mention the agile approach to project management that would be the most appropriate there. Munck af Rosenschöld (2017) proposes organic projects instead of the concept of agile implementation. Baser and Morgan (2008) list three approaches to capacity building: planned (waterfall, sequential), incremental, and emergent (agile), but they do not mention that each of these approaches has long been analyzed in the area of project management. Andrews et al. (2017) provide almost no references to literature and project management achievements. They propose an adaptive and iterative method, while both of these approaches in the field of project management have been known since the 1950s (Larman and Basili, 2003).

The agile approach is related to the interaction of the project target group with the team implementing it and their reactions to the emerging products and effects. When implementing social policies, such contact is essential during the project implementation. On the other hand, projects creating and using software for a long time primarily focused on supporting engineering solutions or other areas where it was easier to define software functions in advance. This was the case when computing was mainly based on mainframe computers. It did not require intensive contact with future users. Only the development of software with extensive user interface functions, supporting frequently used, everyday activities intended for extensive use with personal computers, drew the attention of the software engineering industry to the need for more agile project implementation. This explains why the adaptive approach became popular earlier in public policies than in software projects.

The area of Public Administration and project management would surely benefit from working hand in hand on some topics – like the non-traditional approach to implementation.

1.6 Fragmentary School

The fragmentary one is the last "school" of inappropriate interest in projects. Its representatives are only interested in certain aspects of project management which usually leads them to the wrong conclusions. For example, Öjehag-Pettersson (2017) and Munck af Rosenschöld (2017) ignore project management levels higher than a single project (their programs and portfolios) which leads them to the wrong conclusions about the role of projects in organizations. Asquer and Mele (2018) believe that the management of policy implementation is influenced by, among others, planning, organization, management, staffing, or budgeting. Still, they do not indicate project integration management (PMI, 2017a) as a factor influencing policy implementation. Peters (2000), from ten areas of project management, believes that only personnel management is important. Also, Löffler (2009), Gleeson et al. (2011), deLeon and Steelman (2001), Whitford and Lee (2012), and Ingraham et al. (2003) pay attention only to some areas of management aspects in the context of public policy implementations, for example, communications management or personnel management – not interested in the entire project implementation process. The World Bank Institute has developed a Capacity Development Results Framework (CDRF; Otoo et al., 2009). However, although it describes typical project activities, it does not include important elements of project management – for example, risk management, scope management, or schedule management.

None of these texts contain references to the most important project management area, i.e., integration management (Subchapter 3.12) which is responsible for coordinating all other management processes and, finally for, delivering project products.

The main issue with projects at the Public Administration (and Political Science) side is the lack of understanding of this tool and its role in Public Administration and public policy implementation.

1.7 Different Understanding of Common Concepts

Apart from the issues with public project management on the Public Administration side and those at the project management scientific community side, there is another area that hinders the development of science about public sector project management. This is the area of common concepts, or "interface" between these areas. Some concepts are differently perceived in public management and project management.

An example of such a concept is "governance." It was noted that it could mean different things in different areas (Van Kersbergen and Van Waarden, 2004). This is also the case in Public Administration and project management. One of the main components of governance's definition is its relation to management. The concept of governance in Public Administration has recently been popular – to the extent that the current mainstream thinking is called New Public Governance (NPG; Osborne, 2006; see Subchapter 12.3). According to Kooiman (2010), governance is all interactions associated with solving social problems. Management is part of governability. Governance includes processes and institutions through which citizens articulate their interests and fulfill their obligations (UNDP, 1997). Under (cooperative) governance, programs are managed, and problems are solved (Cepiku, 2017). Hence, governance may be perceived as all processes related to Public Administration, covering, among others, management.

In the project management environment, the perspective on governance is different. Governance is a layer different from managing. Project governance defines goals, determines the means to achieve these goals, and controls (rather than implement) activities that accomplish these goals (Müller, 2009). For Klakegg (2010), the most important governance functions are defining the decision-making process in the project and controlling the quality of the project documents. In project management, governance is one of the processes within a more general management area (Biesenthal and Wilden, 2014; Brunet and Aubry, 2016).

Hence, for the Public Administration environment, management is part of governance, while for project management, on the contrary, governance is a part of management.

1.8 Incompatibility of Conceptual Structures

In commercial organizations (companies) with operational autonomy, project portfolio decisions are primary. Each company independently decides which projects to choose for implementation, i.e., how to construct their project portfolio. In public organizations, the situation is different.

Only the government (central, local) has full autonomy. Subordinate organizations implement government decisions that materialize in public policies. Public policies are implemented by their programs. Depending on the level of autonomy (e.g., culture and science institutions usually have a higher level of autonomy than military policies), subordinated organizations are either given goals and budgets for implementation, or specific projects are defined that must be implemented. Each public organization (like any other) has its project portfolio, but in many organizations, its composition is not a result of the autonomous decisions of the organization's authorities. Unlike in the private sector, in many cases, the initiation, execution, and termination are based on policy requirements or the authorization of higher-level organizations (Baldry, 1998). Using the language of project management, decisions regarding project portfolio building result from budget, policies, and their programs defined at a higher level. Thus, the organization's project portfolios result from the implemented policy programs, i.e., the hierarchy is opposite to that in private companies. The absence of the most important concept of Public Administration in project management standards – public policy – makes evidence of these standards' bias toward the private sector. There are also significant differences between sectors in how the most important decisions are made. In states with a democratic system, decisions concerning implementing public policies and programs belong to the politics sphere; management parameters have only supportive meaning. They result from the distribution of political forces in elected representative bodies. Conservative governments choose different policies and programs for implementation than the leftist ones. In undemocratic countries, the managerial parameters are even less important. These facts are not included in most project management standards, such as those published by PMI (2017a, 2017b, 2017c).

The concept of a public policy portfolio could be introduced in the public sector, but it would have a different meaning than the project portfolio in private companies. The equivalent of the public policy portfolio management process is policy management and models explaining policy-making, for example, the rational choice model (Downs, 1957; Anderson, 2015), Multiple Stream Framework (Kingdon, 2003), punctuated equilibrium (Baumgartner and Jones, 1993).

The knowledge documented in the project management standards relates primarily to private companies. It is not consistent with the conceptual structure of the public sector already at the level of the main concepts. PMI®[2] documents are deeply imbued with the private companies' approach. The differences between the structures of the basic project management concepts in public and private sectors are shown in Figure 1.1.

The discrepancies also concern the purely practical sphere. The project management community (represented by the Project Management Institute or International Project Management Association) and the development community (e.g., World Bank) have different approaches to the very project management. The World Bank uses Logical Framework Approach (Practical Concepts Incorporated, 1979), while the project management community is based on the PMBOK® Guide (PMI, 2017a) or Prince 2® (UK OGC, 2017). LFA deals with what in PMBOK® Guide is called scope management, integration management, and – to a limited extent – risk management. Methodologies derived from the project management community deal with many other areas of management that significantly affect the success of the project, for example, stakeholder management or schedule management. Therefore, it would be difficult for project managers from one of these environments to move to the other.

Knowledge of Public Administration and project management seldom permeates. Researchers usually work separately in their respective areas. Public Administration researchers rarely include project management issues in their areas of analysis. On the project management side, the interest in the public sector and its specific features is smaller than in the private sector. Perhaps this is the

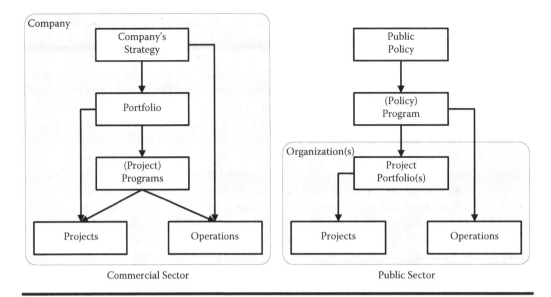

Figure 1.1   Differences between Project Hierarchies in the Private and the Public Sector.

reason for the discrepancies described in this chapter. Both parties: the Public Administration and Project Management community, should work together to overcome these difficulties and develop a systemized common approach to public project implementation.

Project management should support the success of Public Administration. However, the above-described discrepancies make this problematic. The following two parts are devoted to an attempt to bring Public Administration and project management closer together. The second part is devoted to presenting the basic concepts related to project management. After that, I placed a part on basic knowledge about Public Administration.

Notes

1. CMMI® is a registered trademark of Carnegie Mellon University.
2. PMI® is a registered trademark of Project Management Institute.

PROJECT MANAGEMENT FOR PUBLIC ADMINISTRATION

Chapter 2

The Most Important Concepts

2.1 What Is a Project?

Not each work performed by people is a project. At the same time, projects play a vital role in the functioning of any organization, including the public one.

Let's take a look at some works in the public sector.

1. Rescue operation for the victims of a natural disaster
2. Road construction
3. Organization of the event on the occasion of Independence Day
4. Production of a historical film (sponsored by the government)
5. School renovation
6. Reorganization of the health service
7. Implementation of a new type of weapon
8. Implementation of an IT system supporting the staff.

Let's also look at other works:

1. Operation of the IT system supporting human resources
2. The functioning of public transport in the city
3. Treatment of the patient in the hospital
4. Issuing a citizen with a passport in a public organization
5. Protection of the state borders
6. The Ministry of Agriculture (and any other) work
7. Maintaining good relations with neighboring countries
8. Keeping inflation under control.

These works were selected and grouped according to a specific rule. Can you see it at first glance? What is the difference between healthcare reorganization and patient treatment? What about implementing a new type of weaponry and the protection of state borders? Or implementing an IT system and customer service in a public organization? In general, how do the works from the first group differ from those from the second group?

DOI: 10.1201/9781003321606-3

Let us consider the type of goal of the works from both groups and the related time perspective for their implementation. The road construction (first group) was started so that the community could travel on it – once the road is built, the work is over. And public transport (the second group) operates continuously. It will always work as long as the city functions and people move there. Carrying one person or even a large group will not end public transport. Implementing the IT system supporting the staff (the first group) is done to work more effectively, and it will end when the system works throughout the organization. Later, the nature of the work will change, and employees will use it in their daily work. It cannot be said that the system will terminate when someone is hired. The prospect of this system's operation is not determined at the moment of its operation by indicating any event.

All works from the first group have a designated – no later than at the moment of their commencement – result or goal, the achievement of which will result in the completion of this work.

The works listed in the first group are **projects**, while those in the second group are not. So, we can provide the most important information in this chapter: what is a project? A project is a discrete initiative undertaken by an organization to meet a business need. Projects are temporary endeavors.

This definition requires additional comments. Do all the works from the second group have no specific end date? After all, restoring the patient to health ends the treatment. The passport in the citizen's hands ends its service, and the citizen thus achieves the goal – the possibility of traveling around the world. Although people and cities are different in every transport, transporting people between two cities is not considered a project. It is similar to repairing a failure in the electrical system in a building. While each failure may be different and should be dealt with differently, their repairs are not considered projects. These works were not considered projects due to two reasons.

First, their size is relatively small. Using the project management techniques described below would make no sense for work that requires several hours or one working day. In practice, many governments and public organizations specify the minimum planned value (US$ 1000, US$ 10,000, or some other depending on an organization's general budget and size) or the labor volume (5 man-days? 30 man-days?). With such a small amount of work, the expenditure on the use of advanced management methods would not pay off in terms of quality improvement compared to the cost spent.

Second, the process of issuing passports is highly standardized; it requires the performance of several or a dozen or so routine actions and checks. Routine type of activities is another reason why it would not make sense to apply laborious project management techniques separately.

Separation from other works (discreteness) and the effect achieved are more important than non-routine. Building similar buildings or implementing the same IT system in different organizations can become routine after some time, primarily if the same team performs it – but the works are still projects.

Producing a product by some work does not make it a project either. After all, any organized work produces a specific product or effect. The project is not to excavate coal by a mine or permanently feed the poor people.

In practice, both the lower limit of labor intensity and the level of routine work that causes them to be treated as projects are determined in each organization in its way.

The temporariness also requires a comment. After all, everything in the world will end one day, but that doesn't mean everything is a project. If there is a school which the government decides to close after some time, it does not mean that it has become a project in this way. Similarly with taxes, if the government stops collecting a specific tax, it does not mean that collecting the tax has become a project at this point. One of the characteristics of a project is relating the termination of its work to a predetermined pre-planned event, for example, building a stadium or reaching production capacity by an industrial plant. This type of achievement-related time constraint is one of the fundamental factors that characterize projects.

Projects can be classified in many ways, for example, according to the field of application, size, level of complexity, or the organizational unit carrying out the project. But the most important classification is based on the role the project plays in its organization. The project's aim may be to develop the organization (investment project) or obtain direct benefits (operational project). The project of building new facilities, restructuring the organization, and implementing the production of a new product are investment projects. Organizations also carry out projects such as implementing an IT system at the client's (IT companies), organizing a tourist event (tourist companies), or building a monument (culture departments of public organizations). They are related to achieving the statutory goals of the organization. The first two projects have a purely commercial purpose, but this is not the case with the monument construction project. Operational projects may or may not be saleable. It depends on the goals of the organizations implementing them. Basic project types are shown in Figure 2.1.

Public projects, perhaps the most often concept used in this book, should also be defined. I will limit solving this issue to the area of interest of this book, for which I have the goal of providing knowledge about managing projects belonging to a specific set.

A public project is one managed by a public organization.

A public-sector project is a synonym for a public project. Government projects are public-sector projects that are implemented by decisions of the central level government institutions (e.g., parliament, cabinet, particular departments).

You can find out more about public organizations and their types in Chapter 8.

Organizations from other sectors often implement significant components of public projects. Consider building a public road. This project starts from analyzing the transport situation in a given area, defining the route and other road parameters, preparing and performing the tender, signing the contract, etc. They are the tasks of the public organization. The physical building of the road is the task of the company winning the tender – which is also a project from the point of

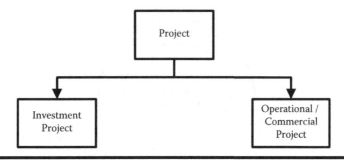

Figure 2.1 Basic Project Types.

view of that company. But this does not change the fact that the public organization is responsible for the entire project and may also define and supervise how the contractor manages the works.

Projects financed by public organizations that are not involved in implementation (a typical situation, for example, in the area of culture) are not of interest to this book. The same concerns philanthropic projects in which private citizens or companies deliver products that are freely accessible to the public. The more, private companies' projects which deliver projects used for fees by the people are not treated as public here.

More on the specificity of public projects and their differences from projects in other sectors can be found in Part III.

2.2 Product Development and Management

Projects are implemented to provide construction, IT system, implement the production of a new car model, develop a new medicine or deliver another product specific to some area of human activity. Each of these areas has produced the knowledge needed to manufacture these products. Builders have been gathering knowledge about building for thousands of years. They know how to create foundations and other building elements. They know the strength of individual building materials and the rules for implementing subsequent construction phases. IT specialists develop the knowledge necessary to produce IT systems. They know programming languages and systems, know computer hardware, know how to create software modules, merge them to create a complete system, and check if the system will work properly. It is the same in any other area of the project application. This knowledge and skills are collected by specialists and documented in the literature of each field and are not transferable between different fields of application. For hundreds and thousands of years, the projects were carried out exclusively by specialists in product manufacturing. The construction sites were led by people who we would be called engineers today.

Over the years, it has been noticed that some activities performed during the implementation of such works (which were not called projects) are similar regardless of the area of application. Each work should have a person who gives orders to other people involved and is responsible for its implementation. Every job should be planned. Since each job requires money, the budget should be part of such a plan. It is good to know when the work will start, end as a whole, and end its component tasks. These are examples of activities that are not part of the production of products but are necessary for the proper organization and successful completion of the work. A set of such activities is collectively called management, and this book is devoted to project management (in the public sector). Hence, project implementation has two areas of activities: product development and project management (Figure 2.2).

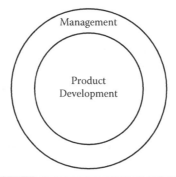

Figure 2.2 Types of Project Works.

Although the goal of any project is not to manage it but to create specific products and achieve its goals, management is a significant factor in the success of any project. The impact of management on the success of a project increases with the size of the project. Some projects are carried out by one person in a short period. For example, a project of a slight modification of an IT system may be performed by one person during a week. Then the management elements are minimal: you need to define the work, and determine who will do it and when it will finish. The amount of management work required in such a project is negligible. On the other hand, the project of sending a lander to another planet involves the work of thousands of people, takes many years, and costs trillions of dollars. It is impossible to succeed in such projects without very efficient management. Management works may consume several percent of the total project budget.

2.3 Components of Projects

In huge works, their division into smaller components facilitates management. The public project to build a new warship can be divided into sub-projects to construct the hull, build a propulsion system, create an armament system, build a navigation system, and construct an IT system supporting the operation of the ship. These works can be carried out relatively independently – although some level of coordination between them is required. Each of these works is called a sub-project and can be managed relatively independently. The sub-projects must establish dependencies regarding the manufactured products and the schedule. The hull must be of a size and load-bearing capacity to install the designed armaments. The propulsion system must have the power to enable the ship to move at the appropriate speed. In turn, time dependencies describe, for example, that at a given moment, the hull must be in a condition that allows the installation of a propulsion system, weapons, equipment for the IT system, and all other necessary elements from which the ship is assembled.

A project may or may not be composed of sub-projects.

2.4 Project Life Cycle: Integration Management

In projects, you can extract the most important parts and points in time that break the project down into these parts. The development of a description of an IT system, its development, testing, and implementation are examples. Or it is identifying the possibilities of building an airport, preparing the technical specification of the airport, building the airport, testing, and operating. The objectives of such partitioning are usually to isolate the main products of the project being implemented and, consequently, the possibility of their verification and making significant decisions on the further implementation of the project or its discontinuation.

In some projects, the decision to initiate it at the beginning is considered unquestionable, and the project is implemented to produce the product, regardless of its usefulness, even if the costs and time increase uncontrollably, and the product has lost its usefulness and it was not possible to achieve the planned benefits. Nobody reads a printed book, a car is produced that people do not buy, training programs are prepared for which no one reports, or weapons are created that, due to technical progress, do not meet the combat requirements. Or the cost exceeds the planned budget in a way that makes it impossible to generate profit. The company implementing the project is losing; society does not benefit. People don't ride the newly built railroad. The cost of building an airport has been exceeded many times over, and airlines do not want to use it. We all know such cases.

To prevent this from happening, the project is divided into **phases**, between which there are so-called **gateways**, i.e., points where the status of the project is verified, primarily in terms of the project's potential intended benefits (PMI, 2017a; UK OGC, 2017; USA GAO, 2020a;[1] Section 20.5.1). Sometimes phases are distinguished for other reasons, such as work logic or schedule. The main gates can also appear, without distinguishing separate phases, immediately after the decision to implement the project or immediately before its completion. The set of major phases and gates is called the **project life cycle**.

2.4.1 Waterfall Project Life Cycle

Historically, the first type of project life cycle was the **waterfall life cycle**, also known as the **sequential** or cascade cycle.

After making the appropriate decision, work on the initiative begins with a Feasibility Study in which the usefulness of the project and the possibility of its successful implementation are assessed. In justified cases, for example, when the project's implementation is required by law, the execution of the Feasibility Study may be omitted. **Initiation** is the phase in which the authorized authorities of the organization analyze the Feasibility Study results to decide on the implementation of the project. The project may not be initiated despite the positive result of the Feasibility Study, for example, if there are plans to implement other, more important projects at the moment. Once initiated, the project enters the **planning phase**. It is necessary to determine what works, when, and by whom will be performed and other conditions for the project implementation. In some projects, planning is not treated as a separate phase – for example, when initiation documents describe how the project will be implemented in sufficient detail. Larger projects are initially planned only in general terms; for example, the most important products and their completion dates are indicated, and the details are settled before the start of each phase. We call such an approach the **rolling wave** planning. The approval of the plan allows the **execution phase** to begin. There must be at least one such phase in each project, and their number depends on the project's specifics. The execution of the plan consists of the implementation of the planned activities. Decisions to continue the project are often conditional, depending on modifying the subsequent phases' detailed plan. The phases need to be designed so that the assessment makes sense; usually, they are about producing some whole, complete products. The **closing phase** is carried out after all products have been delivered by the project and is aimed at the formal closing of work. One of the main activities of closing is disbanding the project team and ending the project manager's job. During the **evaluation**, it is assessed whether the project has achieved its expected results. Closing and evaluation may directly follow one another if the evaluation of the project is possible at the time of its closure, for example, in a road construction project. What will happen after the evaluation depends on its result. A typical evaluation result suggests introducing changes to the processes and products used.

A separate set of activities performed in each project phase is **monitoring and control**. These activities are most visible and intense in the execution phase, where the compliance of the works with the plan must be constantly checked. But activities carried out in other phases also need to be controlled; you have to, for example, check that no action lasts forever.

Figure 2.3 presents a general scheme of project life cycle.

2.4.1.1 Example of Presidential Election

Let's deal with the presidential elections. The initiation of the project is to schedule elections. In this project, the Feasibility Study phase is unnecessary (unless we consider the analysis by the involved political forces regarding the holding of early elections to be the Feasibility Study phase). The ordinance calling elections usually includes the schedule, division into constituencies, the rules for nominating

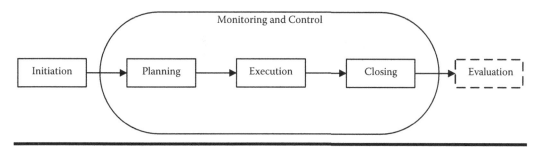

Figure 2.3   General Scheme of Project Life Cycle.

candidates, the method of establishing election commissions, the method of determining the result, etc. (some of these provisions may result from or are contained in other, previously established legal acts). The judiciary verifies the compliance of the decision with the applicable law. If the elections were ordered illegally, their results would be questioned by society and political forces, so the project would not bring the intended benefits. The effect of this verification is usually not presented and commented on publicly, as it does not arouse particular interest in the public, as in the overwhelming number of cases, elections are called legally. Providing the schedule and rules for conducting the elections constitutes planning the project. After the candidates are collected, it is checked whether the number of candidates meeting the required conditions complies with the election rules. If no applications have been submitted or do not meet the requirements, the project is terminated. Then the campaign is carried out, voting takes place, and the results are determined. The judiciary verifies how the elections are held and the elections are over.

The life cycle of an election project is shown in Figure 2.4.

There are two gates separating phases in this project: verifying the legality of the election and checking whether the proposed candidates meet the required conditions. Campaigning, preparing, conducting elections, and calculating the results is one phase of the life cycle – there is no justification for planning significant checks between these activities. Verification of the correctness of the elections (evaluation) takes place immediately after the end of the elections. Depending on the evaluation results, the results are announced, or information about the invalidity of the election is reported. These are the last activities in this project.

2.4.1.2 Example: Implementation of the Paperless Work System in an Organization

An organization is based on a paper document flow. Due to the emergence of cheaper and cheaper means of electronic document circulation, the organization's management board is considering switching to paperless work with documents.

As part of the Feasibility Study, it was assessed whether and to what extent such a project is possible. As a result of the Feasibility Study, the relevant team determined that such a project is viable for legal reasons, the company's necessary financial resources, and the specificity of work procedures.

After receiving the results of this analysis, the organization's management board decided to implement the project. The decision indicates the responsible person, time frames, and budget constraints. As the purchase of software and implementation services will be a high cost for the organization, the management board made the final decision on implementation based on the presentation of a detailed budget which will be part of the project plan. It was also decided that the project will consist of two main phases: a pilot implementation in two organizational units and then a full roll-out in all eight others. The purpose of the pilot implementation is a practical assessment of the implemented solutions and their possible better adaptation to the specificity of

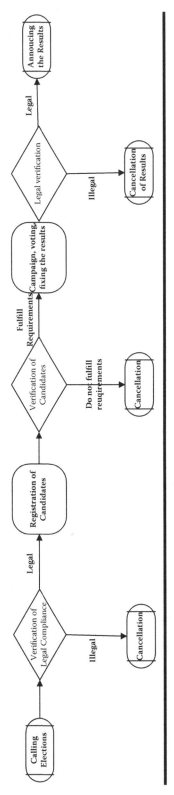

Figure 2.4   Example Project Life Cycle – Presidential Elections.

the organization. After the pilot implementation, the management board assesses whether it is necessary to proceed with the full implementation. The closure of the project takes place when the company is already working in a new way. The desired result of closing the project is blocking the possibility of carrying out any work in this project. In practice, however, it happens pretty often that the closure also results in a list of (minor) corrections to be made to the work results.

The life cycle of this project is shown in Figure 2.5.

This project does not envisage that if new ways of working to be implemented are introduced, the result of the evaluation could be a return to the previous method of working. Evaluation should assess how much better it is to work in a new way. The evaluation result may also be a proposal for implementing improvement works which may take the form of a new project.

The above-presented two life cycles take the form of sets of sequentially performed phases.

The project's products can be delivered at the end in the sequential approach, as shown in the previous examples (the big-bang approach) or during its execution. Consider a project consisting of the construction of several buildings. The construction company has specialized teams to construct foundations and walls, lay installations, and finish (painting walls, laying floors, etc.). Each of these teams performs their work in each building in a logical sequence: construction, installation, finishing. Therefore, the buildings will be put into use successively during the project execution. Such a project scheme is presented in Figure 2.6.

A waterfall (sequential) project can have multiple execution phases. If the production of products in a waterfall project is disjoint in time, it is possible to distinguish phases corresponding to the implementation of individual products. Each of the phases is planned and monitored. This may be the case in an ERP system implementation project. First, the HR module may be implemented, then the production management module, the project management module, and finally the sales support module. After completing the implementation of each module, the project passes through the gate, in which the progress of work and the prospect of achieving the overall goal are assessed, and the project plan is updated, if necessary. The multi-phase waterfall project scheme is presented in Figure 2.7.

Originally, sequential project life cycles were defined for large-scale defense and infrastructure projects with budgets in the billions of dollars. It was believed that the risk of billions of losses was avoided by using sequential, well-documented phases. The introduction of such sequential project management is related to the US Defense Secretary Robert McNamara and his Program Planning and Budgeting System (PPBS) in the early 1960s (Lenfle and Loch, 2010). However, over time, it has been noticed that it is possible to approach the life cycle of projects in other, more effective ways, especially in smaller projects. Recently, non-sequential project life cycles have been used more and more often. Like the waterfall cycle, projects are divided into phases, but their content is planned differently.

An iterative project is one in which the main assumption is to improve the product in consecutive phases until it reaches the desired properties (e.g., quality, efficiency, or business effect). This approach is, in a sense, an extension of the pilot life cycle, but more phases are implemented. An example of iterative projects can be the development of a commercially offered editor and providing users with better and better versions. In subsequent iterations, new functionalities may be added to the product, but this is not the essential characteristic of this type of life cycle.

2.4.2 Incremental Project Life Cycle

An incremental life cycle is the one in which it is planned from the beginning that the goal of consecutive phases is the delivery of new products or functionalities. Assessment of the success of implementing a given phase causes the transition to the next phase.

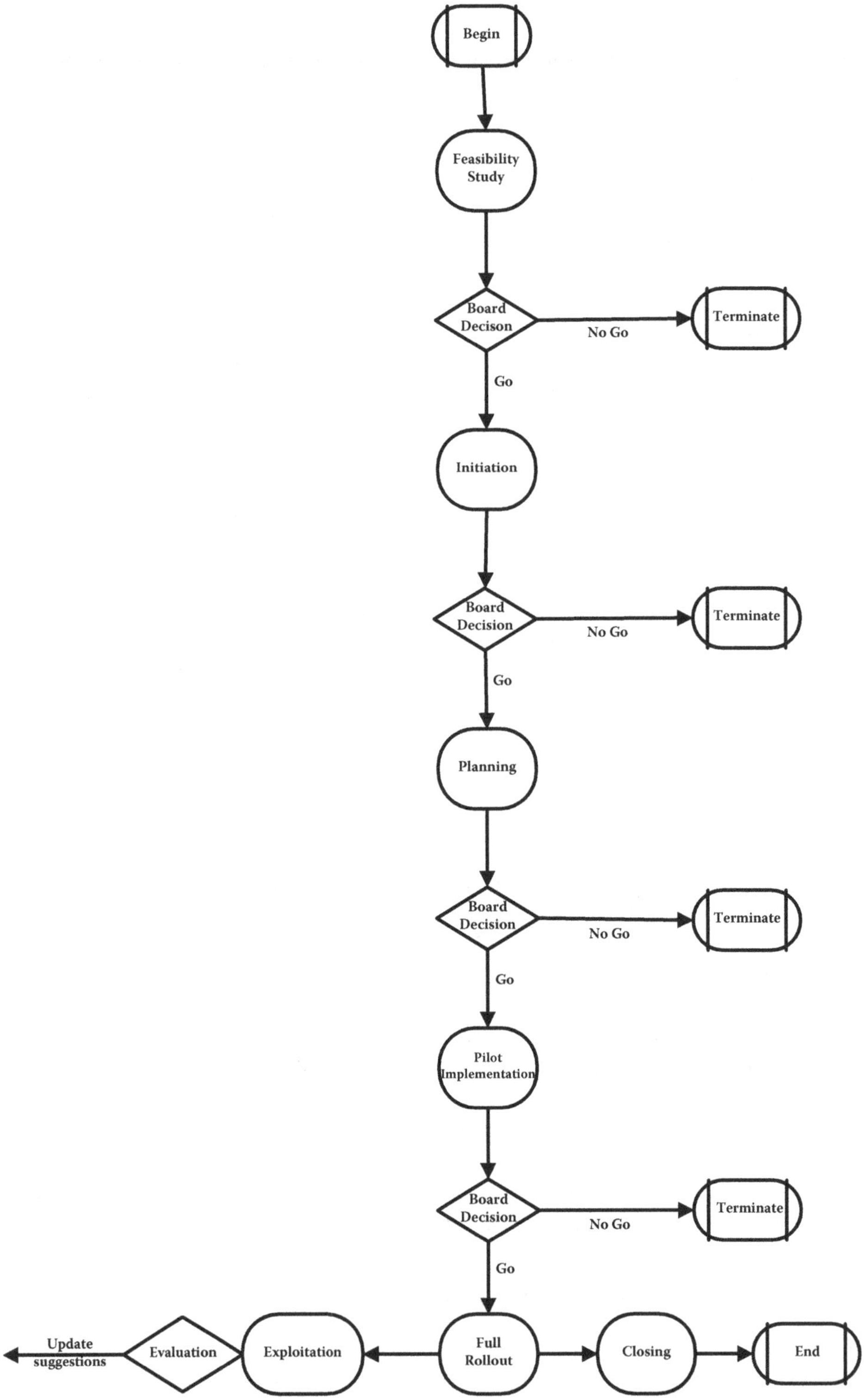

Figure 2.5 Extended Project Life Cycle Example.

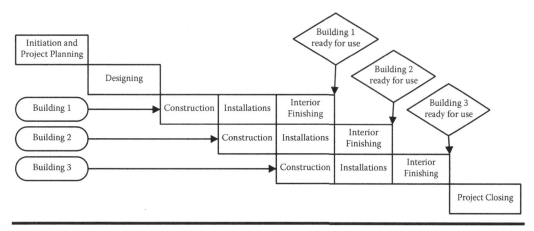

Figure 2.6 Delivery of Products in a Waterfall Project.

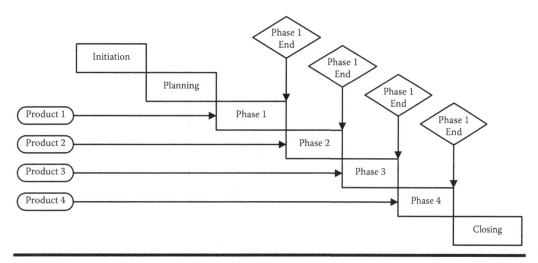

Figure 2.7 Multi-phase Waterfall Project.

Sometimes the concept of an incremental life cycle is used instead of a multi-phase sequential life cycle as new products or functionalities are added in successive phases in both types of life cycles. Some authors distinguish these life cycles by the amount of planning done in transition between phases: more planning in phase results in the incremental life cycle, and less (or no) the multiphase waterfall life cycle. Discussions about the extent of planning in the waterfall or iterative life cycles are rather theoretical with little impact on practice. When moving between phases, it is always necessary to check whether the existing plan can be implemented in its current form (in practice, very rarely), it must be modified or detailed. The name of the life cycle type is not a critical issue. It is essential before starting a project to define how and when the project's product set is defined and modified.

2.4.3 Agile Approach

A particular type of incremental project life cycle is agile (Figure 2.8). The agile approach was introduced to software project management in 2001 (Beck et al., 2001). Project goals and

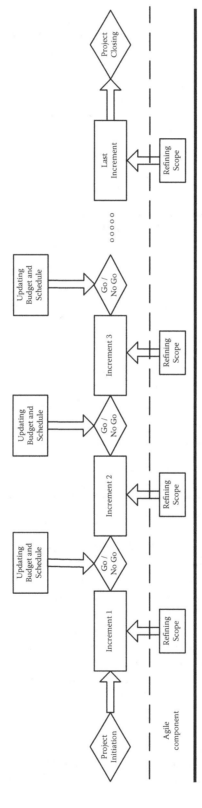

Figure 2.8 Incremental Project Life Cycle and Its Agile Variant.

products are generally defined in advance in the agile life cycle. Still, the detailed requirements and scope of work in increments are detailed between individual pieces of work – and not in advance at the beginning of the project.[2]

The use of the agile approach is justified when large variability of requirements is expected, the modification of these requirements is possible due to environmental or legal conditions, and users may impact the implementation of the project. The organization's employees can be directly involved in implementing an agile project, communicating their needs, and reacting to the already developed product. The agile life cycle can also be used, for example, to develop the capability to do business for a specific social group or inhabitants of a particular locality. Once particular capabilities have been established, it is determined with members of the community concerned whether further work is needed and what will be the extent of the capabilities to be acquired in the next stage. By contrast, this approach is not appropriate, for example, for infrastructure projects where most of the environmental requirements must be agreed upon and approved before the project starts, and the development team must strictly adhere to these provisions and clearances.

2.4.4 Project Product and Its Life Cycle

Projects deliver products or services. People created standards, customs, or best practices for manufacturing products. For example, a proposal is made first for a book, then an outline. The outline is reviewed, and corrections are made. Based on the approved outline, the content is written. The content is reviewed, first by reviewers – specialists in the book's subject. The author responds to the comments and uses them to prepare the final content. After approval, production is ready: an electronic form of the book is created. The book is then printed and bound. The finished book is distributed through appropriate channels, sold, and read.

There is one crucial decision in this project: after the proposal is made, whether the book will be published. So, the life cycle has two phases: the proposition and the book's realization. On the other hand, there are many activities with a book: writing an outline, writing a book, reviewing a book by specialists, reviewing by an editor, preparing an electronic image, printing, binding, etc. The sequence of activities related to producing a project product is called the project **product life cycle**. Project life cycle phases typically fall into product life cycle phases. There is no point in making fundamental decisions about a project if one of its phases is not completed.

Like the project life cycle, the product life cycle can be sequential, iterative, or agile. But they are not unequivocally related. For example, in the sequential project life cycle, products delivered in their various phases can be implemented in an agile manner. When implementing an IT system, it can be assumed in advance that it will be implemented successively in five organizational units. All these implementations together form one phase of the project life cycle: implementation. The implementation methods in each subsequent unit will be determined based on experience from implementations in previous units. That is, the product will be dynamically defined (agile approach) as part of the implementation phase of the project life cycle.

Notes

1. The author of the book was a significant contributor to this document.
2. I use the concept of "piece of work" and not "phase" as work in the agile approach is divided into small periods.

Chapter 3

What Is Managed in Projects?

3.1 A Review of Managerial Areas

Project management is a complex job that requires knowledge, skills, and capabilities from many areas. According to the PMBOK® Guide (PMI, 2017a), there are ten areas of project management.[1] I have selected the most important processes and techniques specific to project management from each area. But projects are performed in the environment of their organizations. Many activities carried out in projects result from the arrangements adopted for all organization activities. For example, a project's budget is based on the overall financial capacity of an organization, and the execution of expenses and financial reporting is based on company-wide procedures. How purchases are made in an organization is usually defined for the entire organization. Project implementation processes are partially determined by the set of all processes of the organizations implementing them and must be integrated with them. For example, employees' evaluation and development path should consider their participation in the implementation of projects, and project risks are part of the organization's risks. Another example of integrating projects with an organization is not releasing funds for a project before their formal start and after their formal closing (which is not strictly controlled in many organizations).

Project **scope management** identifies the set of products that a project delivers and causes the project to produce only the planned products (Subchapter 3.2).

The phases through which the project goes from initiation to completion and evaluation of the results are described in Subchapter 2.4, project **integration management**, i.e., making all processes work well together. The integrated change control is a part of the integration management area (Subchapter 3.12).

Project **schedule management** is the definition of a set of activities leading to the implementation of the project's scope, determining the relationship between them, their duration, which determines the project schedule, and its total implementation time (Subchapter 3.3).

Project **cost management** is determining the project budget and causing the project to be carried out within the approved budget (Subchapter 3.4).

Communications management is making all team members and other entities aware of what is happening in the project. A specific and most important technique for gaining knowledge about the project progress is Earned Value Analysis, described briefly in Subchapter 3.5. The area of **knowledge management** is related to communications management (Subchapter 3.6; Gasik, 2011a, 2011b).

DOI: 10.1201/9781003321606-4

Human resource management is the process of determining the human resources necessary to implement a project and the relationships between them. I discuss the project team and the project management structure in Subchapter 3.7.

Risk management is identifying events that may disrupt the implementation of the project and the implementation of actions influencing the identified actions: counteracting negative events and exploiting positive events (Subchapter 3.8).

Procurement management is the selection of external entities (contractors, suppliers) and making the contracts concluded with them be performed. In this book, I will look at evaluating project subcontractors (Subchapter 3.9).

Stakeholder management is the recognition of who is interested in the project or who may influence the project. Then, actions are taken to ensure support or inclusion in the project of people with a positive attitude toward the project and measures to minimize the impact of people with a negative attitude toward it (Subchapter 3.10).

Project **quality management** is defining the quality standards applicable in the project and ensuring that these standards are met (Subchapter 3.11).

Products developed in the projects must be put into operation. This is often associated with activities such as user training or changing work procedures. Collectively, these activities are called **product transition** (Subchapter 3.13).

3.2 What Should Be Developed in Projects? Scope Management

Projects have their requirements and constraints. Requirements are the conditions set by the project initiator and other stakeholders involved, the achievement of which is deemed to be the project's objective or goal. They define, for example, the length and width of the road, the scope of the material taught in schools after the implementation of the educational program reform, the level of reduction in the incidence of a specific disease, or the number of people who will find employment as a result of the project.

In addition to the requirements, each project has constraints, i.e., conditions that must be met and are beyond the control of the entity defining the project. Typical constraints relate to time (e.g., a project must be completed by 1 January of the following year) and budget (a project cannot be allocated more than US$ 2 million). Constraints may also arise from the implementing organization's environment (a long-term contract to supply materials by a specific organization) or from the law (e.g., a country embargo on purchasing materials from another country).

3.2.1 SMART Requirements

Each project must have an objective. We consider the achievement of the objectives set for the project to be a success. Upon completion of the project, it should be possible to assess whether success has been achieved.

The objective of building a dwelling house is to provide people with housing. The objective of building a new warship is increasing the military capabilities of the state. The curriculum is being changed to more effectively improve students' knowledge. A sports event is organized to disseminate knowledge about the sports association's organizational skills and increase the chances of our players for victory. The book is published to generate profit, spread the ideas of its author, and increase its readers' knowledge.

Do the objectives formulated in this way provide an unambiguous assessment of the project's success? If we had left such objective wording, a discussion would probably have started after the project was completed. Was the book, which was sold in 1,500 copies and brought 22,400 zlotys in profit, successful or not? When should this success be assessed – after six months of distribution or two years? How do you relate to the success of a sports event? Can the number of participants measure it? Or the number of medals won?

If we define the requirements for the project measurably before starting the project, then after its completion, it will be possible to make a more explicit statement about its success.

In the publishing house, the book publishing project's success occurs between the head of the publishing department and the editor responsible for publishing the book. This group will also include a discussion on the effect of the project. The editor persuaded the head of the department to publish a book in English on project scope management.

The SMART model was introduced to enable precise compliance assessment with requirements (Mannion and Keepence, 1995). Each or all of the requirements should be specific, measurable, achievable, reasonable, and time-related.

1. **Specific** means that the requirements unambiguously define their subject. In the book publication project, it was specified that it would be a book on project scope management book in English. For the department head to assess the business effect, this is a sufficient specification, i.e., it really is specific.
2. **Measurable** means that the effect of the project can be measured. Determining that the book will be sold in 1,500 copies and bring PLN 22,400 in profit will allow one to assess whether the project was successful unambiguously.
3. **Achievable** means that the requirements can be met. For the editor to justify the fulfillment of this condition, he should show that he has a potential author and that the company can produce the book and distribute it to bookstores.
4. But not every project we can implement makes sense. **Reasonable** just means that the requirement makes sense. To meet this condition, the managing editor should show that there is a need for such a book and that it is, therefore, possible to achieve the set goal of selling 1,500 books and obtaining a profit of PLN 22,400.
5. The **time-relatedness** condition is met if the moment in time is specified when the verification of the requirement fulfillment will be performed. The required sales volume will be achieved after six months of distribution (you can also indicate a specific date, for example, 31 December this year).

Part of the application of the SMART model is selecting metrics that are intended to demonstrate the project's success. The sporting event's entity determines whether the number of participants will measure the success, the number of medals won, or both.

The head of the department, when considering the editor's proposal, considers business issues. This area includes the sales volume and, above all, the possible income and profit. The perspective of the editor's conversation with the potential author, with whom the order will be placed, is different. The parties must clearly define what the author will provide to the editor to build a contract well. In this relation, the specific condition will be met if the book's content is specified. It will contain an introduction, chapters on creating requirements, building a Work Breakdown Structure, description of delivered products, possible modifications to the project scope and product acceptance, and IT tools supporting scope management. The book written in English will contain a table of contents, illustrations, exercises with solutions, and an index of the most important terms. The measurability condition will be met if the book has between 80,000 and 100,000 words, 40 pictures, and ten exercises in each chapter. Achievability of implementation will be demonstrated if the author presents previous publications and practical experience in scope management. The reasonableness condition will be met if the author in the proposal shows that there is a need for such a book and that the book brings something new compared to other existing on the market (this and the last element will probably be used in the editor's discussion with the head of the department). Eventually, the editor and author should agree on a date to deliver the manuscript to the publishing house.

The book publishing project example also illustrates two concepts: the project **effect** (also called outcome, result, etc.) and its **product**. The project's effect is the benefit obtained due to its implementation. The subject of discussion (and subsequent evaluation) between the head of a department and the editor is the project's effect and the discussion between the editor and the author – the project's product. These concepts should always be distinguished. Projects are not implemented to create a product but to achieve something.

Requirements for projects are usually formulated by both its client and the implementing party. In a house construction project by a commercial company, the future owner wants a comfortable apartment (that materializes in many requirements), and the contractor wants to have a sufficiently high profit. In the project of publishing a book by a private company, the publisher and the author want to earn, the author also wants to gain popularity and support for his ideas, and the reader – to read an interesting book. There is usually one main expectation in commercial companies: that the project will be profitable. Such a requirement may be accompanied by other, but also indirectly related to the commercial side, e.g., obtaining new knowledge, gaining a good reputation which in the future will enable the implementation of a more significant number of more profitable projects. It is different in the public sector. On both the expenditure and the revenue side, finance-related criteria are considered but are not the final selection and evaluation criteria for a project. Such criteria may include, for example, improving the health of citizens, providing education, increasing security, facilitating communication, enabling travel, and eliminating famine. The criteria may also include social justice, equality of citizens, increasing their satisfaction with public services, or increasing support for the ruling party. In addition, the requirements may be inconsistent or contradictory one to another. For example, requiring to work efficiently may conflict with the requirement to reduce unemployment. Requiring treatment for diseases can counter the economy's demands: there are costly treatments to diseases. Requiring the use of the best technology may run counter to the requirement to support local industry. Various methods are used to solve these problems, such as prioritizing and weights for individual requirements or reducing all criteria to one measure, usually financial. Defining and managing complicated and complex requirements is much more complicated in public than in the private sector.

3.2.2 Requirement's Priority – MoSCoW Model

In projects, there are usually many requirements – for example, when implementing a complex IT system or building a new type of weapon, their number can be up to thousands. But not all requirements need to be of equal importance.

Consider the project of building a residential house for a family of four. Someone who wants to order a construction project from a construction company would like the house to have a living room, a kitchen, two bathrooms, four bedrooms, two offices, and a garage for two cars. They would prefer the salon to have a southern exposure. He would like the house to be two-story. This person also has a specified maximum budget for this project. He comes to a construction company with such expectations. And it turns out that his funding, unfortunately, is not sufficient to build their dream house. Therefore, additional arrangements with the construction company are necessary.

The **MoSCoW** model (Clegg and Barker, 1994) helps systematize such a conversation. The individual words in this model indicate the priority of significance for the contracting party:

- *M (Must)*: Without meeting such requirements, the project loses sense.
- *S (Should)*: These requirements will be met if resources (e.g., financial) are found for implementation. But failure to meet them does not make the whole project useless.
- *C (Could)*: The requirements will be met if it does not incur additional costs (yes, there are such requirements!).
- *W (Will not)*: These requirements will not be met.

What can you opt out of, the developer asks? If this house does not have a living room, kitchen, bathroom, one office, three bedrooms, and a garage for even one car, I am not interested in such a construction. In this way, the client defined the M priority requirements.

What about the rest of the requirements? Is the second bathroom, the fourth bedroom, or the second garage more important? I'd like a fourth bedroom and a second garage if this fits my budget. In this way, the client defined the S priority requirements.

What about the southern exposure? You know, the construction cost will be the same regardless of its setting. This is a priority C requirement. Such requirements usually involve choosing one of several options.

During the further discussion, the parties agree that the project will not connect the house to the Internet or install additional protection for doors and windows against burglary. Writing this down in a future contract will reduce the possibility of misunderstandings between the customer and the construction company. Such requirements are marked with the priority W.

Requirements' prioritization arrangements should be made before any project implementation, or part of it, as this is where the most important project parameters, such as cost or duration, are established.

3.2.3 Traceability Matrix

In the book publication example, we had one business requirement and one product to meet the business requirement. There can be many business requirements in real projects, producing many products. A practical issue in projects is to ensure that the set of products created guarantees the fulfillment of all objectives.

Let's deal with the project of implementing an IT system supporting the improvement of public organization management. Before implementation, the following objectives are set (of course, they were much more precisely defined):

O-1: Improving the management of own personnel
O-2: Improving the management of volunteers
O-3: Reducing the duration of case processing
O-4: Reducing the number of errors in issued decisions
O-5: Increasing customer satisfaction
O-6: Improving the service of foreign clients
O-7: Lowering the cost of running the organization
O-8: Streamlining the management of facilities

In this project, the implementation of several specialized modules (project products) was planned:

M-1: Human resource management module
M-2: Financial and accounting module
M-3: Customer service module
M-4: Database of own human resources
M-5: Interface module with nationwide population database

The person responsible for the implementation must determine whether these products (more precisely: their implementation) will meet the expected objectives. A **traceability matrix** (USA NAVY/EC, 1999) can be used for this purpose. Each column in this matrix corresponds to an objective, and each row corresponds to a product (module). At the intersection of the row with the column, I put the √ symbol if the product affects the achievement of a given objective. An example of such a matrix is shown in Table 3.1.

Based on such a matrix, you can check whether you have not forgotten an objective when designing the project's products. If there is no √ symbol in the column of a specific objective (here: O-6), this project objective will definitely not be met. The presence of the √ symbols does not automatically mean that the goal will be achieved. The content of the matrix structures further analyzes meeting the objectives. For example, on its basis, we can pose the question: Is it enough to implement the M-5. Interface module with nationwide population base sufficient to achieve the O-2. Improving volunteer management? Will using standard information for each citizen sufficiently facilitate the work with volunteers?

3.2.4 Work Breakdown Structure

When we know what the project requirements are, we can start describing, in a project-specific way, the products that the project is to deliver.

Let's deal with the project of building a highway from A to B. The distance between these places is approximately 100 km. The construction of this road has been divided into three segments. There will be five gas stations along the road. The regulations require a telephone to be available on the road every 10 km. The road will lead partially through a forest area; hence, building three passages for animals will be necessary. The road will intersect with four other roads; it will be required to build four collision-free intersections. Information is better perceived if presented graphically; the work can be represented as in Figure 3.1.

Such hierarchical representation of the work performed in a project is called **Work Breakdown Structure** (WBS; PMI, 2006a). WBS is most readily represented graphically, but it is also possible to present a text representation where only the codes represent a place in the hierarchy; it is also possible to use indentation.

This level of detail may not be sufficient to accurately describe the scope of work to be performed on the project. Developing a road segment requires building a road, a division zone, side lanes, a green zone, and a fence. In turn, to make a road, you need to prepare the base and the surface. The WBS of this fragment of the scope of work is shown in Figure 3.2.

The actual graphic presentations are in large "canvas," containing hundreds or even thousands of elements. A precise answer to the question of what level the WBS should be refined is beyond the scope of this book. Referring to intuition, it can be said that the WBS element should be written down to the level to which it facilitates the understanding of the whole work (an overly complex hierarchy is not the easiest to understand). Each WBS element

Table 3.1 Traceability Matrix

	O-1. Improving the management of own personnel	O-2. Improving the management of volunteers	O-3. Reducing the duration of case processing	O-4. Reducing the number of errors in issued decisions	O-5. Increasing customer satisfaction	O-6. Improving the service of foreign clients	O-7. Lowering the cost of running the organization	O-8. Streamlining the management of facilities
M-1. Human resource management module	√						√	
M-2. Financial and accounting module							√	√
M-3. Customer service module			√	√	√			
M-4. Database of own human resources	√							
M-5. Interface module with nationwide population database	√	√						

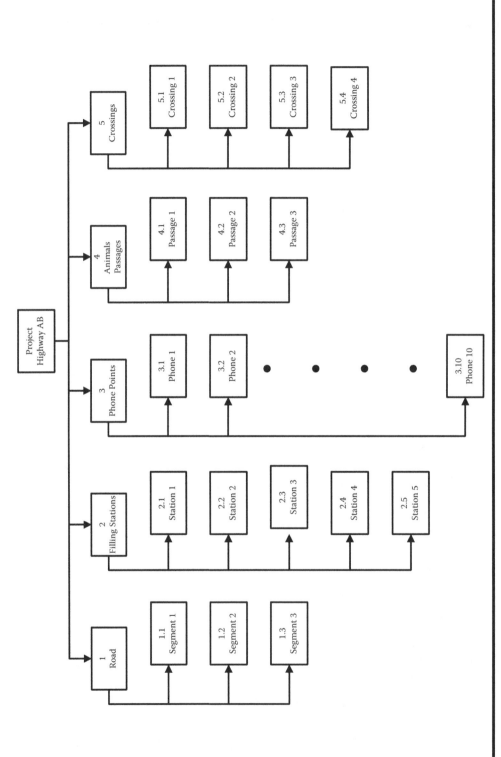

Figure 3.1   Graphical Presentation of Work: Work Breakdown Structure (Part 1).

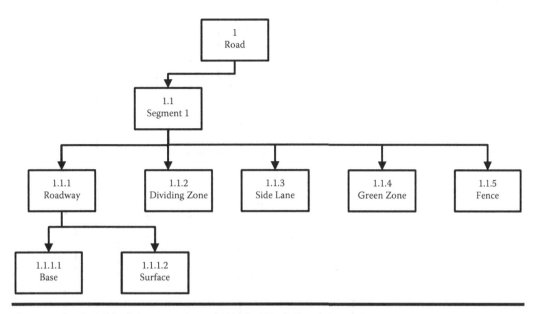

Figure 3.2 Graphical Presentation of Work: Work Breakdown Structure (Part 2).

is described, in addition to the identifier and name, with a lot of information telling what to do, what resources are needed, the duration, the risks associated with it, etc. Each project information is related to some WBS element. For example, the project budget consists of the costs of implementing individual WBS elements; risks are related to WBS elements; each WBS element has quality requirements defined in a way specific to it. You can think of the project description as the root description of the WBS tree. Therefore, **the concept of WBS is the most crucial one in project management**.

In the example above, the WBS components were products delivered by the project. Sometimes it is easier to present the WBS as a hierarchy of works rather than products. The project of conducting a series of training consists of a Feasibility Study, training preparation, and implementation. The WBS for this project is shown in Figure 3.3.

When building a WBS, elements describing products and activities are often combined in more complex projects.

3.3 Project Schedule and the Critical Path: Schedule Management

The two most important questions regarding the project schedule are as follows:

1. How the project activities should be sequenced and organized?
2. What will be the duration of a project?

3.3.1 Example: Building a House

A house is being built. The activities that make up this structure are shown in Table 3.2. The table also indicates which activities must be completed for a specific activity to start (its predecessors) and the planned duration of each activity (in months).

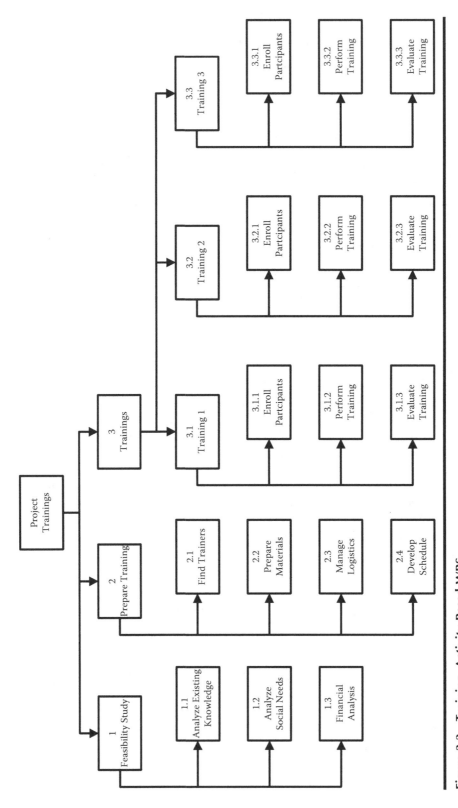

Figure 3.3 Training. Activity-Based WBS.

Table 3.2 Project: Building a Small House

Id.	Activity	Predecessor(s)	Duration (months)
A1.	Building walls and roof	–	5
A2.	Mounting of doors	A1	1
A3.	Mounting of windows	A1	3
A4.	Installing electricity	A2, A3	2
A5.	Installing sanitary	A2, A3	3
A6.	Painting of walls	A4, A5	2
A7.	Fencing the garden	–	3
A8.	Moving to the new house	A6, A7	1

The relationships between activities in this (and any other) project can be represented graphically using a network diagram in Figure 3.4. Activities that play a special role in the project schedule are grayed and connected by wider arrows. These activities are carried out from the beginning to the end of the project, as indicated by the arrows, i.e., the sequence of activities. Let's calculate the duration of all activities marked in red. The sum is 5 + 3 + 3 + 2 + 1 = 14 months. It takes this time to logically follow the activities that make up this path.

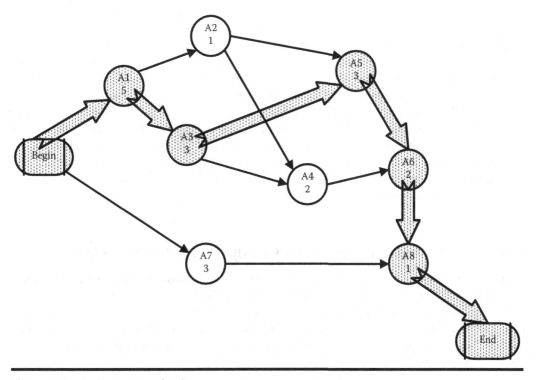

Figure 3.4 Project Network Diagram.

A1. Building walls and roof	5
A2. Mounting of doors	1
A3. Mounting of windows	3
A4. Installing electricity	2
A5. Installing sanitary	3
A6. Painting of walls	2
A7. Fencing the garden	3
A8. Moving to new home	1

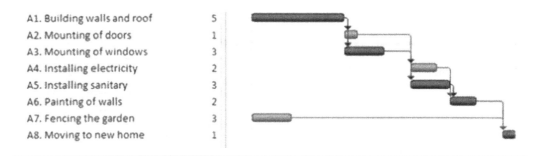

Figure 3.5 Gantt Diagram.

Other possible paths from start to finish include, for example:

- Begin -> A1 -> A2 -> A5 -> A6 -> A8 -> End
  The length of this path (sum of implementation times) is 11 months.
- Begin -> A7 -> A8 -> End
  The length: 8 months.

As other paths have shorter lead times, the lead time for this project is 14 months. This path, and therefore the entire project, cannot be completed faster.

The longest path from the beginning to the end of the project, according to the action dependency logic, is called the **critical path** (Kelley and Walker, 1959).

If any activities not on the critical path are delayed (within certain limits), the project implementation time will not be extended. For example, if the activity A7 (fencing the garden) started in the fourth month (not immediately after the start of the project) and lasted four months (not three), it would take 3 + 4 + 1 = 8 months to follow this path from Begin to End. So, it would not take longer than the house-building work lying on the critical path. On the other hand, increasing the implementation time of activities on the critical path ever increases the implementation time of the entire project. Suppose A3 mounting of windows is delayed by two months and takes five months. Then, to pass through all the critical path, 5 + 5 + 3 + 2 + 1 = 16 months are needed.

Any delay in activities on the critical path causes a delay in implementing the entire project. Therefore, they must be under the special control of the project management team.

The schedule is usually built by specialized software tools and presented in a horizontal bar chart, called the Gantt diagram (Figure 3.5).[2] The critical path is usually marked in red but here, for editorial reasons, is shown in dark gray.

3.4 What Would Be the Project Budget? Cost Management

How much will it cost to build a house? IT system implementation? Designing and organizing the production of a new model of a combat aircraft? This question is asked by everyone who wants to propose implementing a new project and those who plan to implement it. The accuracy of a project cost estimate varies over the life of a project. At first, we know almost nothing about the cost. We have complete knowledge about the cost only after the end of the project when we finally settle it (Figure 3.6).

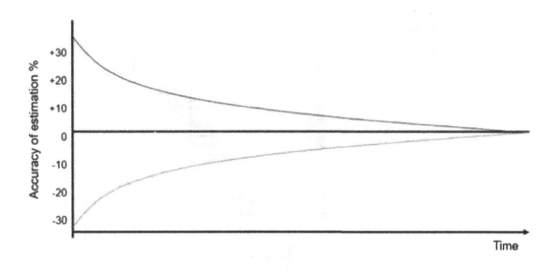

Figure 3.6   The Precision of Project Cost Estimation.

Cost estimates with varying degrees of precision are needed at different phases of the project life cycle.

When considering starting a project, its details are unknown – including the most important artifact, i.e., WBS, which is usually not developed until project planning. Information on previous similar projects may be available. Consider implementing the IT system in organization A costed US$ 2 million, in organization B US$ 2.5 million, and in organization C US$ 2.2 million. So, with a large margin of tolerance, we can assume that implementing a similar system in the next organization will cost around US$ 2.2 million (average of these costs). The construction costs of the four airports were US$ 7.3 billion, US$ 8 billion, US$ 5.4 billion, and US$ 7 billion. To estimate the construction cost of a similar airport, we can assume – again with a huge margin of accuracy – the value of approximately US$ 6.9 billion. Similarly, in the initial phases of the project life cycle, we can estimate the cost of film production or organizational restructuring. This type of cost estimation is called **estimating by analogy**.

Over time, more and more is known about the project. That the stadium under construction is to have 40,000 seats. Or that the motorway section under construction is to be 30 km long. There is often available knowledge about the unit cost of products from previously completed projects. For example, we know from the Internet that the average cost of building one seat in a stadium is approximately US$ 20,000. So, the cost of building a planned stadium with 40,000 seats will be around US$ 800 million. The cost of making 1 km of the highway is approximately US$ 3.2 million, so the cost of our planned 30 km highway should be around US$ 96 million. Estimating project costs depending on product parameters is called **parametric estimation**.

To further increase the precision of cost estimation, you need to have even more parameters. Work Breakdown Structure is used for this, but the artifact is usually carefully developed after the project is initiated – unless the financing entity wishes to have complete WBS in the process of considering project initiation. Then it is possible to estimate the cost of each elementary WBS component and then sum them up that structure (Figure 3.7, note the direction of the arrows). This cost estimation method is the most precise; it is called **bottom-up estimation**.

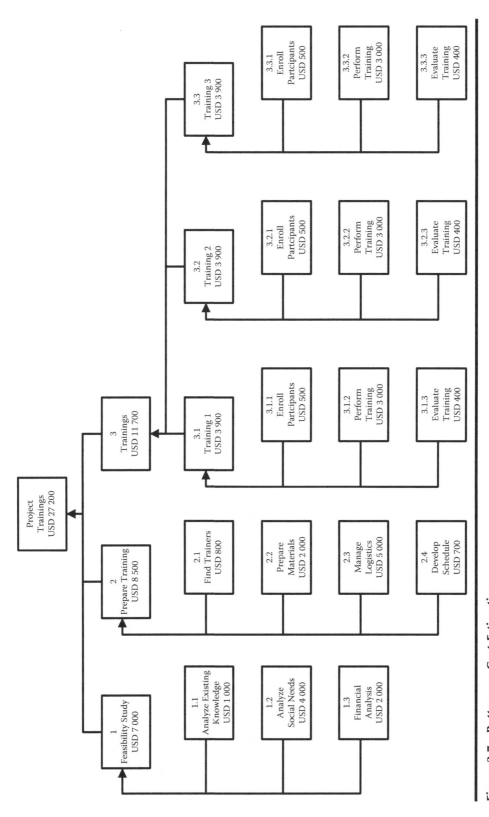

Figure 3.7 Bottom-up Cost Estimation.

3.5 Project Information: Earned Value Management, Communications Management

The most important information about a project is evaluating its performance so far and estimating its most important parameters at the end. The three most important project parameters based on which this answer can be given are: the set of developed products, the implementation schedule, and the budget. So, the answer to the question about the proper implementation of the project should combine these three values. Understanding this led to developing an Earned Value Analysis (EVA) technique (USA DoD, 1967, 1996; Fleming and Koppelman, 2016; PMI, 2019). When describing these fundamental concepts for projects, I must present a more elaborated example.

3.5.1 Example: Laying a New Surface on a Public Road

The objective of the public project was to replace the asphalt on two distant roads, each 40 km long. The company that performs the work has two teams that cannot be moved between these roads due to the distance. It was planned that each team put 1 km of new surface on one day. Replacing 1 km according to the plan costs PLN 300,000. So, the project should end after 40 days (eight weeks) of work and should cost PLN 24,000,000.

After four weeks, in the middle of the planned schedule, the contractor assessed the progress of work: its costs and schedule.

The first team at that time laid a new layer of asphalt on 22 km of road. The cost of these works was PLN 7,000,000.

The second team at that time replaced the surface at 18 km, and the cost was PLN 5,000,000.

Are the works being carried out on schedule and budget? How should the time and cost of completing the project be estimated?

The fundamental values used in EVA are:

> PV – **Planned Value**, the planned cost of product delivery.
> EV – **Earned Value**, cost of delivered products, measured according to the planned budget.
> AC – **Actual Cost**, actually incurred costs.

These values are determined on the day on which the measurements are made. The relationships between these values (usually for educational purposes shown for projects that delay and exceed the cost) are shown in Figure 3.8. This curve also shows that costs at the beginning and end of a project tend to grow more slowly than in its middle parts.

The **unit cost** value can also be used:

The unit cost is the cost of repairing 1 km of the surface:

> UC = PLN 300,000

For the first team, these values were:

> PV_1 = 20 km × PLN 300,000 = PLN 6,000,000 (the team should deliver a product of this value)
> EV_1 = 22 km × PLN 300,000 = PLN 6,600,000 (this is what 22 km of the new surface was supposed to cost)
> AC_1 = PLN 7,000,000 (such costs were borne by the contractor)

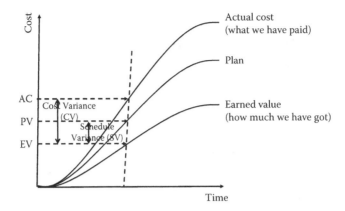

Figure 3.8   The S-Curve. Dependencies between Schedule, Costs, and Products.

And the results of the second team are:

PV_2 = 20 km × PLN 300,000 = PLN 6,000,000
EV_2 = 18 km × PLN 300,000 = PLN 5,400,000
AC_2 = PLN 5,000,000

The Cost Variance formula determines the cost situation:

CV = EV – AC, so much more (less) was spent against the plan.

The formula of the Schedule Variance determines the situation about the schedule:

SV = EV – PV, so much more (less) obtained; a measure of the scheduling situation expressed in money.

For the first group, these values are:

CV_1 = PLN 6,600,000 – PLN 7,000,000 = –PLN 400,000
SV_1 = PLN 6,600,000 – PLN 6,000,000 = PLN 600,000

The corresponding values for the second team are:

CV_2 = PLN 5,400,000 – PLN 5,000,000 = PLN 600,000
SV_2 = PLN 5,400,000 – PLN 6,000,000 = –PLN 400,000

Positive values mean a good situation, and negative values – bad ones in terms of the analyzed parameter (schedule, time).

The first team is ahead of schedule by PLN 600,000 but exceeds the cost by PLN 400,000. The second team is the other way around: it is late by PLN 400,000 but is cheaper than planned by PLN 600,000.

Note that the EVA method's budget and schedule situations are described as monetary units. Schedule delay/acceleration is expressed in the cost of the project's planned products not delivered/excessively delivered. This is very useful for downstream EVA applications.

The progress assessment is used to estimate the time and cost needed to complete the rest of the work and thus to estimate the value of these parameters at the end of the project.

How can we estimate the required cost and completion date for each of these two sub-projects?

Let us introduce two indicators that determine not the variance but the effectiveness of the work carried out in the project.

The **Cost Performance Index (CPI)** tells you how effectively the project spent money:

$$CPI = EV/AC$$

The **Schedule Performance Index (SPI)** shows how efficiently a project is using time:

$$SPI = EV/PV$$

These values for our sub-projects are as follows:

> CPI_1 = PLN 6,600,000/PLN 7,000,000 ~ = 0.94 = 94%
> For each zloty, 94% of what was expected was obtained
> SPI_1 = PLN 6,600,000/PLN 6,000,000 = 1.1 = 110%
> 110% of the expected products were delivered in a unit of time
> CPI_2 = PLN 5,400,000/PLN 5,000,000 = 1.08 = 108%
> For each zloty, 108% of what was expected was obtained.
> SPI_2 = PLN 5,400,000/PLN 6,000,000 = 0.9 = 90%
> Only 90% of the expected results were obtained in a unit of time

Where the variance was less than zero, the performance indices were less than 1 (or than 100%), and where it was positive – they were greater than 1 (or than 100%).

Now we can answer the question: when will these projects end, and how much will they cost? There are three possible options for estimating further work progress:

1. Disruptions that have resulted in the current deviations from the budget and schedule will persist.
2. Disruptions in the remainder of the work will not repeat, and the project will continue as planned.
3. In the remaining part of the works, disturbances will occur other than before.

Two variables define the project parameters for the remaining works:

> ETC – Estimate To Completion, estimated budget needed to complete the work
> ETC = (Budget – EV)/CPI
> TTC – Time To Completion, estimated time needed to complete the work.
> TTC = (Budget – EV)/UC/SPI

On the other hand, the overall values are determined by the formulas:

> EAC – Estimate At Completion, estimated cost of the entire project
> EAC = AC + ETC (total cost will be equal to the sum of expenses incurred and needed to complete)

TAC – Time At Completion, estimated completion date
TAC = Today + TTC (date when the observation was made increased by the time it takes to complete)

■ *Estimating: Option 1*

In the first sub-project, after looking at the work to be done, it was found that in the second part of the work, the situation will be the same: the schedule's efficiency will remain at 110%, and the cost-effectiveness of 94%. Eighteen kilometers of new asphalt remained to be laid.

$ETC_1 = (40 \times PLN\ 300,000 - 22 \times PLN\ 300,000)/94\% = PLN\ 5,727,273$
$EAC_1 = PLN\ 7,000,000 + PLN\ 5,727,273 = PLN\ 12,727,273$
$TTC_1 = [(40 \times PLN\ 300,000 - 22 \times PLN\ 300,000)/PLN\ 300,000]/110\%$
$\qquad = 16.36\ days$
$TAC_1 = 20\ days + 16.36\ days = {\sim}36.36\ days$

So, the company may lose PLN 727,274 on the execution of works on this section, but it will end almost four days earlier, and its resources will be ready to perform other works earlier.

■ *Estimating: Option 2*

In the second section, it was assessed that there would be no disturbances that resulted in deviations from the plan in the first part of the work in the remainder of the works. So, the rest of the project will be carried out as planned. In this situation, the cost should be increased by the budget planned for the second part (PLN 300,000/km), and the time will also be increased by the scheduled number of days. In this situation, the formulas should be substituted with CPI = 1 and SPI = 1.

$ETC = Budget - EV$
$TTC = (Budget - EV)/UC$

And the total values of EAC and TAC are described by the same formulas as in the previous case.
So, in the second sub-project we get the values:

$ETC_2 = PLN\ 12,000,000 - PLN\ 5,400,000 = PLN\ 6,600,000$
EAC_2 = PLN 5,000,000 + PLN 6,600,000 = PLN 11,600,000
$TTC_2 = (PLN\ 12,000,000 - PLN\ 5,400,000)/PLN\ 300,000 = PLN\ 6,600,000/PLN$
$300,000 = 22\ days$
TAC_2 = 20 days + 22 days = 42 days

This section will be completed by PLN 400,000 cheaper, but the work will take two days longer.

■ *Estimating: Option 3*

The third variant of estimating the cost and time needed to complete the project is when we anticipate that the rest of the work will experience disturbances other than the one performed. Then ETC and TTC should be re-estimated to consider the new situation.

The Earned Value Analysis shows that verifying only the cost or only the schedule is not enough to assess the project's status. These parameters should be evaluated jointly, and, on this basis, decisions should be made regarding the further implementation of the project.

3.6 What You Should Know? Knowledge Management

The EVA results exemplify the knowledge needed to implement a project. But the knowledge needed to implement a project is not only the knowledge generated during its implementation. Organizations create, acquire, store, and share the knowledge they need with project teams. Each decision and action in the project can be the basis for generating new knowledge – if they are implemented innovatively or if the encountered circumstances force the project team to develop a new method of operation. Knowledge becomes the most important asset of almost every organization, and thus also of projects. Organizations implementing projects should have repositories to store and share knowledge with project teams.

Knowledge plays an essential role already in the project initiation process. An organization that knows little about car manufacturing should not consider starting a new car model project. After deciding to implement a project, one should understand it thoroughly, that is, determine what knowledge is needed for its implementation. The result of understanding is the division between the knowledge that the organization currently has and the knowledge that it must acquire. On this basis, a Project Knowledge Management Plan should be developed. Its two main components are the description of how to mobilize (acquire) the necessary knowledge and the method of documenting the knowledge generated in the project. Frequently used ways of acquiring the necessary knowledge are training and employing people with such knowledge in an organization or a project. These methods may be supplemented by direct access to sources of knowledge: books, websites, conferences, etc. All these activities ensure that the project has the knowledge needed to carry out its work (Gasik, 2011a, 2011b).

Any project may generate new knowledge. This knowledge can be directly deposited in the project's or organization's repository during production and application. Another way of acquiring and storing knowledge in particular project is through lessons learned reviews. These reviews should be carried out systematically during the project implementation. Leaving such activities for the end of the project may not be effective – for example, due to leaving the project team by the people with the knowledge or due to forgetting.

3.7 Who Will Do It? Human Resource Management

There can be only one answer to the question posed in the headline: people. The **project manager** is in charge of the project. The project team can be divided into sub-teams – this is necessary when the number of people participating in the project exceeds approx. ten people. In larger projects, the project manager's work is supported by a specially delegated team member (often referred to as a **project assistant**) or a team (known as a **project management office**).

The project manager is responsible for everything that happens in the team, in particular for the results of the work of the people entrusted to him. The project manager is also responsible for the money and other resources entrusted to him – such as machines, equipment, and materials. The project manager plays a pivotal role in creating a project schedule and then, once approved, implementing it. The manager should anticipate problems that may arise in the project. Potential problems are called risks (Subchapter 3.8). Allowing the problem to materialize may (but does not have to) mean that the project is not adequately supervised. In managing a project, apart from "hard" skills, such as building a schedule or preparing a budget, soft skills play an essential role: leadership, motivating, conflict resolution – related to the ability to work with people in a team. Soft skills also play an important role in dealing with the external environment, both in one's own and external organizations.

Everyone (except the person or body standing highest in the hierarchy) has their superiors in every organization. This is also the case with projects, represented by the person managing them. The supervisor of the project manager is called the **project sponsor**. This name emphasizes his primary responsibilities: providing resources, in particular financial and personal, for the project. The sponsor is also responsible for the major decisions made in the project, for example, nominating the project manager, approval of the project plan, including the schedule and budget, and possible changes to this plan. The sponsor decides to complete the phase and start the implementation of the gate.

Since the sponsor is responsible for the most critical decisions in the project, he/she is accountable for the project subordinate to him.

The sponsor supervises and controls the work of the project manager. In the initial phase, these persons determine the principles of cooperation, time, and format for progress reporting to the sponsor. The **managerial tolerance** is established, i.e., the maximum level of deviation from the plan within which the project manager may act without obtaining the sponsor's consent each time. If the project is carried out for about three years, deviations from the plan at the level of two weeks can be considered acceptable. The sponsor and the project manager may agree that the managerial tolerance for the schedule for this project will be two weeks. The managerial tolerance can also apply to the budget. If the planned budget is $ 300 million, then variation decisions of $ 500,000 can be considered relatively benign to the project and left to the project manager's discretion. Failure to define the level of managerial tolerance for the project manager would cause the need to disturb the sponsor in the event of the occurrence of even the slightest deviation from the plan.

The sponsor does not work alone – he is the head of the **Steering Committee**. Apart from the sponsor, the Steering Committee should include a representative of the organization implementing the project (**supplier representative**) and the organization for which the project is being carried out (**client representative**). For example, if a project is implemented within the organization by its own IT department for the sales department, duly authorized representatives from both departments should be on the Steering Committee. The sponsor often invites other people to participate in the work of the Steering Committee as needed, such as lawyers. The project structure is pictured in Figure 3.9.

Regardless of the composition of the Steering Committee, as the sponsor is responsible for the project resources, the decisions made by the sponsor are final.

The construction of the Steering Committee, its subordination, and its powers is an element of the internal governance of the project (more on governance in Chapter 9).

During the project implementation, each participant should know well the expectations toward him concerning the implementation of each activity. There are four main types of people involved in activities:

1. R – Responsible, performs the activity directly
2. A – Accountable, approves and is accountable for the performance of the activity
3. C – Consulted, provides the information needed to implement the project
4. I – Informed, is informed about the implementation of the activity

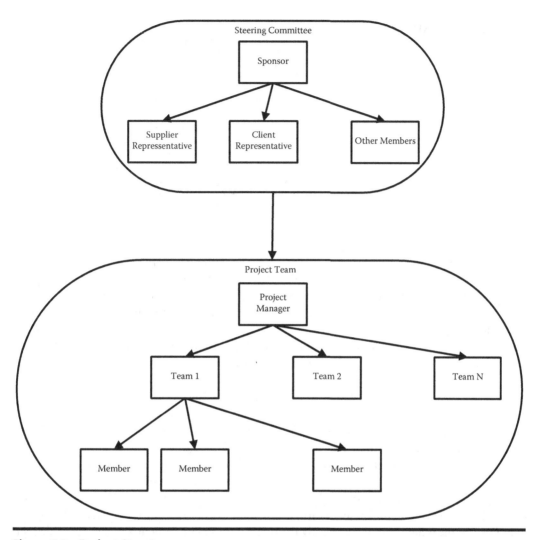

Figure 3.9   Project Structure.

Writing this information in tabular form creates a two-dimensional matrix, called the **RACI matrix** (Andersen et al., 1987; Smith and Erwin, n.d.). A sample RACI matrix describing a project of Martha's birthday party is shown in Table 3.3. Each activity must have precisely one accountable person and at least one responsible person.

3.8 What Are We Afraid of? Risk Management

Projects are very often non-routine activities. If it's the first time we do a particular job, we may not know all the circumstances surrounding it. The first wedding, trip to distant countries, or the film's first production can always bring surprises. Anyway, each subsequent job of the previously per-formed type is not identical to the previous one and may also bring some dangers. It may happen that the tour is canceled due to an epidemic, or the film's production is delayed due to the failure to

Table 3.3 RACI Matrix for a Project of Martha's Birthday Party

	Grand Mom	Mother	Father	Martha	Chris	Guests
Inviting the guests	I	A	I	R	I	I
Defining the menu	A	R	C	C	C	–
Buying products	I	C	A	I	R	–
Cooking	A	R	C	C	I	–
Eating	R	A/R	R	R	R	R
Cleaning	–	A	R	I	R	–

build the scenery at the scheduled time. Surprises can be not only negative but also positive. After planning a trip to distant countries, you may find that the exchange rate of his currency will drop. In some projects, it may turn out that we managed to recruit a more qualified person than we assumed.

Such potential surprises in projects are called **risks**. Negative are **threats,** and positive are **opportunities**. As far as it is possible and reasonable, the project management team should prepare for risks. The reasonableness of preparing for risk is the result of estimating two parameters: **probability** and **effect** (positive or negative) and the resulting **exposure** of the project to this risk:

$$Exposure = Probability \times Effect$$

One should respond to risks that have a sufficiently high exposure (it is determined differently in each project). The set of resources that we can devote to also affects whether or not to deal with a particular risk. If the result of the risk materialization would be an immediate, unplanned project interruption, it is necessary to respond to such threats. An example of such a risk may be the abandonment of the project team by key team members. A response to such a risk should be prepared regardless of the probability of its occurrence – even if we estimate that its probability is only 1%. On the other hand, if we identify the risk of a one-week delay in a project scheduled for five years, we may not deal with preventing it, regardless of the likelihood of this event.

Under the guidance of its manager, the team decides how to respond to risk. The four main types of response to threats are:

■ *Avoidance*

Causing the probability to drop to 0
Example: Early purchase of the currency for which we will purchase the things needed in the project ultimately minimizes the exchange rate risk.

■ *Mitigation*

Reducing the probability or effect of a risk
Example: Having (less qualified) substitutes for people needed in the project

■ *Transfer*

Transferring responsibility to another entity
Example: Signing a contract with a subcontractor with contractual penalties for delays covering the liabilities for delays imposed in our contract with our customer.

■ *Acceptance*

Taking no action on risk
Example: Any risk that is not prevented.

■ *Contingency plan*

A specific type of acceptance. Not minimizing the risk and preparing actions in case of its materialization
Example: Releasing software that may have bugs and maintaining a team to fix these bugs.

The reaction types for the opportunities are respectively:

■ Exploiting
■ Escalation
■ Sharing
■ Acceptance

3.9 Subcontractors' Projects: Procurement Management

Many organizations do not have the resources or skills to complete projects on their own. There is no need for an automotive company to have an extensive construction department if the need to build new facilities occurs every few years. It is similar to IT systems: their implementation takes place every few years, so in a retail company, there will be a team maintaining IT infrastructure and applications, and developing implementation skills is not necessary. Therefore, many projects are carried out under contracts by external companies. The skills of obtaining products and services from the outside are being developed, not abilities to produce them separately.

Procurement management in the public sector is pretty well analyzed and documented; legal regulations describe how public organizations make purchases in many countries. But the purchase that is carried out through the project requires special skills.

When selecting an external contractor for a project, many factors are considered in the selection criteria. The price and scope of the offered works are always important which must be consistent with the documented needs of the buyer. In addition, the selection criteria include the implementation time, the range of optional functionalities, the quality of the offered product, and many other criteria specific to a given area of operation.

Completing an order from the vendor's point of view is a project. The question that should also always be asked when choosing a project contractor is: is this company able to efficiently implement its project and meet our requirements? Often, the contracting authority tries to get an answer to this question by requesting to demonstrate the implementation of similar projects in the past, having appropriate resources or references. A specific way to ascertain the feasibility of a project is to evaluate potential contractors through a project **maturity model**. The maturity model classifies organizations according to their ability to carry out specific tasks – projects in this case. The more mature the organization, the more likely it is that it will be able to implement the project following the accepted requirements. Carnegie Mellon University developed the first widely used maturity model for the US Department of Defense order. This was the Capability Maturity Model (CMM®, Paulk et al., 1993, later Capability Maturity Model®[3] Integration, CMMI®, SEI, 2010).

More on maturity models you may find in Chapter 24.

3.10 Who Will Help Us? Who Will Disturb? Stakeholder Management

In a highway construction project, drivers may be well-disposed toward it, while residents or environmentalists may be reluctant. Some employees may have a negative attitude toward an organizational restructuring project, for example, due to a perceived threat of job loss or the need to learn and follow new procedures. In the project of organizing a sporting event, athletes, owners of sports facilities, or fans may be positive, and there is a risk that the inhabitants of the town where the event is to take place, especially in the case of large events, will be hostile – due to the fear of severe disturbances of quiet.

The people who may be influenced by the project and those who influence the project are called **stakeholders**, as in the case of permanent organizations. Stakeholders are primarily Project team members, the Steering Committee, the project implementing organization, subcontractors, and customers – but these basic stakeholder categories are dealt with in other areas of project knowledge management. Usually, when you write and think about project stakeholders, you think about people and their groups that are not formally part of the project structure.

Project stakeholders should be identified as early as possible. The most important of them should be known when deciding to initiate the project. In this phase, the question is: are there no people or groups who can put the project at risk of failure? And on the other hand, who will want to get involved in the project if the resources and abilities of his team are not enough?

The three most important features that characterize the project's stakeholders are:

1. Attitude
2. Strength
3. Influence

The stakeholder attitude to the project can be negative, indifferent, and positive. Strength speaks of position in the project delivery environment. Influence determines the level of possibility to influence the project. Strength differs from the ability to influence: not everyone who has a strong position in the environment can significantly affect a project. For example, the opposition party, even the strongest one, cannot usually stop implementing a government project. However, it may create an unfavorable atmosphere toward the project, which may hinder its implementation.

The techniques of managing project stakeholders do not differ significantly from the techniques of influencing the stakeholders of other organizations. An example of such a technique is the Stakeholder Analysis Matrix (Figure 3.10).

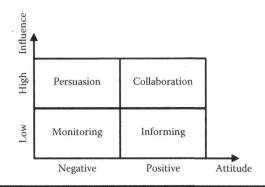

Figure 3.10   Stakeholder Analysis Matrix.

If a stakeholder has little influence, i.e., he cannot help us too much but is positive about the project, we should inform him about the course of the project. Failure to inform can create an impression of disrespect and the risk that this stakeholder will lose his positive attitude. If the stakeholder is positive and can do a lot, you should encourage cooperation.

We treat negative stakeholders differently. If they have a significant impact and can hurt us a lot, the best solution is to convince him (if he wants to contact us at all). Negative stakeholders, who do not have much influence, should be monitored to see if they are getting too strong. Sometimes individual stakeholders do not have much power, and once an influence group is created, their influence dramatically increases. We should observe and, if possible, counteract such proceedings.

3.11 How to Deliver Good Projects? Quality Management

Project products should meet stakeholders' expectations, and the project should be efficiently implemented.

The project should have defined its product requirements. The most important are those defined explicitly by its stakeholders. In the case of the construction of a public building, it can be virtually any of its characteristics: cubature, location, communication accessibility, and specific functionalities. Some of the quality characteristics are classified into categories. For example, regarding performance – a public building should have an appropriate number of rooms, a kitchen and dining room, heating, and elevators. The characteristics called **features** are close to performance which are not in the primary functional scope but make it easier to use – for example, automatically lowered curtains and automatic temperature control. The product should comply with standards – for instance, in terms of thermal insulation or earthquake resistance. Other qualities include durability, aesthetics, and affordability. You should strive to document all project requirements, but this is not always possible.

The causes of problems with project outputs and effects can be complex and multi-dimensional. One of the techniques for analyzing problems in projects is the so-called Ishikawa diagram, also called the cause-and-effect diagram or fishbone diagram (Ishikawa, 1976). This technique works by finding problems and their causes from the main to the most detailed. Problems identified are written in a fishbone-like form (Figure 3.11). They can be more complex, with multiple levels of "bones."

The quality of the project's products results from its implementation processes. This is why project management areas and processes are defined so that an efficiently implemented project

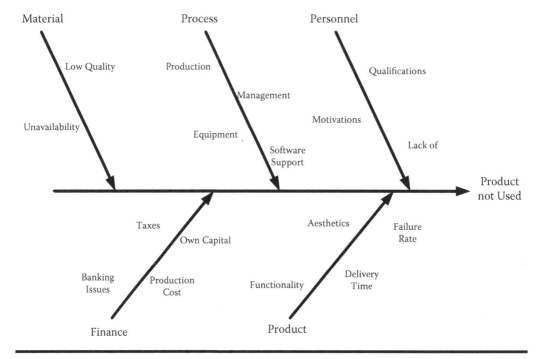

Figure 3.11 Ishikawa Diagram.

produces the expected products within a specified time and budget. Each project implementation process affects quality, and therefore is subject to quality management. Both the form and content of the processes are of interest. The structure of a process is its correct, unambiguous definition, for example, defining the functions performed in it, the conditions for their initiation, the roles involved in its implementation, the inputs and outputs created, and the criteria that these objects must meet. When we talk about content, we mean the overall suitability of processes to achieve their purpose. Examples include using the suitable materials to build a house, the qualifications specified for roles, or the correct sequence of actions in the process. Usually, they are specific to the area of project implementation.

The activities related to the quality include both defining the characteristics of products and processes and checking whether the developed products and applied processes meet the previously defined requirements and expectations.

Checking project processes and the project as a whole to make sure that the project will achieve its goals is called **quality assurance** (ISO, 2021; PMI, 2017a). The quality of the project products also depends on the project management processes – that is why they are needed – and therefore they are also subject to quality assurance. The main methods of implementing assurance are reviews and audits. These techniques involve getting acquainted with the current progress of works in the project, determining compliance with the accepted requirements and standards, and formulating recommendations for a possible repair or improvement of the project implementation. The **review** is carried out with the active participation of the project team (possibly only by them), while the **audit** is carried out by external entities outside the project team with only passive participation of team members (although this nomenclature is not always used uniformly). The ultimate goal of both types of activities is to assess the ability of the project to achieve its goals and suggest improvements, if adequate. In practice, I have not met any review or audit which have not formulated any improvements.

A concept similar to quality assurance, not always distinguished, is the **project assurance**. Usually, organizations implement similar activities under the label of quality assurance or project assurance. When it comes to project assurance, the external context of the project is also taken into account, for example, business conditions, which are often beyond the influence of the project, but knowledge about them is needed to make the most important decisions by authorized bodies. When talking about project assurance, the independence of the assurance team is more strongly stressed than in the case of quality assurance.

Checking the quality of the project's products is called **quality control**. The ultimate goal of quality control is to determine whether the product is usable by the user (if it is the project's final product) or whether it can be used in the next steps of the project or subsequent processes. All important project products and artifacts, such as requirements or design, are subject to quality control. The most important and popular technique of quality control is verification. It checks, one by one, whether the product meets the previously formulated requirements. For example, checking that the IT system performs its functions.

3.12 Plans and Changes: Integration Management Again

The arrangements for the elements of the project described in the previous sections should be analyzed and documented in the relevant plans before starting the project. Each project must have its project plan. The plan includes not only a schedule and budget but also describes how to proceed and expected results in other areas of management. For example, a human resource management plan should describe the processes related to the recruitment, development, and work termination of project team members, but also their expected qualifications. The procurement plan should include the procurement processes and the characteristics of products purchased from outside the organization.

The sub-plans of the project are as follows:

1. *Scope management plan*: What the project is supposed to deliver (and work processes)
2. *Schedule management plan*: Schedule (and schedule-related processes)
3. *Cost management plan*: Budget (and budget-related processes)
4. *Quality management plan*: How to achieve the required quality
5. *Communications management plan*: Processes of information distribution
6. *Knowledge management plan*: Acquisition of needed and storing of generated knowledge
7. *Human resource management plan*: Required staff (and staff-related processes)
8. *Risk management plan*: Risks (and risk-related processes)
9. *Procurement management plan*: Acquired products (and procurement-related processes)
10. *Stakeholder management plan*: Project stakeholders (and processes related to them)

Collectively, these plans form a **project management plan**. Usually, the project manager is responsible for developing these plans. Project planning should be tailored to its specifics. Particularly a lot of attention and time for careful planning should be devoted to complex projects at high risk (Zwikael, 2020). Plans, including the overall project management plan, are approved by the project sponsor. From the moment of approval, the project manager (and the entire team reporting to him) is obliged to implement the adopted plans (according to managerial tolerance). The sponsor-approved plan on which the project works is called the **baseline**. The project manager reports to the sponsor on the progress of the work in relation to the baseline.

It is not always possible and rational to plan a project right away for the entire project duration. If the project is to last, for example, five years, when starting it, it does not make sense to specify who will be specifically involved in carrying out its work in the last six months. Even the set of works in this period is difficult to predict. The solution to this problem is the **rolling wave** scheduling. In this approach, an overall plan is prepared for the entire duration of the project, and its details are defined immediately before the start of the next phase of the project or its calendar period (e.g., for the next three or six months).

What to do when the values of a specific parameter exceed the adopted tolerance thresholds in the reporting period? The two main types of action in this situation are:

1. Corrective action
2. Changing your baseline

Corrective action is work bringing a project back to implementing as planned. Suppose the disturbance is related to the schedule (probably the most common type of deviation). In that case, you can, for example, move employees and other resources from timely work to delayed work, implement new working methods or ask the sponsor for new resources.

If corrective actions do not bring the project back to the baseline boundaries, request a change from the baseline must be submitted to the sponsor. Project changes typically affect more than one management area. Changing the schedule may cause changes in the scope of work, required staff, budget, or any other project element. Therefore, the change should be integrated; its request should include changes in all component project areas. As a result of the approval of the change request (or part of it), a new baseline is created, according to which the project manager is obliged to implement the rest of the project.

3.13 Transition of Project Products

The development of the project's products is not the same as putting them into operation. Products must be transferred from the laboratory environment to the environment where they will ultimately function. These activities concern both the project product and the target environment. These activities are called product's **transition**.[4]

At the product side, validation is performed. This is the process of checking whether the product can be used in the environment for which it was developed. As part of the validation, it may turn out, for instance, that although the system has been developed following its requirements, it is not able to process the required data volume, and therefore it is useless. In this case, the set of requirements should be supplemented with a requirement of processing a sufficiently large data set. Enacting new laws by the legislature may be another exemplary reason for the incompatibility of project products with their target operational environment. Negative validation results usually indicate deficiencies in the definition of product requirements. A kind of validation technique is pilot transition – i.e., its transition to limited organizational space. A pilot transition usually identifies deficiencies in the product and then remedies them.

The target environment changes due to the adoption of the project's products. After the initiation of exploiting the online IT system for citizens, old on-site service procedures are no longer needed and must be replaced with new ones. New work procedures must be established when the health services legislation is changed. Part of the transition of products into the operating environment is a change in the procedures and perhaps also the organizational structure of the public organization. The transition of new solutions may result in the need to change the staffing: parting with some employees, acquiring new ones with different qualifications, or training employees in new work processes. Information campaigns for people using new services or services delivered in a new way may be necessary. For example, drivers need to learn what behavior is expected when traffic regulations change.

Notes

1. The management areas and practices presented in this subchapter are mainly described for the management of projects, but they have their equivalents for other management levels: management of portfolio or management of project programs. Therefore, the general concept of project management is used and not any of the component concepts.
2. Such a diagram was originally invented in 1896 by Karol Adamiecki, a Polish researcher working in Russia and called "harmonigram" (Marsh, 1976; Morris, 1994). Henry Gantt re-invented this concept later in 1919 (Gantt, 1919).
3. CMM®, CMMI®, and Capability Maturity Model® are registered trademarks of Carnegie Mellon University.
4. In a project management environment, they are often called implementation, but this word in this book is primarily used for public policy instruments and policies.

Chapter 4

Preparing Organization for Project Implementation

The processes from all just described areas must be jointly executed and synchronized by the integration management processes in each project. But it is too late to start thinking about projects the moment each project starts. Organizations should prepare for project implementation before its starting. First, they should know what the projects are for and what their goals are. I discuss these issues in Chapter 11 on public policies – because in public institutions they are used to achieve policies' goals. Among typical organizational activities, I will deal with two issues:

1. Linking projects and permanent organization
2. Storing, retrieving, and using knowledge

4.1 Permanent Structures and Projects

Each project is performed by a specific organization. Employees of this organization are delegated to work in the project and may have a supervisor on both the line organization and the project side. This raises the issue of the relationship between the project team members and the parent organization. The three main approaches to solving this problem are as follows:

1. Functional organization
2. Project organization
3. Matrix organization

A **functional organization** is one in which employees belonging to their functional units (e.g., bricklayers, mechanics, equipment operators) are delegated to projects. Reporting to functional managers is more important than reporting to the project manager. Then the project manager plays the role of coordinator.

A **project organization** is one in which the primary organizational units are projects. No organizational units are gathering functional specialists in a pure project organization.

DOI: 10.1201/9781003321606-5

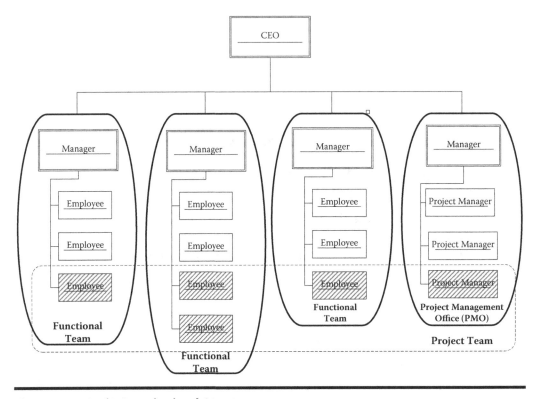

Figure 4.1 Matrix Organizational Structure.

The main disadvantage of the functional organization is the small power of the project manager and of the project organization – limited ability to retain specialists in the company between the periods of work on projects. The type of organization that tries to solve both of these problems is the matrix organization. In such an organization, functional units employ and prepare appropriate specialists for work. After working in their parent organizational units, specialists can be delegated to projects. After completing their project work, they return to these organizational units. During the project implementation, they are primarily subordinate to the project manager. The matrix organization is schematically presented in Figure 4.1.

A specific functional unit (which can be present in any type of organization) is the unit responsible for project management. Such a unit is called Project Management Offices (PMO).[1] PMO is responsible for delivering project managers and their working methods. PMO can also supervise the work of individual project managers. The primary task of PMO is to define and maintain project management methodology. Part of this work may be implementing and maintaining software applications supporting project management. PMO may also employ employees to support project implementation, such as work-related to handling project documents or communication within any project or between projects.

Another type of mixed functional and project structure is embedded project structures. These are functionally organized organizations at higher organizational levels, while projects (or project programs) are organizational units at the lowest level. For example, there are road, rail, and aviation divisions in the infrastructure ministry. In the last division, an organizational unit may be created whose task is to build a new airport. Or an organizational unit responsible for constructing a nuclear power plant is established in this ministry.

4.2 Project Knowledge in Organization

Like all human activities, projects need knowledge. This knowledge applies to both management and the area of project implementation. Project knowledge is managed at the micro and macro level (Gasik, 2011a). Micro-level management concerns the knowledge needed to solve a single problem (or part of it). Knowledge management at the macro level is the assessment of all knowledge possessed by the organization and taking actions to eliminate knowledge gaps or cause the development of knowledge where the organization can achieve an advantage over the competition.

The basis of knowledge management at the micro level in an organization is a knowledge repository. Any member of the project team who produces or acquires an item of knowledge potentially useful in other projects (or processes) should be able to report it to the repository managers for the assessment of its suitability (Subchapter 3.6). The knowledge repository is also supplied by units responsible for project implementation (like PMO) and other units involved in project implementation. Knowledge should be indexed in a way that allows easy access to it and updated. An extension of the concept of a knowledge repository is the concept of a set of organizational resources, including, for example, document templates, management support applications, procedures, etc.

Knowledge management at the macro level begins with defining the area and goal that the organization is to achieve through the proper use of knowledge in the project environment. Will it be to become a global leader in the production of vaccines against infectious diseases, or to ensure timely, cost-effective construction of local roads. Based on the identified gaps (and the existing opportunities), the organization develops a plan for acquiring such knowledge. The methods of obtaining knowledge include, for example, training, conducting own research works, conferences, acquiring and using literature, acquiring specialists from the market or the above-mentioned maintenance of a knowledge repository.

Note

1. PMOs may have other names, like project office, portfolio management office, program management office, etc.

Chapter 5

Managing Multiple Projects

5.1 Is One Project Enough? Project Programs

In one country, the structure of the railway network did not allow for the efficient transport of goods and people between different regions. Residents and administration knew that a comprehensive modernization had to be carried out. And as part of this modernization, it will be necessary to build many rail connections between individual cities, i.e., implementing many projects. Each new episode improves communication between the cities it connects. But a comprehensive improvement will only come about when the entire network of connections takes a new look.

The Apollo program. Its goal was "landing a man on the Moon and returning him safely to the Earth" (President Kennedy, 1961). To achieve this goal, it was necessary to carry out many projects. The most important components of the Apollo program were the projects of the consecutive spacecraft missions numbered from Apollo 1 to Apollo 17.

Rich crude oil and natural gas deposits have been discovered in yet another country. It was necessary to build many shafts exploiting these resources to manage them properly. It was also needed to make oil and natural gas processing plants and pipelines to transport oil and gas. Each such work was a project that brought its effects.

In each of these three cases, implementing a single project did not solve the problem. It was necessary to implement many projects to achieve the required effect. A well-thought-out design of the new rail transport system will have a better effect than if each province planned its railways. Achieving the goal by the Apollo program was equivalent to the deliberate, consistent implementation of many projects. This brings us to the concept of the program (PMI, 2017b). To distinguish between programs that mainly consist of projects and programs that implement policies (Subchapter 11.4), which are essential in the area of public administration, we will use the name "project program."

> A **project program** is a collection of projects and other activities[1] undertaken to achieve an effect that would not be possible if these projects were carried out separately.

The structure of the project program is shown schematically in Figure 5.1.

DOI: 10.1201/9781003321606-6

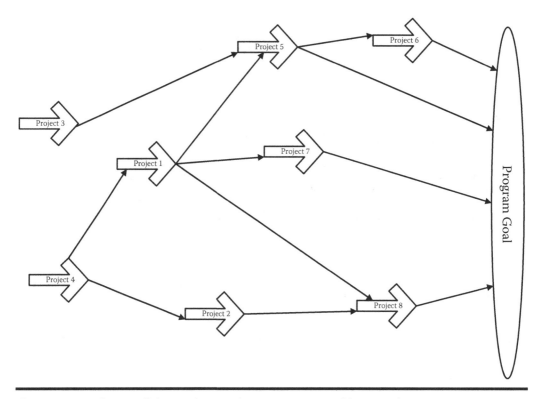

Figure 5.1 Projects Collaborate in a Project Program to Achieve Goals.

The effect should be understood as the **operational effect**, i.e., related to the interaction of component products of the projects. If built under the project program, the railway network will serve citizens better than if the construction projects were not coordinated. Therefore, when managing a project program, attention is paid to factors other than managing a single project. As the program's component projects emphasize their products and effects, the project program is oriented toward the benefits being obtained. In the railway network project construction program, due to the construction of certain sections of railways, the communication between cities X, Z, and C will improve in the first five years, and the communication between cities A, F, and K in the following years. In addition, the project program manages areas similar to the areas of project management – e.g., scope management, communication, or stakeholder management – but these areas apply to the project program as a whole. The project program may also define the management methods applicable to the component projects.

5.2 Which Projects to Implement? Project Portfolios

A set of projects implemented in a given organization (or its component unit) is called the **project portfolio** (PMI, 2017c). If an organization allocates funds for separately managed types of projects – for example, IT or development projects – we are talking about a portfolio of projects of this type: IT project portfolio or development project portfolio. Then the considerations presented next refer to such defined areas of the organization's activity. How are organizations' project portfolios constructed and maintained?

Each organization has a finite set of resources: finances, people, equipment that can be used in its work. Some of these resources may be allocated to the implementation of projects. Each organization also needs to implement many projects. Project initiatives can arise independently of each other in each organizational unit. It very often happens that their reported number and demand for resources exceed their capacity. Then a decision should be made to select a specific set of projects to be implemented. Ideas not selected for implementation may eventually be rejected or designated for implementation later.

In addition to the organization's resource pool, there are other criteria for selecting projects. The selection of a project depends on its compliance with the scope of the organization's activities and its strategic goals. If the organization is involved in the production of cars, the proposed project of implementing aircraft production will not be accepted. Suppose the company's strategic goal accepted by the shareholders is an expansion to the Asian market. In that case, the project proposal that creates the basis for expansion into South America will also be rejected.

Another important factor that can be considered a type of the previously mentioned factor is the level of achievement of strategic goals. In a commercial company, if two projects will cost US$ 5 million each, but one will generate income of US$ 5.5 million and the other at the same time US$ 5.7 million, it is reasonable to choose the latter project. In companies whose goal is not to maximize profit, the criteria for achieving strategic goals usually are much more complicated.

As a given type of work usually carried out for the first time by an organization, projects are subject to risk. Risk is another factor that should be considered when selecting projects for implementation. Suppose two projects are proposed, each costing US$ 3 million and each promising income of US$ 3.5 million. Which one should you choose then? The one whose implementation is unchallenged, i.e., carries less risk. If it is assessed that the implementation of the first project is almost certain, while the success of the second one depends on the favorable market conditions (i.e., it is burdened with high risk), then the former should be chosen.

Hence, the main factors determining the selection of projects for implementation are as follows:

1. Compliance with the organization's strategy
2. Resource demand
3. Level of proposed benefits
4. Level of risk

Apart from this classification, there are projects that the organization must perform. Suppose the new law requires IT systems to be adapted to new regulations, for example, regarding accounting or personal data protection. In that case, they are mandatory and are not subject to analysis, taking into account the above criteria. The organization must find resources to implement them. If it doesn't, it will lose its ability to operate.

The organization may also have to implement other projects. If it does not have complete autonomy, for example, because it is part of a larger group, the projects to be executed may be defined by the superior bodies. It makes sense to talk about project portfolio management if the organization has at least partial autonomy in selecting projects to be implemented. That is why only one portfolio of the entire organization is referred to in some organizations. In others where departments have operational autonomy, one can speak of managing the portfolio of individual departments.

Portfolio management does not end when projects are qualified for implementation. It may happen that as a result of an inefficient implementation (e.g., exceeding the schedule or budget, failure to deliver planned products) or a change in the implementation environment (another company launches a very competitive product on the market), the project loses its sense. Then it is possible to redefine the methods or goals of the project implementation and ultimately make a decision at the portfolio level to terminate the project implementation. If an organization has never made a decision of this type, there is a suspicion (which must be further analyzed) that it does not actually manage its project portfolio.

Apart from a set of all projects of a specific organizational unit, the concept of a project portfolio has one more meaning in practice. This is a group of projects and project program meeting some conditions. One may talk of a portfolio of projects implemented for a particular client or a portfolio of the largest projects – for example, with a budget of over US$ 10 million. A frequently used concept is a portfolio of the most important (major) projects. You can also talk about a portfolio of projects implemented in a specific year. The purpose of defining such portfolios is to focus attention or gain knowledge of specific projects and not necessarily decide on the continuation or killing of particular projects. If manager X's project portfolio is poorly managed, it is possible to change manager to person Y and not kill all these projects. If a familiar name is needed for a set of projects and does not meet the definition of a project program or family, then the word portfolio is most often used. Sometimes it is referred to as the managerial portfolio (all implemented projects) and auxiliary, analytical portfolios – any set of projects that is interesting to someone from any point of view.

5.3 How to Analyze Projects? Project Families

The company carries out a project in which it creates an IT system. Then, projects are carried out that implement this system in other organizations.

Stadium construction projects are underway. These stadiums are needed for the football championship project to be organized.

A construction company is implementing a project to implement methods of building high-speed railway lines. Then, it carries out projects to construct such railways in various countries.

There is a logical link between the projects in each of these examples. Specific projects are carried out so that another project can be implemented. Without the project of creating an IT system, this system cannot be implemented. Without building the necessary stadiums, the championship cannot be played. If high-speed rail construction is not mastered, high-speed rail construction projects cannot be implemented.

A project carried out earlier, which enables the implementation of subsequent projects, is called an **ancestor** (or **parent**). A project that can be implemented thanks to it is called a **descendant** (or **child**). The relationship that binds these projects together is called **parenthood**. And all related projects are called **project family** (Figure 5.2; Gasik, 2008a).

If the project is part of the family, it cannot be assessed in isolation from other family members. Suppose an organization decides to develop a commercial information system, then evaluate this initiative. In that case, one must consider the effects obtained thanks to the subsequent implementation of this system – that is, child projects. To assess the financial aspect of a football championship project, it is necessary to consider not only the championship itself but also all projects (e.g., stadium constructions) that made it possible to implement it. In the first of these cases, when we analyze an ancestor project that was an investment enabling the implementation of subsequent projects, we speak of an **investment family**. In the second case, when we want to evaluate the target-descendant project together with other projects that made it possible to implement it, we evaluate its final effect; we are talking about a **commercial family**.

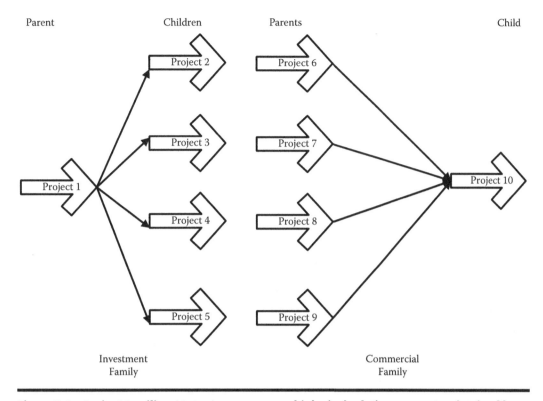

Parent   Children   Parents   Child

Investment   Commercial
Family   Family

Figure 5.2   Project Families. *Note:* **As you can see, biological relations are not maintained here.**

The links between projects are multi-level. For example, before building an IT system, it might be necessary to master a new technology or implement a procurement project that supplies the necessary equipment and recruitment projects.

The purpose of introducing the concept of project families is to enable the analysis of project effects.

Note

1. It is about management activities needed to manage the whole program, e.g., controlling the implementation of component projects.

Chapter 6

Other Approaches to Project Management

The content of the previous subchapters was based mainly on the content of the PMBOK® Guide, a document developed by the Project Management Institute (PMI), an organization for project management professionals. Number of PMBOK® Guide copies distributed (over 5,000,000), number of PMI members (612,245) and affiliates (314, in 214 countries and territories), number of PMP®[1] certificates based on PMBOK® Guide (1,036,367; PMI, 2020) makes PMBOK® Guide the most important project management standard which justifies the orientation of this chapter around PMBOK® Guide.

PMBOK® Guide is an example of project management methodology.[2] A **methodology** is a systemized set of structured practices, guidelines, processes, etc. describing ways of performing some work (PMI, 2018a; PWN, 2006). The methodologies are available in both publication and implementation versions. No publisher or project management organization has anticipated the situation you have in your institution – because each institution is unique in its own way. Therefore, if you are reading a project management methodology and you think that it is too complex, too much, or too little detailed, then know that this is the case in every organization. Your task is to choose what you really need from the publication version and transform it into running processes. Then you have the implementation version of the methodology.

It is different when an external organization (e.g., superior or donor) requires you to use some methodology. This means that just they took on the transformation of the publication version into the implementation version or otherwise built the project management methodology.

A methodology can be a **standard** if it instructs to perform activities in a specific way and a **norm** if used to assess activities' compliance with the provisions of this document.

PMBOK® Guide is not the only document that tries to systematize project management. Other widely used project management standardization documents are the Logical Framework Approach (LFA; Practical Concepts Incorporated, 1979), the ISO 21500 standard (ISO, 2012), and SCRUM. The reason for listing LFA here is that this methodology is used in World Bank projects. I mention ISO 21500 because it is a project management publication issued by the world's most important ISO standardization organization. We devote a few words to SCRUM because it is probably the most popular of the agile methodologies.

DOI: 10.1201/9781003321606-7

Table 6.1   Logical Framework – Example

	Narrative summary	Indicators	Verification	Risks/ Assumptions
Goal	Poverty reduction	Increasing life expectancy to 65 years	Statistical office calculations	The non-agricultural factors influencing poverty will not deteriorate
Objective	Increase in rice production	2,000 kg/ hectare	Agricultural agency calculation	The average weather will not change over the period observed
Outputs	Trained farmers	200 trained farmers	Final test	
Activities	Training on effective methods of rice cultivation	20 trainings	Verification of documentation by representatives of the financing organization.	Acquisition of four instructors who know the local realities and the LFA approach

6.1 Logical Framework Approach

Logical Framework Approach (LFA) is a methodology used by the World Bank for the implementation of projects, in particular, development ones. The dependence of the financing of development projects on this methodology obviously influenced its popularity.

In this methodology, the most important object is a two-dimensional array. Its lines are artifacts created during the project, and its columns are the properties of these artifacts. An example of a straightforward LFA table is presented in Table 6.1.

The purpose of the LFA table is to embed and justify the project's validity in its implementation environment. The content of each higher row is derived from the content of the lower row. Activities, i.e., the content of the project, lead to the achievement of results, leading to accomplishing the objectives and goals. Usually, a goal is achieved by realizing more than one objective; achieving the objective consists of more than one output. Sometimes you have to do many things to get the output. The set of activities described by one such table is a project.

The hierarchy of main concepts ensures the execution of logically consistent actions to achieve a specific goal.

6.2 ISO 21500 Standard

The ISO 21500 standard was created due to an attempt to standardize project management on a global scale. The main elements of this standard are management processes. Processes are grouped and classified according to subjects.

Process groups are:

- Initiating
- Planning
- Executing
- Controlling
- Closing

Subjects are:

- Integration
- Stakeholder
- Scope
- Resource
- Time
- Cost
- Risk
- Quality
- Procurement
- Communications

Process groups and subjects are analogous to PMBOK® Guide's content. Basing the ISO standard on the PMBOK® Guide processes confirms this document's suitability for managing all kinds of projects. The ISO 21500 standard has not diminished the popularity of PMBOK® Guide or Prince2® so far.

6.3 SCRUM

The agile approach is gaining more and more popularity also in public administration (PMI and NAPA, 2020). Recently, SCRUM (www.scrum.org; Schwaber and Sutherland, 2017) has been a very popular agile approach, especially in IT projects.

Before starting the project and during the implementation of works, requirements are created and placed in the Product Backlog. A team of a few developers and product owner work closely together to develop the product. Physical collocation of team members is strongly recommended. The project is divided into "sprints," which usually last no more than two weeks. The scope of the sprint's work is determined and documented immediately before its implementation by selecting the appropriate number of requirements from the Product Backlog. At the beginning of the sprint, work is allocated to the individual team members as part of the planning. Every day, during the sprint, team meetings of no longer than 15 minutes are held to evaluate the work from the previous day and the scope of work for a given day. The sprint is never extended (such an approach to scheduling is called *time-boxing*). This means that if the workload of implementation is wrongly estimated, some of the requirements for a given sprint may not be delivered and may be returned to the Product Backlog. During the sprint implementation, the developers closely cooperate with the product owner, allowing them to resolve issues on an ongoing basis. Some sprints (called a "release") result in a usable product being released. Releases or their groups generally correspond to the concept of phase in traditionally managed projects.

Notes

1. PMP® is a registered trademark of the Project Management Institute.
2. PMI avoids naming the PMBOK® Guide a "methodology," but this is not in line with the definition of "methodology" accepted in this book.

PUBLIC ADMINISTRATION FOR PROJECT MANAGERS

After presenting the basic project management concepts, I will move on to the description of concepts related to Public Administration and its operating (further we will learn that it is called "public policy," Chapter 11). In addition to explaining these concepts, I will show how project management can be useful for Public Administration.

Thanks to this knowledge, project management specialists will have a common language with the people for whom these projects are implemented: Public Administration employees.

DOI: 10.1201/9781003321606-8

Chapter 7

State, Its Structure, and Development Phases

The highest hierarchically object related to public administration is state. But to be able to say what a state is, the concept of country must first be defined. Everyone can indicate which country they live in, what are the neighboring countries and probably many other countries. The countries are, for instance, Poland, the United States, and Senegal. Each country has three basic characteristics. First, it has a defined territory (down to the area of contention). The country also has its citizens – that's why, for example, Antarctica, although it has a well-defined territory, is not a country. Finally, the country is sovereign, that is, it has no supreme authority over it. Usually, whether a certain territory is a country as we speak here depends on its recognition by other countries. There are several territories in the world that I will not mention here, so as not to expose myself to criticism of their inhabitants, who aspire to be an independent country, but are not recognized by anyone.

According to the CIA World Factbook (CIA, 2020), there are 195 countries in the world, the largest in terms of population is China (about 1.4 billion people), and the smallest is Vatican (about 1,000 people). In terms of territory, Russia is the largest (over 17 million km^2) and the Vatican is again the smallest one (0.44 km^2).

The concept of the country is related to the concept of the state. According to the definition of Weber (1947), the state is an entity with potentially supreme authority over all activities carried out in the country. The scope of activities on which the state actually exercises more or less direct influence depends on the state's decision. The state is the country's infrastructure of operating. The state functions to ensure that citizens can live and organizations (including enterprises) can achieve their goals. Just like the transport infrastructure serves to ensure that people can move between different places on previously constructed roads, leading along a specific route. Drivers live their lives mostly off the road. Citizens also have their own areas of operation that do not relate to the state. On the roads, the order is ensured by the police, and in the state by the legal system. The scope of the state's functioning, like the network of communication infrastructure, may have a different scope and quality. How many roads are needed and how many state institutions should exist depend on the decisions shaping the state or the communication system. Roads may be wanted by some and undesirable for others, for example, due to noise or the

DOI: 10.1201/9781003321606-9

necessity to cross highways in remote places. State mechanisms can be a major obstacle to the functioning of society – for example, in undemocratic, autocratic, or tyrannical states. Sometimes roads are built to make life easier for individuals or groups of people. Likewise, it can be with a state – it can better serve these or other groups in society.

The main task of the state is to maintain its sovereignty over its own territory (Bäck and Hadenius, 2008). To perform this and other tasks, the state has a monopoly on two functions: the use of force and the compulsory collection of taxes. State power can be used on its own territory to enforce certain behaviors of its citizens and also outside it to wage wars and participate in other military activities.

States may be associated in **supranational organizations**. The best known and developed organization is the European Union. Under the Treaty on European Union and its subsequent modifications, EU member states transferred some of their powers – for example, those relating to travel or capital transfer – to the EU. States may also belong to specialized organizations, such as the International Telecommunication Union which defines rules for the management of radio frequencies common to the Member States, thanks to which, for example, in border areas it is possible to receive radio broadcasts without interference. Membership in such organizations does not infringe the sovereignty of the member states, as both accession and possible withdrawal (as recently happened with the United Kingdom from the European Union) are based on the sovereign decisions of independent states.

States can be unitary and federal. They can also have associated or subordinate territories. A **unitary state** is the one in which all important decisions are reserved to the central government. The existing constituent administrative units do not have the possibility to legislate or establish the structures governing them. Consequently, all these units operate the same way. If there are differences between the constituent units, they must be approved by the central government. The central authority may define the scope and rules of functioning of the constituent units to the level of detail it deems appropriate. Examples of unitary states are France, Uruguay, and Japan.

The **federal state** is composed of more or less independent subordinate units. We will call them "states" – this is probably the most common name (e.g., USA, Brazil, Australia, India), although in different countries they are called differently, for example, provinces in Argentina or Canada, lands in Germany or cantons in Switzerland. It happens those different components of a federal state have different names, for example, in Russia the components are oblasts, republics, and krai (country).

The component states have varying degrees of autonomy as defined by constitution. Usually, foreign policy, defense, and the monetary policy are reserved to the central (federal) level. Responsibility for other sectors can be differently distributed. For example, transportation, natural resources, and health issues can be spread across central and component state levels. Education, culture, and welfare are often the responsibilities of component states. The states of the federally structured countries can autonomously function within the limits of their powers and therefore are the subject of our considerations equal to entire countries. The general scheme of lower-tier government is the same as that of central government: legislature, executive, and judiciary. The relationship between central and local government is beyond the scope of our book. What is important to us are the structures and processes of project implementation, not their source: local or imposed by central law.

There are two levels of government in federal states: federal and of component states. Each of them has specific areas of operating and ways of linking between levels of government. Perhaps the most common type of relations is the financial one: the central governments finance areas of component governments, even if they are reserved to the responsibility of the lower components.

In this way, federal governments subsidize, for example, the development of infrastructure, education, or health care for which state governments are responsible. But central governments do not usually dictate how these processes state governments must be implemented – doing so would be contrary to the principles of federalism and autonomy. One of such areas of operation is the implementation of projects. Usually, state governments have autonomy in defining (or not) them. Therefore, e.g., in the United States or Canada, we can talk separately about the processes and systems for implementing federal government projects (relating to projects in areas reserved for projects at this level of government) and state government projects.

The word "state" has two main meanings: the country governing apparatus and a component of a larger country; its meaning depends on the context. Since this book deals with administration, we will use the word "state" more often in the former sense.

Federal states are usually formed in one of two ways: bottom-up or top-down. We deal with the formation of a bottom-up federation when independent states recognize that they will form a common state organism. Perhaps the best-known example of such a creation of a federal state is the USA which was created as a result of the creation of the union by thirteen British colonies and then other states joined them. We deal with top-down creation of a federal state when, due to difficulties in the central management of a country, units of a lower level are separated from it. In this way, states in Brazil were distinguished, where they were initially separated on the basis of the decision of the king of Portugal, conferring power over certain territories to certain people. In bottom-up federal states, the level of autonomy is usually greater than in top-down states.

States go through three major phases of development (Rolland and Roness, 2009; Rose, 1976; Premfors, 1999):

1. Protection of existence
2. Infrastructure development
3. Ensuring the well-being of citizens

In the first phase, the primary task is to ensure the existence of the state. Its territorial scope is defined, often through conflicts with neighboring countries or with the one from which the given country separates. The ways of exercising power and the institutions through which the state will operate are established. In particular, an administrative system, army, and internal security forces are being formed.

After consolidation, the state moves to the phase of infrastructure development. In this phase, the main tasks of the state are the development of the infrastructure of all sectors necessary for the functioning, for example, housing, transport, energy, food, and other things necessary for the life of its citizens, mining, basic tasks of the health service and education.

When all these areas are working reasonably well, states begin to focus on well-being of citizens. The state tries to involve as many citizens as possible in using the generated goods. Mechanisms of economic protection of citizens are created. The processes of providing state services to citizens are being improved. The task of the health service is not only to intervene in emergency situations (diseases, accidents) but also to ensure a long healthy life for citizens. The state develops or stimulates the development of cultural, entertainment, and sport organizations.

Each phase of state development has specific types of implemented public policies and projects (Section 11.11.4).

Chapter 8

Government and Public Administration

8.1 Organizations and Institutions

To perform any task for which a group of more than one person is needed (and certainly when hundreds or thousands of people are required), you need to define the purpose of the work, define the methods of work and the relationships between these people (or subgroups, if it is necessary), and make the group work together. Such a group of people is called an **organization** – a group of people working together to pursue a goal (Rainey, 2014; George and Jones, 2012; Daft, 2010, and many others). The organization is a store where four people work (including the store's owner-manager), and the state mentioned above, whose tasks can be performed by hundreds of thousands or even millions of people (almost two million civilian employees work for the American government; USA OPM, 2021). Organizations often consist of smaller units that can be arranged hierarchically, also meeting the definition of an organization.

There are organizations whose primary goal is to implement projects – for example, construction and IT companies. Construction companies perform building construction projects, and IT companies – projects for developing and implementing IT systems. This type includes public organizations responsible for projects implementing specific public policies, for example, the Ministry of Transport accountable for creating transport infrastructure. A large part of the projects is needed by organizations dealing with natural resources – building or licensing mines is one of their main activities. It is similar to defense ministries that implement new weapons types not to lose the arms race. Institutions responsible for culture implement projects to build theaters or museums or organize cultural events. Organizations in which projects play are the main tool for achieving their goals are called **project-based organizations** (Ekstedt et al., 1999; Soderholm, 2008; Hobday, 2000).

There are also (public) organizations that, at first glance, are not interested in projects – for example, institutions, including ministries of education or justice. Schools continuously teach children and adolescents. Due to strictly defined criminal procedures, courts process cases that are more of routine work than projects. It is similar to the health service or telecommunications. The health service performs treatments and procedures related to ensuring patients'

DOI: 10.1201/9781003321606-10

health. The task of the telecommunications sector (and the relevant ministry or supervisory office) is to provide continuous, reliable telecommunications between citizens and organizations operating in the country. This is a typical operating, routine activity. Organizations whose statutory goals are achieved by performing routine operations are called **operation-based organizations**.

Consider the role of projects in operation-based organizations. The telecommunications sector is undergoing development. Several decades ago, the cellular communication system was introduced. These were massive works related to the allocation of frequencies to telecommunications companies. The allocation was made based on conducted tenders. Each such tender was a project: it was a non-routine activity with a well-defined result: managed frequencies. The fifth generation of mobile communications is currently being rolled out in many countries. The implementation of these works is also a project. The health service had to organize a way to fight COVID-19. This work is a project and has resulted in vaccination processes and other virus-fighting methods. Courts are implementing a new IT system. Such action applies to hardware and software and usually reorganizes the entire work. These are time-limited works with a well-defined effect: new court work procedures. Project. Another public organization finds that its structure hampers the effective achievement of its goals and therefore decentralizes it, granting more powers to local branches. Finite-time activities have an effect – a new way of operating the organization is also a project. Every mode of operation, every process, if it is working, means that it was once implemented. From the point of view of project implementation, such organizations are called "project-dependent" because the implementation of their activities depends on the efficient implementation of projects that shape their methods of operation. Most if not all operation-based organizations are project-dependent. Of course, a project-based organization also depends on projects making changes and modifying their processes.

Public organizations, like organizations from other sectors, implement projects to:

1. Directly achieve their goals
2. Make changes necessary to achieve new goals
3. Modify and improve their activities

Hence, they are divided into two types:

1. **Project-based** in which the primary goals are achieved through the implementation of projects (point 1).
2. **Project-dependent** in which the achievement of goals depends on the implementation of projects (points 2 and 3).

Every organization, also public, belongs to one of these groups. There are no public organizations that would not need projects – to achieve their statutory goals, to develop or improve operations.

In everyday language, the concept of institution is often equated with that of an organization constituting a state (Hill and Hupe, 2014). The government itself is an institution. The institution is the court. But organizations exist to do activities. Consistency is essential for the activities of state-owned organizations. An institution is some pattern of behavior in the functioning of society (Ostrom, 2008), and this concept is also applied to processes performed by the state (and by citizens like money borrowing, but this is of less importance in this book). An institution is an election process, a lawsuit, a census, power structures and processes, or the process of getting married. Understanding the concept of an institution depends on the context; it may refer to both the organizations and the (institutionalized) processes. Still, in this book, most often, we use this concept to denote organizations.

Due to their one-off nature, most public projects are not institutions. Building a specific motorway or restructuring an organization is not institution. On the other hand, some projects may be treated as institutions – perhaps the best-known example of an institution-project is elections. On the other hand, an institution, a particular implemented pattern of behavior, may be the process of implementing a public project, from submitting an initiative to evaluating its effects.

8.2 State and Public Administration

The state comprises more or less dependent organizations, each of which has a specific goal and implements particular activities. The most important organization in the state is the government which has supreme power over the country. The government enacts laws and is responsible for implementing and enforcing them. The government solves or stimulates the solution of the most important problems facing the state's territorial integrity state and its citizens (Bauhr and Nasiritousi, 2012). According to Lascoumes and Le Gales (2007), the government deals with governing, i.e., making decisions, resolving conflicts, creating public goods, coordinating the behavior of non-state actors, regulating markets, sourcing, and allocating resources. According to yet another definition, the only thing a government does is implementing public policies (Dye, 2013; Colebatch, 1998).

The government has three main components to accomplish these tasks: the legislature, the executive branch, and the judiciary. The legislature establishes laws, i.e., the most important rules for the functioning of the state and society. And among the laws, the most important is the constitution,[1] which, among other things, defines the most important rules governing the functioning of the country and tells how the law is to be enacted. The task of the executive branch is to implement and enforce the laws established by the legislature and ensure the functioning of the state. The judiciary's mission is to assess the compliance of the behavior of entities operating in the state with the applicable law and settle disputes that may arise between these entities.

For the subject of our interest, the legislature and the executive branch are most important. Legislation is essential because it enacts laws relating to the functioning of the state and the implementation of public policies by governments. Laws and regulations may apply to the implementation of projects. But I will not detail the legislative process (although it is also project-based) because only its effect is important to us. The judiciary may draw conclusions about the shape of the law, but it is a marginal part of its activity in quantitative terms, and I will not deal with it.

The executive branch is the most developed and most important for our analysis part of the government. It coordinates and integrates sources of information and advice and uses well-thought-out processes to ensure the coherence of policy issues and priorities (Tiernan, 2012). The cabinet heads the executive branch – a team of essential officials – ministers – responsible for individual ministries. It performs its tasks within the framework of the powers resulting from the binding constitution and laws specified by the legislature. These tasks are performed directly by a government-subordinated set of organizations having a hierarchical structure. For these constituent elements of the state, i.e., sub-organizations, names such as department, organizational unit, agency, ministry, or other, specific for the area of the organization's operation, are used. For example, in the army, we have regiments, battalions, or platoons, just to name some of the names used.

The collection of these organizations is called the **government administration**. A more extensive definition (UNDP, 2003) says that government administration is the organizational structures, personnel, and processes involved in implementing policies, laws, and other rules established by governments. The government administration is responsible for the government's interactions with other public stakeholders.

Apart from the central government administration, there is also a local government administration in the state, the activities of which are subject to the laws established by the legislature. The level of operational subordination of local government administration to government administration results from the regulations adopted in a given country. The law may define requirements for implementing projects in local government units.

The central government administration and local administration together constitute the **public administration**.

For example, in the Australian state of New South Wales, municipalities are governed by the Local Government Act (NSW Parliament, 1993). Under this law, councils of communities have the authorization and duty to provide services and facilities for the community on behalf of themselves and higher government levels. Many activity areas are left to council responsibility, such as health, culture, public transport, industry development and support, and tourism. They must also develop a strategic plan for their area, a delivery plan (e.g., listing projects planned to be implemented in a given council term), and an operational plan. Within the state government, there is a Department of Local Government that oversees the functioning of local governments. Local councils must notify, report, or obtain approval for some of their activities, including projects, from the Department's Director General.

8.3 Types of Governmental Organizations

There are several main types of government organizations. I describe them in the following sections.

8.3.1 Central Institutions

The most important type of public organization is the ministry (known in many countries as departments). The ministry is the state's highest-level organization that deals with a specific area of the state's functioning, headed by a member of the cabinet (minister). For example, the Ministry of Agriculture may deal with issues related to the production and processing of food, and the Ministry of Defense can deal with the state's territorial integrity using military means. From our point of view, an interesting ministry dealing with the most important programs for the country is the Indian Ministry of Statistics and Programme Implementation (India MoSPI, 2021). Central governmental organizations may have their local representations established to perform tasks more efficiently in a specific territorial unit. Field representations may cover all government tasks in a given area (e.g., in Sweden, counties are headed by governors subordinated to the central government) or a part of these tasks – for example, education or employment. Government field representations should not be confused with local governments that usually have complete responsibility for some administration parts.

8.3.2 Government Agencies

The government (each branch) may also establish specialized agencies that perform strictly defined functions. They may be subordinated directly to the cabinet, parliament, or higher-level government institutions. One of the world's most famous US government agencies, the Central Intelligence Agency, is tasked with providing the intelligence needed to keep the United States functioning. The task of the National Health Fund (Poland NFZ, 2021) in Poland is to finance health services and medicines. In Botswana, the Citizen Entrepreneurial Development Agency (Botswana CEDA, 2020) is responsible for supporting the development of entrepreneurship of citizens. Project agencies have been set up in some countries. In the United Kingdom, this is done by the Infrastructure and Project Authority (UK IPA, 2021). In Uzbekistan, there is the National Agency for Project Management (Uzbekistan NAPM, 2021). Government agencies can also support private projects – for example, in Australia, they are facilitated by the Major Projects Facilitation Agency (Australian Government, 2021).

8.3.3 Governmental Enterprises

But ministries and agencies are not sufficient to keep the state running smoothly. For territorial and functional reasons, enterprises are created that meet the goals set for governments. They can be established by the government or private entities and perform tasks assigned to the government. In this situation, we are dealing with the so-called QUANGO (quasi non-governmental organization; Wettenhall, 2003). These companies may be subordinated to agencies or ministries. Examples of state-owned enterprises include Gail India Limited, which produces and distributes natural gas, British Business Bank, financing small and medium-sized enterprises, and the Bavarian Hofbräuhaus Arizona Brewery. An enterprise in any market sector may be a state enterprise because governments are market actors. The pool of enterprises owned by a government is primarily determined by the operating philosophy of a given government. The government may wholly own state-owned enterprises, or they may be public companies, where the government has a direct or indirect interest. The classification of organizations related to public administration is not unambiguous. For an overview of these classifications, see Wettenhall (ibid).

State-owned enterprises and government agencies are sometimes collectively referred to as parastatals.

8.3.4 Project Implementing Agencies

Governments set up special organizations to deliver large-scale projects or project programs. They are more or less independent or under government influence, often partly privately owned. Two facts make this book's focus: the establishment by government decision and the purpose of the activity – implementing a project or project program. Such organizations may be called agencies or companies.

For example, in Australia, Snowy Hydro (n.d.) was created to manage hydropower development in the Snowy Mountains area. In India, National High-Speed Rail Corporation Ltd (India NHSRCL, n.d.), jointly owned by the federal and state governments, was established to build high-speed railways. The British Government established HS2 Limited to build a high-speed railway from London to Birmingham (www.hs2.org.uk). Together with the governments of Berlin and Brandenburg, the German federal government established an agency building the Berlin Brandenburg Airport (Wikipedia, 2022). Specialized governmental organizations are also created to organize major sporting events, such as the Olympic Games or world championships in popular sports.

8.3.5 Committees

Governments can also set up committees: permanent (i.e., operating throughout the government's term, usually dealing with legislation in a specific area) or temporary (ad hoc), set up to deal with a specific topic. There are several dozen standing committees in the United States Congress (https://www.congress.gov/committees), for example, the US Senate Committee on the Budget or the House Transportation and Infrastructure Committee. Committees can also deal with other issues, such as ethics in relation to members of parliament (Committee on Ethics). In India, there is a committee for the MP Local Area Development Scheme, which deals with projects carried out at the request of individual members of parliament (India MoSPI, 2014).

An example of a committee dealing with a particular issue in the United Kingdom was a team led by Lord Esher, who published his report in 1904 UK Brett, 1904), which analyzed the entirety of the British army. Another British team, led by Sir Peter Gershon, dealt with the efficiency and expenditure of the public administration. His report was published in 2004 (UK Gershon, 2004). The effectiveness of project implementation in the Netherlands was dealt with by the Elverding commission (The Netherlands Elverding, 2008). An example of a team set up to tackle a specific problem in Australia is a defense procurement review by a team led by Malcolm Kinnaird established in 2002 (Kinnaird, 2003). Such teams are important for project implementation for three reasons. First, their work is project-like: they have a specific goal and time perspective to perform the work. Second, they can deal with projects directly or indirectly – such as the Kinnaird report or the Latham report. Moreover, the result of the work of such teams is usually the introduction of changes in the functioning of a specific area, and the implementation of changes is project work.

All organizations that are components of public administration – central or local – and state-owned or in which the state has a majority stake are collectively referred to as the **public sector**.

The set of governmental organizations is shown schematically in Figure 8.1. The diagram is illustrative; it does not show any actually functioning administration – each of them has historical

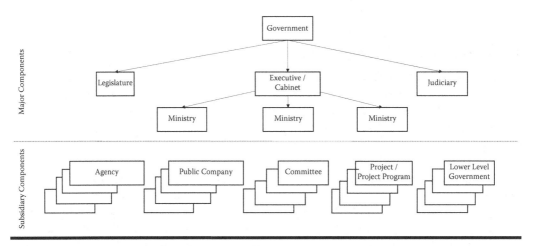

Figure 8.1 The Structure of Public Administration. *Note:* **Be aware that the diagram does not show the containment relation.**

conditions and selects and adjusts its elements to its needs. The diagram also does not show the relationship between the primary and additional elements, as they may be differently defined in each country. For example, the implementation of public projects and project programs may be overseen by agencies or ministries rather than by specially created organizations.

From our perspective, it is important to look at organizations from the point of view of the ability of the government to exert influence on their operation, in particular, on the methods of project management. The government has less and less direct control over successive tiers of these organizations.

Most public administration organizations implement projects to achieve their statutory goals or facilitate their development – perhaps the only exception is committees. Hence, **the government as a whole is a project-dependent organization**.

8.4 Autonomy and Centralization

Two concepts are important for organizations operating within complex structures: autonomy and centralization. The concept of autonomy refers to a single organization, and it is the scope of an organization's authority to make decisions without the consent of other external entities (Brock, 2003). On the other hand, centralization is the level of concentration of decisions in one place and refers to the entire structure containing many organizations. Government organizations may have varying degrees of autonomy (Brock, ibid; Bach, 2014, 2016). Flinders (2006) compared the state structure to concentric circles, increasingly distant from the center, with an increasing level of autonomy. The level of autonomy is influenced by the regulations governing the organization and relations with other entities – superior and parallel ones. For example, decisions about road construction in an agency responsible for transport can be influenced by the department responsible for the agency, transport law, and decisions of the department responsible for environmental protection. From the standpoint of autonomy, individual functions of the organization can also be considered – for example, strategic management or defining work procedures. In organizations that are elements of hierarchical structures such as governments, strategic management usually has little autonomy; it is subordinated to the strategic plans of hierarchically higher organizations. This approach is called cascading strategic goals. On the other hand, work procedures, mainly those specific to a given organization, are usually autonomously defined by a given organization. Work procedures not specific to a given organization may be determined in a non-autonomous manner by higher organizations. For example, the Ministry of Finance usually sets out procedures related to the financial operations of all government organizations. Procedures for a given area can be defined in three ways: by higher organizational units, by the organizational unit in which they are performed, or they can also be undefined – then whenever work of this type is performed, the work process is defined from scratch.

Public sector project management processes can also be defined in these ways. It is related to the level of maturity of the public administration in project management (for more, see Chapter 25).

8.5 Public Services

Public services are activities performed for the benefit of society or its subsets. Decisions about their implementation do not depend on other forces, such as the market, and the state is responsible for them (Spicker, 2009). A public service, for example, is keeping order. Education and transport can also be a public service to a specific, basic extent. Public services in free-market economies do not usually include, for example, tourism, car production, or book publishing. The activities of religious associations, in particular their provision for confessors, are also not public services.

Public services may be provided by organizations within the state structure or by private companies. For example, population registration services are usually provided by state institutions and health services by state or private healthcare units. The provision of public services may be financed by the state or, wholly or partially, by the public. City public transport is often funded through the city budget and, at the same time, from ticket fees.

Public services can be the result of operational activities (e.g., issuing all kinds of certificates, lawsuits) or the implementation of projects (e.g., construction of public buildings, organization of sports events). How public services are delivered may be entirely defined by the government or shaped by the citizens for whom the services are provided. For example, parent representatives can be involved in shaping the teaching practices in schools.

The quality of public services is influenced by the intentions and plans of the state (as we will soon learn, called "public policies," Chapter 11), the availability of resources, and management (Boyne, 2003b). When assessing public services, the most important criterion is the level of satisfaction of the needs and the recipient's satisfaction. The economic criteria do not have to play a decisive role (Boyne, ibid). However, all benefits must be included in the state budget. One of the evaluation criteria may be the number of services provided – the scope of some services may be limited due to state capabilities and their specificity (e.g., education at the academic level or participation in cultural events).

8.6 Civil Service

The public administration consists of politicians and civil servants. Politicians are elected and civil servants hired. The task of civil servants may be to support politicians and their work processes or to perform activities related to the provision of public services, or enable the provision of these services. The appointment of a civil servant may entail specific obligations (e.g., availability, high level of service) and privileges (e.g., exceptional durability of employment). There is a significant variation in the concept of civil servants between countries. Local government employees may or may not be considered civil servants. In some countries, civil servant status is granted only to those employees who meet the relevant requirements, such as education or seniority. Therefore, the concept of a civil servant does not have to be equivalent to that of a public sector worker.

All civil servants or their distinguished subset constitute the Civil Service.

Civil servants have general values related to their functioning (Spicker, 2009) (some of which are shared with workers of other sectors). These people should be trusted – the person applying for service should be convinced that his case will be appropriately dealt with. They should be fair – the citizen cannot be deceived. They should have a sense of service – the necessity to carry out their duties no matter the circumstances. The service is associated with a sense of altruism – valuing the values and needs of the citizen no less than his own. Civil servants should present the integrity of beliefs and behaviors, that is, profess and implement the same values. Civil servants should have deep substantive knowledge of their area of activity. In particular, they should be immune to corruption.

Specialized civil servants dealing with project management in public organizations with no PMO (Subchapter 20.1) are sometimes called Project Management Agents.

8.7 What Is Important in the Public Sector?

Usually, a private organization's primary goal is profit; the main goal of private sector organizations is to maximize the financial benefits for their shareholders. It is also natural to ask what is the primary goal of a public sector organization, i.e., what is the purpose of their operating. Here, the situation is unclear; it is difficult to give one goal. In public organizations, profit is either not a goal or one of many goals. At the project level, Wirick (2009) and Pūlmanis (2015) also note that public projects do not have a single goal of profit maximization as it is in private projects.

For example, hospitals operate to fight disease, schools to prepare young people for living in society, and the police to ensure the safety of citizens and enforce the law. The aim of the public sector organization may be to improve health, provide cultural goods, increase the safety of citizens, reduce unemployment or win sports competitions by the country's representation. Examples of broadly defined goals of public sector projects are economic and social development or poverty reduction (Moe and Pathranaraku, 2006).

There are also more general criteria for evaluating public organizations (Kuipers et al., 2014), for example, effectiveness (level of achievement of goals), efficiency (effort required to achieve the goal). Other general goals can also be, e.g., public accountability, honesty, openness, responsiveness to policy, fairness, due process, social equality, the criteria for the distribution of manufactured goods, and correct moral behavior (Boyne, 1998; Rainey, 2014).

In general, what is expected by society from public organizations is called **public value** (Moore, 1995; Crawford and Helm, 2009). The public value may be the products and effects of activities and the ways of working of an organization. The public value in terms of outputs and outcomes is determined in ways specific to the nature of the given organization. Just as private sector organizations are rated for their profit, public organizations are rated for their achievement of public value. A public organization is as good for society as it achieves public values, and more precisely, how society perceives its achievement (Boyne, 2003a). Also, public projects should deliver public value.

However, the goals of public organizations do not have to be considered constant, depending on the level of country development (Chapter 7) and the way of thinking about the functioning of the state (Chapter 12).

The goals of public organizations may be perceived differently in different societies. For example, the primary goal of natural resource organizations in one country may be to maximize the volume of resources harvested and, in another, to protect the environment. The main goal of organizations responsible for physical education may be to raise health or ensure domination in international sports competitions. In one country, the value may be to provide access to weapons for every citizen, while in another, it may minimize that access.

Because different groups in society may have different views on specific issues, the assessment of public organizations may be ambiguous. Choosing the main goals of governments is often the subject of election campaigns. The minor goals may be defined in other ways, such as through public discussions or public opinion polls, and others result from current needs, culture, or history. The assessment of the quality of public organization may change over time (Boyne, ibid), for example, due to changes in the economic and social or political situation. The tasks of labor offices may differ depending on the level of unemployment.

Conflicts between different public values are possible (Fottler, 1981; Rainey, 2014). For example, a requirement of efficiency may conflict with a requirement for public oversight or employment equality.

Achieving almost all public sector values, both specific, such as the mining industry's efficiency, the level of education, and general, horizontal, such as efficiency and effectiveness, is related to implementing projects. You can read more about the quality of government and public administration in Chapter 13.

Note

1. Not all states have a constitution. The United Kingdom is an example of a country that does not have such a document.

Chapter 9

Governance

9.1 What Is Governance?

Each organization operates according to a specific – more or less structured and documented – order. Each component of the organization has, above all, specific goals, tasks, and methods of their implementation, as well as relations with other components. And tasks and relationships can refer to other components of this organization and entities outside the organization. The company's production department, also from the public sector, has primarily internal tasks, and the marketing department – primarily external tasks.

> Determining an organization's goals, values, structure, relations between organizational units and main roles, and the high-level manner of achieving its goals is called organization **governance**.

The set of defined relationships, goals, and methods of achieving them and the level of detail of their definition is different for each organization. However, it is generally accepted that the details of performing actions are not an element of governance. Indicating specific people who perform tasks or technological processes is usually not an element of governance.

Hence, we have two main levels in the management of any organization. The management board of each organization acts at the governance level and makes the most important decisions regarding the functioning of its organization. Making such decisions may also be delegated by the management board to persons and units located lower in the organizational hierarchy. These decisions may result from environmental conditions, for example, from ownership relations or subordination in larger structures. But every other employee of the organization also makes decisions related to the work performed by him/her, within the framework of his/her powers and responsibilities defined by the organization's governance. This is called the management level.

In the following text on governance, first, I will discuss some topics related to the state level, then I will proceed to the project governance, and, finally, I will try to describe the results of merging these two areas resulting in the public project governance.

9.2 State Governance

A state, like any other organization, has its specific governance mechanism. The purpose of **state governance** (also known as **public governance**) is to ensure its coordinated and coherent operation taking into account its individual components' different purposes and objectives (Pierre, 2000). The essential elements of state governance are usually described in the constitution. The lower level makes the system of law in a given country. Other less formal arrangements define how public decisions are made and how public activities are performed from the perspective of preserving the state's constitutional values in the context of changing problems, actors, and the environment (OECD, 2005). An example may be the practice of consensual making of essential decisions in some countries. Another example of an informal element of state governance may be the role of the leader of a political group, even if he/she is not formally part of government structures. For Klakegg et al. (2008), state governance is the use of institutions, structures of power, and cooperation in allocating resources and coordinating activities in society or the economy. According to Ramesh and Howlett (2017), state governance primarily considers hierarchical structures of state power, market mechanisms, and citizens' network influence and associations. State governance can be understood as a way of organizing society to make decisions (CIoG, 2002). These methods arise and are determined as a result of the clash of interactions between actors in society (Kooiman, 1993). The governance of a country can be influenced not only by internal structures but also by supranational organizations to which the state belongs.

The definition of governance and the understanding of the state are closely related. In one country, some drug use is regulated by the state, while in others, it is not. Drug use is a problem in some countries and not in others. So, drugs and related activities may be subject to governance but may not be. If we assumed that governance should formally cover the entire functioning of the state, then we could speak of empty governance in relation to soft drugs. In a country with no armed forces (e.g., Costa Rica, Iceland), their use is not subject to governance. In the state environment, governance is the exercise of political power to manage the nation's affairs (World Bank, 1989, 1993). Abbott and Snidal (2001) limit the scope of governance to problems facing the state, but this definition ignores the trouble-free actions and opportunities facing society.

The essential elements of state governance are selecting, monitoring, and changing governments, the ability to form and implement policies, and the respect of citizens for such defined governance (Kaufmann et al., 2010). One of the main elements of state governance should also be the long-term strategic plans of a state. Since public policies cover the entirety of state action (Subchapter 11.1), state governance can be considered the way policies are formulated and implemented (Ramesh and Howlett, 2017; Mainwaring and Scully, 2008; Kaufmann et al., 2010).

State governance is implemented using various methods. Historically, coercive methods have been used most often: commands, orders, punishments, etc., characteristic of undemocratic forms of government. Nowadays, incentive and persuasion-based methods are increasingly used (Subchapter 10.2). Citizens are included in governance – then we are dealing with participatory governance, which emphasizes the participation of citizens and their associations in governance (Munshi, 2004), particularly in the implementation of policies (Cepiku, 2017a). Citizens can group to influence any area of the state's functioning: the judiciary, health service, education, foreign or other policy. Network dependencies between civic groups may also be formed without the involvement of governments (Kooiman, 2010). Important in democratic state governance are the principles of participation, accountability, transparency, the rule of law, separation of powers, subsidiarity, equality, and freedom of the press (United Nations, 2006).

The state governance is subject to change. The rules for making changes in governance are called **meta-governance** (Peters, 2010). The most important element of state meta-governance is the way of establishing the state's constitution. In some countries, a specific parliamentary majority is sufficient to change it, while in others, a nationwide referendum is also required. But meta-governance is also about changes in lower levels of power. Government typically has the authority to define and implement governance structures and processes.

Improper governance arrangements can lead to conflicts. History knows many cases where, for instance, inappropriate power distribution between ethnic groups in a country has led to such conflicts. And good state governance ensures productive specialization and peaceful social cooperation (Boettke, 2018) and improves political, social, and economic well-being (Mainwaring and Scully, 2008).

Different states have different governance abilities or **governability** (Kooiman, 2010). In one country, the public is more likely to follow applicable rules and standards, and in others, it is less so. The level and structure of governability stem from rules of the power-trust-control triangle (Bachmann, 2001; Heiskanen et al., 2008), where there is less trust between society and government for historical or current reasons the government must use more rigid methods of enforcing regulations. Conversely, where the public feels that the government is acting in its interest rather than its own, regulation can be more general, leaving more freedom to members of society and their groups. An excellent example of the influence of traditions on governmentality is Botswana, where the traditions of the direct impact of community members on pre-colonial leaders survived the colonial period and are now exemplified in the current direct assessment of government by members of the public (Botlhale, 2017; Acemoglu and Robinson, 2008).

9.3 Project Governance

The two-level decision-making structure also applies to projects. The governance level covers the main principles and decision-making processes shaping the entire project implementation system (or a single project) in an organization is called **project governance**. The processes, methods, techniques, or tools subordinate to them, affecting projects to a lesser extent, are called **project management**. The implementation of the project management processes, e.g., scope management, risk management, and communications management, takes place within the higher-level framework established by the overall governance system of the organization (Too and Weaver, 2014). The dividing line between governance and other management processes may be set differently in each implementation, at the government, subordinated organizations, project program, or project level.

The project governance concerns several major areas (Müller, 2009). The first is defining the **role of projects** in the functioning of an organization. The most generally defined roles of projects are enabling the development of an organization or implementation of its business (or both). In many IT or construction companies, projects are the primary tool for achieving their business goals. In other companies like schools or hospitals, projects are implemented mainly for the goal of organizations development. Project governance consists of structures and processes (Du and Yin, 2010). It must be consistent with the overall governance of the organization.

Having defined the role of projects in its organization, the management board should make the most important decisions regarding project management. This area of project governance is called **governance of project management** (Müller, 2009). And the way to implement project management may be, for instance, the decision to create a PMO and develop project management methods and tools by this organization. Related to this should be decisions on training or recruiting project management specialists. The number of specialists and management methods should be adjusted to the needs resulting from the governance of the organization. Suppose the organization's main task is to carry out routine processes, for example, retail commerce. In that case, the project management needs are smaller than in an organization whose primary business is construction projects.

According to Klakegg et al. (2008), project governance is the definition of the structures through which project objectives are established, the means of achieving these goals, and the means of monitoring their implementation. In Müller's hierarchy, this is closest to the governance of project management.

The next area of project governance is the main principles of project implementation (**governance of projects**), of project program implementation (**governance of project programs**), and of implementation of the project portfolio (**governance of project portfolios**) (Müller, 2009) (Figure 9.1). These three areas of governance operate within the framework defined by the role of projects and governance of project management. The successful implementation of projects, project programs, and project portfolios depends on their governance and the governance of project management (McPhee 2008; Shiferaw et al., 2012; Gasik, 2016).

Governance of projects must ensure the implementation of the target vision of the project (Shiferaw and Klakegg, 2012). Garland (2009) defines governance of projects as the framework for making decisions leading to the project implementation within which the major decisions are made. Projects must have a current, published implementation plan (Barkley, 2011). Each project must have a specific methodology and a method of making significant decisions, for example, the method of forming the Steering Committee, the rules of staffing the position of the project manager, the main gates through which the project should pass, the way of reporting the project manager to the Steering Committee and the sponsor to the senior supervisor. Governance of projects must clearly define project responsibilities (Shiferaw and Klakegg, ibid). There must be a project manager (Barkley, ibid). All other roles and responsibilities regarding the project should be clearly defined and assigned (Crawford and Helm, 2009; Barkley, 2011; Too and Weaver, 2014). The project should have a specific set of business goals agreed upon by all stakeholders. A collection of project products should be approved (Barkley, ibid). The governance of projects

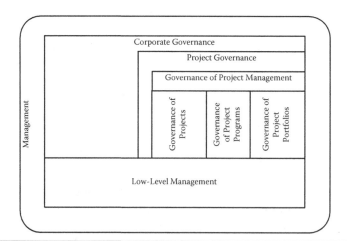

Figure 9.1 Organization's Governance. (After Müller, 2009.)

determines how resources and risks are allocated to stakeholders (Klakegg et al., 2008; Barkley, ibid). Governance guarantees the identification of all stakeholders and defines communication methods with them (Barkley, ibid). Governance must be prepared to respond to external stimuli (Shiferaw and Klakegg, ibid).

Tools of governance of projects (also project programs, project portfolios) include reporting, audits and reviews, and project sponsorship by the top management of relevant organizations. Governance also includes controlling and ensuring project compliance with objectives and regulations. Supervision is concerned with the schedule and budget, the services provided and their value, and customer satisfaction. Compliance with targets is guaranteed through audits and reviews (Barkley, ibid; Crawford and Helm, 2009). Governance documents should be reviewed. An element of governance implemented after the completion of the project is the assessment of its compliance with the original goals (Barkley, ibid). In the governance of project programs, the ways of identifying component projects should also be added to these principles. In the governance of project portfolios, the key governance decisions are the rules for qualifying project initiatives for implementation based on compliance with the organization's strategy and the continuous control of the suitability of the portfolio components.

Governance is a different, higher level of management. Specific ways of creating a schedule, identifying risks, and carrying out individual works are the responsibility of the project manager and the members of the project team. This is the project (project program, project portfolio) management level. Management roles and responsibilities must be kept separate from governance roles and responsibilities. The director of the company or department should not say what the process of commissioning or controlling a team member's work should look like; this should be the project manager's responsibility.

Governance consists of structural elements (entities involved in project implementation, e.g., PMO) and processes (e.g., project selection process, project supervision process) (Hall et al., 2003). An organization uses structures to achieve its goals. Governance ensures that management works within defined structures (Too and Weaver, 2014).

According to Klakegg et al. (2008), the governance systems of public projects vary in terms of the set of stakeholders involved in decision making. In the United States, United Kingdom, and Canada, this system only includes project shareholders. In Germany, Norway, or Japan, there is a social system that also considers non-shareholder stakeholders. The first of these approaches facilitates the implementation of projects, while the second – greater stakeholder satisfaction and ultimately a higher level of success.

Bekker (2014) distinguishes three ways of thinking about project governance. The school of one company is rules for intra-organizational projects and their application at the technical level. The school of many companies concerns projects in which two or more companies are involved based on contracts. Governance rules apply to the technical and strategic levels. The Big Capital School includes projects carried out by temporary organizations. The rules concern the formation of these organizations and institutional rules binding in these organizations. The situation is similar in the public sector, where internal projects are implemented, projects using contracts and large projects and project programs, implemented based on special laws (e.g., construction of mega-airports or building entire cities, such as Canberra, the capital of Australia). The level of governance complexity increases with the level of the school thus defined.

In turn, Too and Weaver (2014) define two governance schools. According to the first approach, there is one governance process in an organization that covers the entirety of its functioning. The second school requires the creation of various types of governance in multiple areas of the organization's operation, e.g., IT governance, knowledge governance, public governance, or project governance. Each area of governance is distinguished from the others. The entire governance system is responsible for the integrated functioning of the constituent governance systems.

The importance of governance processes means that the main problems with project implementation relate to its governance (Sargeant, 2010). Bad governance is one of the reasons for the failure of public sector projects, i.e., failure to deliver the expected benefits (Shiferaw and Klakegg, 2012). Patanakul (2014) also points to governance as one of the problem areas of public sector projects. These problems may include improper start, the lack of suitability of public sector projects to the needs of the implementing countries, and the lack of durability of products (Klakegg, 2009).

A problem related to the governance system of public sector projects is the lack of coherent integration of policies and strategies. Public administration and parliament should play a strategic role in this system (Jänicke et al., 2001).

The governance process of public sector projects can be adversely affected by government structure, lack of experience, lack of knowledge, information, ways and means of measuring facts, and pressure from decision-makers (Shiferaw and Klakegg, 2012).

Governance ensures efficiency and effectiveness of management (Hall et al., 2003), influences transparency, accountability and relates to the interests of project stakeholders. There are two directions for improving the effectiveness of public sector projects: improving management and improving governance. Influencing governance has a more significant effect (Hall et al., ibid). Without an established governance framework, the productivity of project managers would be at risk (Barkley, 2011).

Public projects often involve many people (e.g., community members and employees of organizations for which projects are implemented, public sector management staff, politicians, project shareholders), organizations (e.g., funding organizations, legislators, media, the hierarchy of executive agencies, public agencies other than the organization implementing the project) and contractors (including subcontractors and sub-suppliers). Many decisions in projects must be accepted at many levels of the organizational hierarchy. Governance processes introduce order in decision-making processes involving all these entities, ensure that optimal decisions are made in such a complex environment (Williams et al., 2010). In particular, the proper involvement of sponsors in key decisions increases the likelihood of success for public sector projects.

There are usually many stakeholders in public project programs and strong political and legal empowerments. One of the most important success factors is the precise definition, documentation, and enforcement of their responsibilities and authorizations.

9.4 Public Project Governance

Since the state is an organization, some of the knowledge about organization governance may be applied to state governance, particularly to project implementation. The most important role in the public project governance system is played by governments (the legislative or executive branch), which have supreme power over the entire jurisdiction; they can also make decisions regarding projects and their implementation. The first group of decisions concerns project governance, while the second – individual, usually the most important, projects. Governments allocate budgets to them, thus confirming the decision to initiate them. During implementation, governments can accept reports and decide on the continuation, modification, or termination of projects. Governments usually obtain information on project implementation through specialized bodies such as audit chambers or the Governmental PMO (Section 20.1.1).

The state governance covers the main principles shaping project implementation in the state as this is one of the most important areas of government activity. The implementation of public policies takes place through the implementation of their projects and project programs, so the public governance includes, among other things, mechanisms of their selection and implementation. The project component of state governance creates and enforces the link between the state development strategy and project objectives (Shiferaw and Klakegg, 2012). Governance during project selection ensures that appropriate, sustainable alternatives are selected (Klakegg et al., 2008).

Project governance in a state covers the area from the government through its institutions to individual projects. These formal and informal arrangements determine how public decisions about projects are made and how public sector projects are executed (Williams et al., 2010). For public investment projects at the national level, governance consists of two subsystems: political (decision making and prioritization) and administrative (Klakegg et al., 2008). These areas should be separated from each other and have clearly defined responsibilities (Williams et al., 2010).[1] Public sector project management processes operate within established and formally applied governance structures (Kwak et al., 2014). For Barkley (2011), governance of public projects is about identifying the relationships between actors involved, describing the flow of information between all stakeholders, providing proper reviews of issues in each project, and ensuring that appropriate approvals and guidance are issued for each phase of each project. Since the government can be considered a "state board," government structures are part of the governance of public sector projects (Klakegg et al., 2008). Public sector project managers need to know which government entities have the authority to define and amend the specification of project delivery rules (Barkley, 2011). They must have a specific process for reducing the inconsistencies between government entities if and when they do occur (Brunetto and Farr-Wharton, 2003).

As in any organization, the public project governance consists of two main areas (Müller, 2009). The first is the role of projects in the state, and the second – governance of projects (project portfolios, project programs). Both of these types of governance should be embedded in the governance environment of the state. The set of public policies covers the entirety of state action. Policy programs implement them. Policies (and policy programs) can be operation-based and project-based. Projects are needed to implement both types of policies (Section 11.4.2). Therefore, the role of projects in the state is enabling (for operation-based policies) or implementing (for project-based policies) the public policy programs.

> The state could not achieve most of its goals without efficient and effective project management.

According to the previously introduced classification (Subchapter 8.1), **the state as a whole is a project-dependent organization**.

For projects to deliver the expected results, they should have specific rules for their implementation. The government defines such rules, i.e., governance of projects. Projects may be components of project programs.

There are two types of programs to be distinguished in the public sector: policy programs and project programs (Section 11.4.1). The goal of policy programs is to implement policies. The goal of project programs is to group projects to allow them to be efficiently managed. Project programs may be included in policy programs. In the case of project-based policies, one project program

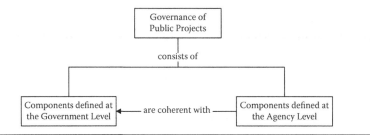

Figure 9.2 Components of Public Project Governance.

may be sufficient to implement its program. The management of a public organization's project portfolio is instrumental in implementing the projects necessary to achieve the goal of a government policy (and policy program). Therefore, I will analyze together governance of projects, governance of project programs, and governance of project portfolio as one area under the label of governance of projects.

Governance of projects at the government level is the definition and enforcement of the manner of implementation of activities in the field of project delivery (Chapter 20), support (Chapter 21), and development (Chapter 22).

The extent of government-regulated processes may vary from country to country. Governments do not have to decide on all aspects of the governance of projects in all their reporting agencies. The management of each agency has a certain level of autonomy (Subchapter 8.4) in defining the processes for implementing their projects. Therefore, some elements of governance of projects are determined by the organizations implementing the projects themselves. Leaving the determination of the governance of project or its elements to the level of the organization may result from two reasons. Governments may not be aware of the importance of project management for state development, i.e., they unconsciously leave the definition of project governance to agencies. Governments that are more mature in project management may deliberately leave some elements of governance of projects to their agencies. For example, determining how projects will collaborate between the organizational components of a government agency is usually set at the agency level.

Governance of projects consists of arrangements for projects defined at the government level and defined at the level of component organizations (Figure 9.2). Hence, this is multi-level governance (Marks et al., 1996). The elements of governance of projects defined at the organizational level must be coherent with those defined at the government level.

Larsen et al. (2021) show an interesting comparison of the combination of governance of the central level and the level of state-owned enterprises in Norway. There are the same main decision gates at the enterprise level as in the government's Project State Model, but different processes lead to their implementation.

Note

1. This is consistent with Woodrow Wilson's understanding of state structure (Subchapter 4.9).

Chapter 10

Instruments – What Governments Can Do?

10.1 What Is a Policy Instrument?

The most important concept in the area of functioning of governments is the public policy. Before I define and discuss it, however, I will deal with its components – the tools of the government operation, i.e., the instruments of public policies.

Examples of instruments include building a hospital network, levying taxes, establishing a new ministry, introducing new civil law, introducing new healthcare procedures, changing political relations with a particular country, or introducing an order to publish certain information – for example, on the progress of public projects.

A public policy instrument is a government action used to achieve a specific goal (Van Nispen, 2011; Canada TBoCS, 2007). Other definitions enrich the understanding of this concept. Howlett and Rayner (2007) define it as a technique or method of governance that involves using state resources or their conscious limitations to achieve public policy goals. On the other hand, Fontaine (2015) believes that it is an artifact used to introduce an action, measure, or set of measures to achieve a specific goal. According to Le Gales (2011), the instrument of policy is a technical and social device that organizes social relations between the state and those affected by them in accordance with the representation and meaning carried by them. It is a specific type of institution to bring a particular notion of policy/society relations and sustained by the concept of regulation. Since the state has potentially full power over its jurisdiction, policy instruments can apply to every element of the country's functioning, not only to public administration – they often refer to, for example, market mechanisms (Vargas and Restrepo, 2019) or how citizens organize. For Musgrave (1994), instruments are devices through which public goods and services are provided, income and goods are redistributed, the economy is stabilized, and social, economic, and environmental sustainability are achieved. The instruments have a generic scope – the same type of instrument can be used for various problems (Lascoumes and Le Gales, 2007). The sets of instruments increase as countries develop (Musgrave, ibid).

Instruments are described by laws, decrees, ordinances, regulations, contracts, circulars, and other forms of regulation specific to a given country (Musgrave, ibid). But these documents

DOI: 10.1201/9781003321606-12

themselves are not instruments – they are their descriptions which should not be confused. An instrument such as establishing a government-level Project Management Office may arise both due to adopting the relevant law and a decree by the prime minister – but in both cases, it is the same public policy instrument.

Lascoumes and Le Gales (ibid) distinguish instruments (e.g., the organization of the Olympic Games), techniques (e.g., the collection of facilities necessary for the implementation of the Games), and tools (facility construction projects, competition implementation processes, and accompanying activities).

The set of instruments may be limited or modified due to the existing traditions or governance structures. For example, the instrument of requiring different wages for both genders may not be feasible in patriarchal societies. The quantitative control of project implementation instruments will probably not be adequately implemented if there are no project management procedures in a given country.

Many authors have listed exemplary policy instruments (Thynne, 1998; Salamon, 2002; Blomkamp, 2018; Howlett, 2009, 2019; Mees et al., 2014; Linder and Peters, 1989; Musgrave, 1994; Le Gales, 2011; de Bruijn and ten Heuvelhof, 1991; Canada TBoCS, 2007; Lascoumes and Le Gales, 2007) and tried to classify them (de Bruijn and ten Heuvelhof, ibid; Lascoumes and Le Gales, ibid; Peters, 2000; Anderson, 2015; Linder and Peters, 1989; Howlett, 2009, 2019; Mees et al., ibid; Van Nispen, 2011; Pal, 2014; Jann and Wegrich, 2007; Vargas and Restrepo, 2019; Juerges and Hansjürgens, 2018; McDonnell and Elmore, 1987). The number of attempts to systematize the instruments shows the multitude and diversity of possible ways governments operate. These classifications can be divided into those relating to the implementation process and those relating to the policy effects.

10.2 Different Ways of Instruments Implementation

The most important difference between the methods of implementing the instruments is doing something or doing nothing (Pal, 2014). Doing nothing can be further broken down into two important categories. First, the state may not be interested in a specific area of society's functioning, such as religion or culture. The second type of doing nothing concerns areas that the government is interested in, but in a given period it does not take action for various reasons – e.g., acceptance of the status quo, lack of resources. If schools work well, keep it that way. A kind of supplement to this classification is the level of the directness of the instrument (Van Nispen, 2011), describing the level of involvement of public entities in its implementation. **Direct instruments** act directly on citizens, and the state is only involved in the supervision of its implementation. For example, introducing criminal law provisions, where citizens must directly comply with the provisions, and the state's participation is only the control of compliance with the introduced provisions. **Indirect instruments** are implemented through state institutions. An example of an indirect instrument is the introduction by the government of a new process of any authorization issuance (e.g., for construction or commercial) by its institutions (Van Nispen, ibid). Instruments can also be divided into **internal** instruments of the state and **external** instruments concerning society. An example of an internal instrument could be the reorganization of a specific ministry. An external instrument is, for example, the construction of a motorway or the introduction of regulations regarding the cleanliness of inland waters. Together, both these Van Nispen classifications show a continuum of instruments from internal, which involve only state institutions, to those in which the state's participation is negligible, and the entire instrument is implemented by society. A similar division concerning the sphere of implementation was given by Howlett (2019), who divided the instruments into market-based (e.g., the introduction of tax incentives or subsidies) and state-based (e.g., implementation of new elements of criminal law).

From the managerial point of view, the instruments are divided into **project-based** and **operation-based** instruments. A project-based instrument is one which essential part is the implementation of one or more projects. An operation-based instrument is one that is implemented through the usually long-term performance of specific operations and processes. An example of a project-based instrument could be constructing a motorway or organizing a sporting event. Another project-based instrument is using a particular weapon – such as an atomic bomb – in a war. An example of an operation-based instrument may be the introduction and implementation of a specific method of tax collection – e.g., the introduction of a VAT rate of 17%. An example of an operation-based instrument is also the introduction of computer science education in state primary schools. This classification, very important for the implementation of instruments and policies, we will call the **management type classification.**

The division into automatic and non-automatic instruments is somewhat related to the type of management (Van Nispen, 2011). Automatic instruments are those that do not require the creation of new organizational structures – that is, to implement them, extensive organizational projects do not have to be implemented. An example of an automatic instrument is the change of curriculum in schools. And a non-automatic instrument is a change in the structure of education, for instance, from a two-level to a three-level system.

The role of the state in defining and implementing instruments is changing. There are two generations of them (de Bruijn and ten Heuvelhof, 1991; Lascoumes and Le Gales, 2007):

1. *First-generation, hard instruments*: Instruments based on directing and giving orders. For example, introducing new taxes and military instruments: fighting a war and introducing a new weapon. For instance, legal instruments introduce the new criminal law code.
2. *Second-generation, soft instruments*: Instruments based on cooperation, information, incentives, persuasion, and partnership. For example, introducing incentives for the use of non-traditional energy sources; establishing forums for collaboration between the government and specific groups of people, public-private partnership.

This division is important due to countries' needs at different levels of their development.

The instruments of the second generation do not replace but complement the instruments of the first generation. It is difficult to imagine a deliberative (i.e., determined due to discussions with interested parties) way of managing a war campaign. First-generation instruments are well managed in a top-down hierarchy-based Weberian style (Subchapter 12.1). In contrast, second-generation instruments are well-managed in an alternative, humanistic style where more emphasis is placed on collaboration and other relationships between all stakeholders involved (Peters, 2000). Peters also divides instruments into capital-intensive and work-intensive. Capital-intensive instruments, such as the co-financing of failing workplaces, do not require extensive implementation processes. Work-intensive instruments may or may not be complex. For example, sending a man to the moon, a labor-intensive instrument, was at the same time very complex. On the other hand, increasing the extraction of a particular mineral by a state-owned enterprise may only require increasing the number of employees.

10.3 Types of Policy Instruments' Results

Each instrument has its specific effect.

A popular division distinguishes between **procedural** instruments and **substantial** instruments (Howlett, 2009; Blomkamp, 2018; Anderson, 2015). A procedural instrument prescribes the implementation of specific processes. A substantial instrument is one which essence is to dispose of a particular resource. Examples of procedural instruments may be ordering the implementation of a particular process in implementing projects or determining the method of access to public information. An example of a substantial instrument could be building a power plant or paying a certain amount for each child in a country.

A breakdown of public instruments according to the subject of intervention is often used (Howlett, 2009; Mees et al., 2014; Le Gales, 2011; Jann and Wegrich, 2007; Vargas and Restrepo, 2019; Lascoumes and Le Gales, 2007; Juerges and Hansjürgens, 2018). In this classification, the following types of instruments are distinguished (I give examples under the name):

- *Political, related to power*: Introduction of a new electoral threshold. Introducing the new administrative division of the country.
- *Legal and regulatory*: Probably the most comprehensive set. Modification of the labor code. Defining the rules of access to public waters. Establish a new road traffic law.
- *Managerial*: Implementation of the Governmental Project Implementation System (Chapter 23). Introducing the obligation for company crew members to participate in their management boards.
- *Economic*: Increase or decrease in taxes. Subsidies for a specific type of production. Introducing payments to children under 18 years of age.
- *Related to communication and information*: Publication of acts in the *Journal of Laws*. Reporting the progress of work of the largest projects. Carrying out a census.
- *Organizational*: Establish a social dialogue council. Liquidation of the Ministry of Infrastructure. Establishment of the Analysis Department at the Ministry of Finance. Some of the researchers mentioned above add still other instruments to their classifications.
- *Normalizing (standards and best practices)*: Introduction of a standard for permissible air pollution by car engines. Introducing the best practices of performing audits by the audit chamber. Implementation of the ISO 9000 series standards.
- *Based on collaboration, agreement, and incentives*: Subsidies for replacing coal stoves with gas ones. Progressive income tax. Establishing a cooperation council between the government, employers, and employees.
- *Informal*: Informal meetings, networking, selective presentation of information.

One of the classifications described by Van Nispen (ibid) is the breakdown of instruments into prohibitions (what must not be done), orders (what must be done mandatory), and information tools. In turn, McDonnell and Elmore (1987) built an interesting classification in which, like Van Nispen, they distinguish orders. Two of their other categories are capacity-building (see Chapter 14) – very important from the point of view of state development, and changing the state governance system, also related to state development.

Van Nispen (ibid) also proposes to classify instruments according to the level of their visibility in government activities (i.e., showing them in plans and reports). McDonnell and Elmore (ibid) propose as features of public policy instruments also their general nature (applicability in various situations), social intrusiveness (the level of permanence of presence), the cost of implementation, and the precision of the group to which the instrument is directed. The introduction of an order to maintain social distancing is a very intrusive instrument of the epidemic prevention policy for society and the demand to register exotic animals – not so much. An example of an instrument with a precisely defined target group is the introduction of free medicines for people over 75 years of age. The policy of promoting innovation by introducing a new mode of public purchasing has a less precisely defined target group – it is not known how many companies will decide to use this mode.

This chapter shows that the classification and research of public instruments are at a low level of sophistication and, indirectly, that it is a very complex area. Therefore, it is important to develop a classification that can be the basis for further analytical work. The most important of the described classifications to understand the essence of the instruments seems to be:

- Division into project-based and operation-based instruments
- Division into hard and soft instruments, the first and second generations

Instruments are components of public policies that we will now deal with.

Chapter 11

Public Policies

11.1 What Is a Public Policy?

A single instrument does not necessarily lead to the desired goal. Instruments produce products, not effects. To increase immigration, it is not enough to change the right to cross the border and stay in the country. It may also be necessary to introduce incentives and preferences for the desired professions, and introduce rules for recognizing foreign diplomas or activities related to the integration of foreigners into one's society. Only the consistent performance of all these activities – a set of instruments – leads to the desired effect.

In this way, we are getting closer to the fundamental for the entire Public Administration concept of public policy. For the administration's actions not to be contradictory and lead to specific goals, the government must deliberately and consistently define its intentions and activities in each area.[1]

It is not easy to state what the concept of public policy means. Many authors have made such attempts, each contributing to some extent to increasing the knowledge on the subject. There are many such attempts to define it. Trying to determine the understanding of this concept, I reviewed many publications; among them, I chose those that explicitly try to answer the question: what is public policy (Table 11.1). There are probably many more definitions of public policies, but this review already allows for a particular understanding of this concept.

11.1.1 The Problem; Policy Area

First, there must be a problem, a social interest, an opportunity to do something better, a chance to improve the quality of life in society (Anderson, 1975, 2003; Hogwood and Gunn, 1984; Kilpatrick, 2000). For example, the level of innovation in the country is low. Or the poor condition of the roads makes transport difficult. Either there is a shortage of drinking water. Or inflation is too high. We will generally call it a problem or a policy area. The problem may be in any sphere. It need not be – and in most cases is not – a sphere of activity reserved to the government. For example, it may relate to bank commissions (banks are usually part of the private sector), how citizens' associations are registered, or information that must appear on the packaging of products sold in shops.

DOI: 10.1201/9781003321606-13

Table 11.1 Components of Public Policy Definitions

	Anderson (1975)	Anderson (2003)	Cochran et al. (2010)	Cochran and Malone (1995)	Dye (2013)	Hill and Varone (2017)	Hogwood and Gunn (1984)	Jenkins (1978)	Kilpatrick (2000)	Lane (2000)
Government actions	√	√	√	√	√			√	√	
Way to achieve goals								√		
Implementation of programs				√						
Resource allocation										√
Income redistribution										√
Funding Priorities									√	
Laws									√	
Regulatory measures	√	√	√	√					√	√
Intentions, purposes	√	√	√	√				√	√	
Attitude						√				
A pattern of decisions made							√			
Stability		√								
Specific topic/subject of interest	√								√	
Problem		√								

(Continued)

Table 11.1 (Continued)   Components of Public Policy Definitions

	Anderson (1975)	Anderson (2003)	Cochran et al. (2010)	Cochran and Malone (1995)	Dye (2013)	Hill and Varone (2017)	Hogwood and Gunn (1984)	Jenkins (1978)	Kilpatrick (2000)	Lane (2000)
A specific situation							√	√		
Political decisions				√				√		
Political actor								√	√	
The personal, group, and organizational influence							√			
Subjective perception							√			

11.1.2 The Vision of the Target State

For a specific problem or an area, the government should have a vision of the target state (Althaus et al., 2013; Anderson, 1975, 2003; Jenkins, 1978; Parsons, 1995). For example, it is reducing inflation to 5% annually, achieving a place in the top ten innovation classification of European Union countries, or providing commuting to work within half an hour for every employee in the country.

11.1.3 Political Decisions

To start working toward a specific public goal, a political decision must be made (Jenkins, 1978; Parsons, 1995), i.e., one in which political actors participate. Non-political entities: individuals, groups of people, or organizations can also have an impact (Hogwood and Gunn, 1984). Implementing policies requires making many decisions, not only about starting to deal with a specific topic but also about how to act in the course of policy delivery (Althaus et al., 2013; Colebatch, 2006; Parsons, 1995). For example, the government has to decide whether to base transport on road or rail in the country. It also should choose whether to focus on its food production or its import from abroad. These are the decisions that are essential for the functioning of the state. The decision is a component of the concept of public policy.

The goal and decisions can be supported by a hypothesis (Althaus et al., 2013) showing how measures and actions will change the situation. For example, allocating 2% of the state budget to science will ensure reaching the place in the top ten of the most innovative EU countries. Or reducing the speed limit in cities by 10 km/h will reduce the number of fatal accidents by 12%.

11.1.4 Actions

The leading actor in public policies is the government. Most authors explicitly indicate that government actions are a component of public policy. We know from the previous chapter that they are called public policy instruments. Dye (2013) argues that everything government does is public policy. The actions of the government as a component of public policy – but without limiting the actions of governments only to policies – are also seen, for example, by Anderson (1975, 2003), Cochran et al. (2010), and Jenkins (1978). Some authors point to specific types of action used by governments. For Cochran and Malone (1995), policy objectives are achieved through policy program implementation. The government may also have a specific attitude toward a problem that shapes its way of operation (Hill and Varone, 2017). An example is making decisions that take into account the interests of the consumer and not the producer, which was the attitude of the Margaret Thatcher government. Public policies concern all entities within the government's jurisdiction; they are actively or passively involved in their implementation.

This short overview makes it possible to attempt formulating (yet another) definition of public policy.

The public policy addresses a specific area of concern or issue and indicates a direction or goal there. It is implemented or facilitated by the government through particular instruments. The set of public policies covers all government activities.

11.2 Structures of Public Policies

The government distinguishes between areas that are the subject of its active interest. The government can contribute to cultural activities in one country, while it cannot in another. In landlocked countries, the government usually does not deal with maritime affairs. Not every country has to be interested in the production of cars, and if it is, the state may interfere (or not) in this production in various ways. The activities of government and Public Administration are exhaustively broken down into areas. As the areas of interest of the state have a (multidimensional) hierarchical structure, public policies also have a hierarchical structure. For example, in the area of foreign policy, one can distinguish the area of European policy and any other territory of the world, and in the area of Asian policy – policy toward China. The area of transport infrastructure consists of the area of road, rail, air, and water infrastructure. In turn, in the area of road communication, communication areas can be distinguished in individual regions of the country. Another dimension of the hierarchy of areas of activity results from the state's administrative structure. If the state has a federal structure, then each central policy should have its counterpart in the policy of the constituent units. Healthcare in the United States is the sum of the approaches to health care in all component states.

Public policies are well separated at the level of their definition – due to the existing structure of Public Administration. It should be known what the state wants to achieve in the area of transport, employment, economy, etc. From this point of view, these are areas of public policy (Burnstein, 1991). These goals and policies may be the responsibility of the Ministry of Transport, the Ministry of Employment, and the Ministry of Economy. However, at the level of implementation, public policies are often not separate. For example, the construction of an expressway can be considered from the point of view of meeting transport needs, enabling the creation of new enterprises, approach to project implementation, using new technologies, and reducing unemployment. In each of these areas, the government may have specific policies. For example, the issue of using the latest technology, which is largely automated, and reducing unemployment should be addressed – and this would be facilitated by the use of more people in construction. At the implementation level, the policy domains are not mutually exclusive. This is an additional complication in both their formulation and implementation. Public policies in common areas should be coherent (May et al., 2006).

The relationships between policies, instruments, and their domains are shown in Figure 11.1.

11.3 Strategy and Policy

The concept of strategy is akin to policy. The strategy usually applies to the entire organization. The state is a hierarchical set of organizations, each of which, including the state as a whole, has its strategies. The strategy defines long-term goals to be achieved and the most important ways to achieve them (Joyce, 2017). State strategies are sometimes called multi-annual plans (e.g., the Indian Five-Year Plan, India Planning Commission, 2013; the Kenyan Plan, Kenya Government, 2013). State agencies must also have strategies in some countries (e.g., the United States, under the GPRA, USA Congress, 1993). There are different views on the relationship between strategy and policy. According to Archibald (2008), the policy is a more general concept, and strategies are used to implement policies. In countries where strategies are the primary documents determining the functioning of the state, the opposite is true: policies are part of the strategy because there are no entities above the strategy. And according to Joyce (ibid), the terms policy and strategy are

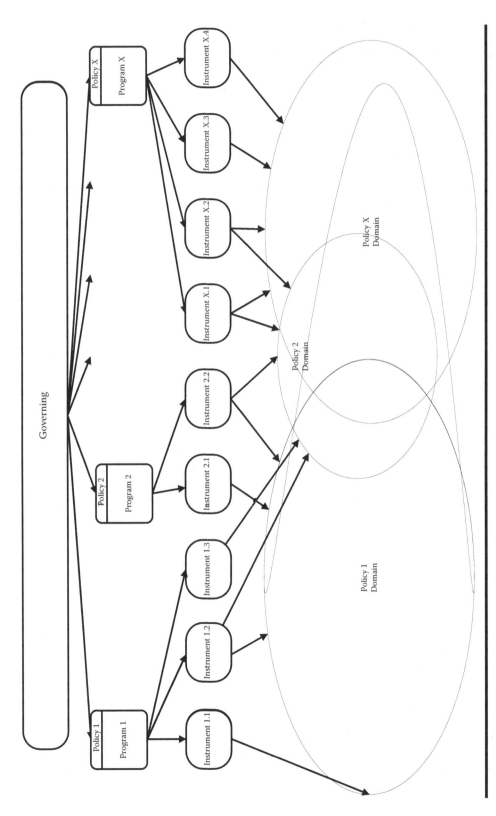

Figure 11.1 Policies, Instruments, and Domains.

used interchangeably. For this book, I will adopt, as it is in practice, among others in the cited countries, that the state's strategy covers the entirety of its functioning, especially all public policies. Otherwise, it can be said that the strategy is a top-level policy. The overall state strategy may consist of strategies relating to particular areas, for example, defense strategy, environmental protection strategy, or communication strategy. Each component (or main ones) strategy may contain many public policies.

11.4 Policy Programs

Policies are implemented by their instruments. The set of policy instruments is called the **instrument mix** (Mees et al., 2014; Howlett, 2014; Howlett and Rayner, 2013). Activities that implement policies' instrument mixes are called **programs**.

11.4.1 Policy Programs and Project Programs

We will call programs described here **policy programs** to distinguish them from the **project programs** described in Subchapter 5.1. The concept of the program means something else in the area of Public Administration and in the area of project management. In the Public Administration environment, "programs are what governments do" (McConell, 2010, p. 350). On the other hand, in the project management community, programs are only groups of projects ending when all projects have delivered their outcomes (PMI, 2017b). Policy programs may contain projects as well as non-project instruments. Part of the policy program to combat COVID-19 is social distancing and vaccination. Fiscal policies encourage specific behavior by requiring you to pay a certain amount of taxes. Consumption of more fish or vegetables may be part of food policy. Project programs do not include the use of project products, while policy programs do. Hence, knowledge and practice developed by the project management community only partially may support policy programs. Attempts to apply project program management knowledge to each policy program cannot be effective and discourage the Public Administration environment from using it.

11.4.2 Project-Based and Operation-Based Policies and Programs

Researchers point to the differentiation of public policies depending on the level of use of projects in them (Sjöblom, 2006; Sjöblom et al., 2013; Jałocha & Prawelska-Skrzypek, 2017). From the point of view of using projects in public policies, two types of policies can be distinguished: **project-based policies** (Karaburun, 2009) and **operation-based policies**. Accordingly, there are also two types of policy programs: **project-based policy programs** and **operation-based policy programs**.

Project-based policies are those whose effects are almost identical to the project implementation's effects. Examples include infrastructure policies or many military policies. Infrastructure policies are implemented through projects that create this infrastructure – for example, the construction of roads, mines, or airports. Military policies are often implemented by producing new types of weapons, which is an example of purely project work.

Operation-based policies are those whose effects are obtained through performing continuous processes (operations). For example, tax policies, health policies, educational policies – all of them – initially create some changes in operations (e.g., establishing new tax rules, implementing new rules for access to health services, or forming new teaching bases), but their results depend on the long-term implementation of these changed operations (e.g., collecting taxes, treating patients, or teaching pupils).

In most operation-based public policies, project management techniques are required. For example, tax policy: although its final effect depends on tax collection processes, it must start with implementing appropriate changes in these processes, which have the nature of project-type work. However, the reverse is not valid. There are project-based policies in which the routine, operational element is negligible or nonexistent. For example, the policy of achieving US military advantage over Japan during World War II, in which one of the main elements was the Manhattan project of producing and using the atomic bomb, lasted about six years and had no routine component.

> Projects are necessary for implementing any new or changing existing public policy.

Lack of introducing changes in a specific area of government activity means a continuation of operations or conscious decision not to intervene there (Tinbergen, 1998). Only these types of public policy do not require any project to be performed. The continuation of previously implemented policies is the most frequently implemented type of policy. Probably no state would bear introducing changes to too many public policies simultaneously. But it is also the type that perhaps rightly attracts the least attention of public policy researchers.

The role of projects in implementing various types of policies is shown in Figure 11.2 (this is a refinement of a part of Figure 1.1).

The classification of public policies that has just been introduced helps us address differences between policy programs (as understood by the Public Administration and Political Sciences communities) and project programs (a component of project management; more in Subchapter 5.1). Project program management knowledge developed in a project management environment is the primary tool for implementing project-based policies and their programs. But project management is also indispensable for operation-based policies. In this case, projects are applied for implementing changes needed to change the way operations are performed – or to implement new operational processes.

An attempt to compare policy programs and project programs is described in Table 11.2.

Understanding the term "program" is different in Public Administration and Project Management environments. Due to its essential importance and long-term use, it cannot be replaced by another word in either of these two areas. However, failure to address this issue may cause further misunderstandings between these communities. Hence, I propose to distinguish between *policy program* and *project program* concepts.

11.5 Public Policy Life Cycles

The public sector implementation cycle relates to a specific area that the government influences. It starts with identifying the area of activity in which public organizations or communities are interested. The area of interest may be the use of the Internet, road transport, or any other area necessary for the functioning of society. Some areas can be managed centrally (e.g., motorway

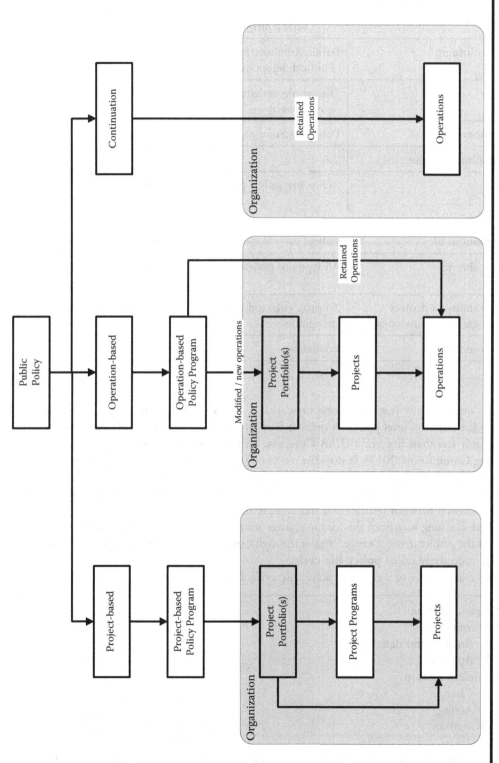

Figure 11.2 Role of Projects in Public Policy Implementation.

Table 11.2 Comparison of Policy Programs and Project Programs

	Policy program	*Project program*
Area of usage	Public Administration/ Political Sciences	Project Management
Scope	The whole activity of the government	Selected projects
Components	Policy instruments	Projects
Operations (product use)	Yes	No
Goal	Achieving policy goal	Improving results of managing component project
Component of	Policy	Project portfolio
Applicable to	All types of policies	Mainly project-based policies
Applicability of project management knowledge	To project-based policy programs	To all project programs
Governed by	Politicians	Managers

system), others locally (e.g., telephone network). A special body can be created to manage it. A strategy (i.e., the top-level policy) is defined for that area. The strategy implementation period is usually not less than five years (USA Congress, 1993; Kenya Government, 2013, 2013; India Planning Commission, 2013). It may last ten years (NSW Parliament, 1993) or a more extended period (e.g., Poland Rada Ministrów, 2017). Based on the strategy, policies are developed that are defined by the area of operation, intentions, and actions (policy programs). A general diagram of the relationship between policies and strategies is shown in Figure 11.3.

There is a long way from the idea to change something in the functioning of the state to the effect of the implemented change. It goes through certain phases to bring results to the state and society, collectively called **policy life cycle**.

The main phases of the public policy life cycle are:

- Agenda-setting
- Forming
  - Analysis and design
  - Approval
- Implementation
  - Change
  - Acting
- Evaluation

In the following subchapters, individual phases of the life cycle of public policies are discussed.

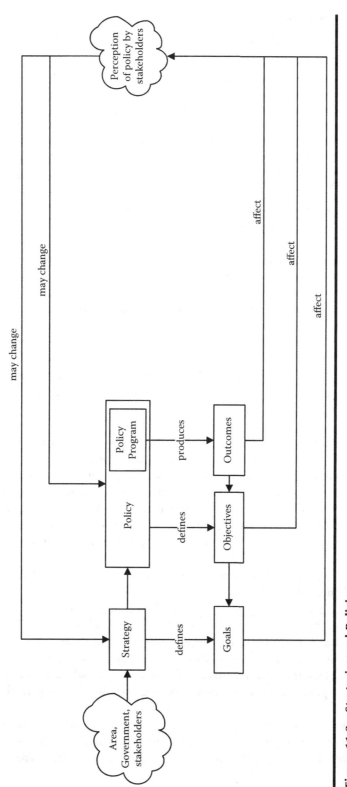

Figure 11.3 Strategies and Policies.

11.6 Agenda Setting

For the state to act in a specific area, first, there must be awareness in society or government of the existence of a problem that should be solved, or of an opportunity to use which could improve the functioning of the state or improve the lives of its inhabitants. In society, it may turn out that the level of education of people living outside the cities is too low. Or that the time it takes for residents to travel to work is too long. Or that the organization of the Olympic Games could promote the city and the state. Or that the effectiveness of missile defense is too low. Environmentalists believe that the state is not doing enough to protect the environment. The government recognizes that a specific issue needs to be addressed. The government itself may take this decision, or it may be consulted with the public, or the public or its groups may propose it.

This is the first phase of the public policy life cycle – including an issue in the set of issues of interest to the state, i.e., setting the agenda. The agenda lists the areas where state interventions are to be carried out and the reasons why the interventions should be carried out. For the picture to be complete – as I wrote earlier, the policies cover the entirety of state action – it must be remembered that earlier defined policies are implemented in areas not covered by the current agenda.

The agenda-setting process is not a simple, one-dimensional, rational process. As in the whole life cycle, political forces are at work in it, driven by various motivations. Some groups may be interested in maintaining the status quo, others in changing their functioning due to possible better social effects after the change, and others – in benefits, for example, financial, related to a possible change. Professional groups currently operating in a given area are often interested in maintaining the status quo. But these are not the only motivations in the agenda-setting process. These motivations are determined differently in each case.

Let us consider the policy of transition from classic coal-based energy to nuclear energy. For example, miners and employees of traditional power plants may be against implementing this policy. Ecological environments can support this policy for essential reasons as it reduces environmental pollution. They may also be the opposite due to the risk of a nuclear power failure and ecological contamination. In turn, construction communities may support the construction of new types of power plants due to the opportunity to use their production capacity. And energy specialists support the development of nuclear energy due to the limited resources of hard coal and the need to provide more power to the developing economy.

Introducing the issue to the government's agenda is the first fundamental, crucial step toward defining and implementing public policy.

Many researchers have analyzed the processes of including specific areas and issues on the agenda. For example, Kingdon talks about the policy window when a favorable environment is created for addressing a particular issue (Kingdon, 2003). Baumgartner and Jones believe that the government deals with a specific issue when an event undermines the rightness of the policy currently being implemented (Baumgartner and Jones, 1993). Sabatier believed that policies are placed on an agenda when a sufficiently strong coalition to support them is formed (Sabatier, 1988).

In the language of project management, we say that a stakeholder analysis should be performed in the agenda-setting phase (Subchapter 3.10). The stakeholder analysis is all the more important, the more directly the policy concerns non-governmental stakeholders.

11.7 Policy Forming

Once you have decided to address an issue in a specific area, you need to clarify the main policy parameters. The main questions that need to be answered are:

1. What are the phenomena, and how are they interacting in the policy area?
2. What do we want to achieve?
3. What means do we have at our disposal?
4. How will we achieve this?

The process of answering the first two questions is called **policy analysis,** and the process of answering the last two questions – **policy design**. In practice, these activities are not disjoint. They can be performed simultaneously; answering these questions often brings you back to some previous question. The mix of available resources can influence the goals you achieve. Understanding the nature of phenomena can change how these goals are achieved.

11.7.1 Analysis, What Is Going On?

Policy analysis is the search for the best solutions to existing public problems (Anderson, 2015) using the available knowledge (Howlett, 2019). The task of politicians is to shape public policies following the preferences of the groups they represent. Still, in most cases, politicians follow rational criteria when making decisions (Howlett, ibid) within the framework defined by their political preferences. As each problem relates to a specific area (or areas) of society, it isn't easy to formulate a single method of analysis that would apply to all policies. Transport problems are solved differently, differently military, and differently, problems related to the welfare of the society.

However, some general recommendations can be made (e.g., Patton and Sawicki, 1993), although these authors explicitly advise against using the entire pre-defined policy analysis approach. Perhaps the most important recommendation is to involve stakeholders in the analysis process. The inclusion of stakeholders is all the more important, the more directly the policy influences them. Stakeholder involvement is essential for two reasons. First, they know the subject of the analysis. Second, they usually have a specific attitude toward the analyzed issue. The most general division of public policy stakeholders divides them, according to the types of possible influence, into three groups (Ramesh and Howlett, 2017; Orchard, 2014):

1. *Political stakeholders*: The government, non-governmental political groups seeking to participate in the power process
2. *Market stakeholders, business*: entities involved in business activities aiming at maximizing commercial benefits
3. *Society*: Individual citizens and their organizations and associations pursuing civic goals

In addition, as a separate group, you can indicate:

■ *Non-national stakeholders*: For example, other countries, supranational organizations

The issues related to the stakeholders are governance structures: both the entities (who take part in the decision-making process) and the type of relationship (e.g., decision-makers, contractors, advisers, consultants). Action roles are well described using the RACI matrix (Subchapter 3.7). Taking these factors into account is essential in analyzing public policies.

Analytical work begins with an inventory of the current situation – describing the entities and processes in a given area. In business analysis, this stage is called "as-is" analysis. As part of this analysis, it is necessary to determine a set of stakeholders, their scope of influence, and who is the most important, i.e., has the power to make decisions. If the analysis is performed on a commission from a non-major stakeholder (e.g., trade unions, social associations), their understanding of policy success should also be identified. In the energy policy analysis, one of such criteria for the government may be the non-reduction of employment, not purely energy parameters. This success criterion should be the axis of the analytical work. From the area of business analysis, the Ishikawa diagrams (Subchapter 3.11) are well suited here, showing hierarchical cause-and-effect relationships relating to the selected criterion. Other techniques that can be used in policy analysis are process analysis (e.g., Recker et al., 2009; Weske et al., 2004) or data analysis (e.g., Lewis-Beck, 1995). Artificial intelligence techniques can be used to analyze social phenomena that are massive in nature, for example, to recognize the results of people's responses to specific stimuli (machine learning, neural algorithms, Big Data techniques; their description is beyond the scope of this book).

The ways of defining policies, depending on their orientation to the objective knowledge and needs of the society and interested communities, can be divided into four groups:

1. Based on citizens' needs and objective facts
2. Created by independent experts
3. Created by government experts
4. Defined directly by decision-makers

The most objective and community-oriented approach, based on objective facts, starts with identifying problems perceived by citizens and data about them by the agency responsible for the area of operation (USA Congress, 2018). If citizens complain about shortages in the number of educated mathematics teachers, then objective data should be collected after writing down this problem. For example, teachers' age, duration of employment, data on this issue in different states, etc.

The second approach is to outsource analysis to independent teams or individual specialists entrusted with creating a solution proposal to the government. Their independence increases the likelihood of obtaining an effective solution that is not oriented toward the current political preferences of the ruling party. Such a method has long been used, for example, in the United Kingdom and also in Australia. Sir Michael Latham's team was set up to solve problems in the UK construction industry (UK Latham, 1994). The UK government has twice commissioned Mr. Peter Gershon to analyze issues related to public procurement. As a result of the first of these reports (UK Gershon, 1999), among others, the Office of Government Commerce was established. Similar to this is the deliberative approach in which the solution to the policy problem is developed by a group of representatives of the community or society (often randomly selected).

The third method of solving public problems is the creation of strategic analysis teams in government structures. Creating such teams ensures knowledge-based decision-making but carries the risk of biasing policies toward the needs of the government rather than the public. For example, the Government Strategic Analysis Center operates in the Lithuanian government (Lithuania Government, 2021) or the Government Strategy Department in the Finnish government (Finland Prime Minister's Office, 2021). Strategic units can also be located at individual ministries, e.g., the Policy and Strategy Analysis Team, the US DoT (USA DoT, 2021), or the Strategic Research and Analysis Division of the Department of Transport Ireland (Ireland DoT, 2021).

The least advanced way to create a strategy is direct strategic decision-making by top decision-makers. With such solutions, the impact of knowledge on Policy-making is minimal, and there is a high probability that the policies will be geared to the needs of the ruling team. This method of determining strategic directions is characteristic of autocratic systems, although it may also appear in other political solutions.

11.7.2 Policy Goals

The goals of public policies can be defined in various ways. Some policies do not have a specific time perspective or indicator values that will cease to be implemented when achieved. An example is a policy of increasing the real incomes of citizens. If a policy does not have specific targets, it is advisable to have sub-targets for specific points in time. For example, your per capita income in 2025 will be US$ 35,000. Other policies have a specific target for certain indicators, but their achievement does not lead to policy abandonment. An example is the policy of ensuring equal income for both genders. After this equality is achieved, the implementation of policies should be continued so that no differences arise again. There may also be policies with a specific time perspective. Typically, such policies also have specific target indicators. For example, due to the policy of improving competitive sports, our national team will win 15 gold medals at the Olympic Games in 2028. Both for the time perspective and the values of indicators, not specific values, but ranges of target values may be determined.

The way of defining the policy should enable its evaluation, i.e., it should meet the SMART requirements (Section 3.2.1).

11.7.3 Policy Design

Policy design aims to define the mix of policy-making instruments and establish policy program. The policy program estimates activities duration and their dates. Policy design begins with the mapping and evaluating available instruments (Howlett, 2019). In conjunction with the knowledge obtained during the analysis, this knowledge determines the space of policy design (Howlett, 2014), in other words – from which set of policy instruments future policy actions can be selected. Design links knowledge and action (Hill and Hupe, 2014). The state of domain knowledge, the available instruments, and the knowledge of the stakeholders make it possible to define the policy goals. Public policies tend to have more than one goal – or in other words, a factor that must remain under control. A typical combination is a relationship between economic policies and employment (e.g., the length of roads built and the number of people employed in construction). Poverty reduction policy may target the maximum number of people with incomes below a certain threshold, the minimum number of new businesses created, and the minimum number of people who will find employment. The set of instruments leading to them should be indicated for specific purposes. Tinbergen's law (cit. after Mees at al., 2014) is in force here which says that for each policy goal, there must be at least one instrument that implements it. In social public policies, the task of which is to shape the behavior of the society, the participation of people who constitute the policy's goal in their development is very appropriate. In this situation, the policy-making process is primarily political in nature. It is a clash of political forces (Winter, 2012). The situation in infrastructural public policies or military policies is different.

In democratic countries, society primarily indirectly influences these policies through the election mechanism. In 2007, the Australian Labor Party came to power in Australia. One of the items on its political agenda was to create Infrastructure Australia, the body overseeing infrastructure development in that country (Infrastructure Australia, n.d.).

A way of designing policies very closely related to stakeholders is co-design (Blomkamp, 2018). Negative stakeholders are not involved in co-design. This way of designing is usually iterative: developed, in cooperation with some or all stakeholders, some versions of the policies are then presented to all stakeholders. In the next phase, the stakeholders express their views on the developed version of the policy and work out corrections. Co-design can increase policy complexity and a loss of state control over policy.

When designing a policy, in addition to the purpose and instrument mix, it is important to define the management method. For example, the Weberian model of management is appropriate for infrastructural instruments. More flexible, iterative management methods based on co-management are appropriate for purely social instruments to activate specific social groups.

The policy forming process identifies several possible policy implementation programs in the linear version of the decision-making process (USA GAO, 2020a). Each policy program may have different goals. The process of analysis and forming begins with the formulation of policy evaluation criteria. For a health policy related to a specific disease (e.g., COVID-19), these may be the cost of implementation, the likelihood of obtaining a cure, the number of fatalities, the level of economic losses, and the time taken to control an epidemic. The second phase identifies alternatives that are then assessed against the adopted criteria. The work results are documented and passed on to the next stage – approval.

11.7.4 Approval

Approval is the decision to start implementing a specific policy program. While in policies geared toward society, all stakeholders played an important role in earlier steps, government organizations play the major role in the approval. Due to their role in the state and society, they assume responsibility for the future of public policy. If the linear model is used, the policy parameters identified in the previous step, the level of political support, and available government resources are considered. The final decision is made by the minister, cabinet, or parliament, following the governance rules in force in a given country. According to the classification of Jann and Wegrich (2007), decision-makers are divided into those who make decisions based on knowledge (technocrats, rational model) and decisionists who primarily take into account political relationships (model of government policies).

Allison (1971), analyzing the decisions made by the US government during the Cuban crisis in 1962, distinguished three ways of making decisions. A rational model is when the government decides based on complete information about the existing situation and possible action effects (Jann and Wegrich, 2007). But governments do not always make decisions that way. The construction of unnecessary infrastructure facilities, excessive armaments, and missed propaganda programs are examples known to everyone. In the organizational behavior model, a problem is assigned to existing organizational units for analysis which usually deal with it separately. Organizations and organizational units have their action processes that run for decisions. Decision-makers consider only some of the obtained analysis results and options for action. Usually, they run the one that ensures the achievement of short-term effects. In this model, the decision is seen as the result of an organizational process. In the third model of government policies, they are seen as an effect of interactions (not necessarily formal) between political forces and, above all, between their leaders. Leader's advisers have different ways of acting, different backgrounds, different knowledge, different levels of charisma. They also have the power to prevent the policy from being implemented if they do not accept it. Therefore, the leader must seek consensus amongst his team.

Yet another interesting model of decision-making regarding public policies was developed by Kingdon (2003) based on the general garbage can model (Cohen et al., 1972). This model deals with a chaotic and dynamic situation known as "organized anarchy." The participants are loosely related to each other. The proposals change during the negotiations, the participants do not have a chance to participate fully, they have limited time and energy. Many things happen simultaneously. In this confusion, participants try to understand it all, especially their role. Problems, solutions, and participants appear and change randomly. Based on these three streams, suggestions for solutions also appear semi-randomly. Sometimes decisions are made, and sometimes they are not. It happens that problems other than those initially analyzed are solved.

Decision-making models – or work models more broadly – extend between the garbage can and GAO's highly structured model.

In summary, policy decisions consider the substance of the issue, the processes of policy development, and political relationships.

11.8 Implementation

Historically, in the old days, it was considered that issuing appropriate laws and regulations was equivalent to implementation; that is, post-approval actions were not of interest to practitioners and scientists. This situation changed due to the publication of the book Pressman and Wildavsky (1973), devoted to the implementation process, dealing with the Economic Development Administration (EDA) policy established by the USA Congress in 1965. This policy aimed to provide permanent employment to minorities through economic development. Pressman and Wildavsky described an EDA implementation in Oakland, California.

After the decision is made, the implementation phase begins. Implementation is turning policy into action. It is about making policy decisions that structure the implementation process (Hill and Hupe, 2014). Such a definition does not include the operation of the implemented instruments in the implementation process. In line with a different approach (Cochran et al., 2010), implementation also includes outputs and outcomes. This means that the operation and exploitation of the implemented instruments are also part of the implementation. Action leading to the achievement of the intended effects is undoubtedly part of the life cycle of public policy. Therefore, to distinguish a separate phase of public policy action, we recognize that the action is part of the implementation. An additional argument for such an approach is the possibility of introducing changes, i.e., elements of its implementation, in the policy during its operation.

A new policy is always a change. The only type of policy that can be pursued unchanged is the continuation of a policy previously followed in a given area – and in this section, such an approach is not of our interest. Two steps are therefore important in implementing a policy:

1. Change implementation
2. Operation of the policy

This division is related to reservations that have been formulated concerning each phase model of the life cycle of public policies, and above all, concerning the possible blurring of their boundaries.

New policies rarely wholly redefine the way of operation in a particular area, so their policy programs also include some previously implemented operational processes. For example, a new HR policy that promotes equality in all dimensions can use existing recruitment processes by changing only specific parameters. A similar comment applies to the salary payment process. However, new processes are usually at the heart of implemented policies. The implementation of new reconnaissance techniques can fundamentally change intelligence policies.

Implementation can involve all kinds of organizations operating in a given country – governmental, non-governmental, and private (Jann and Wegrich, 2007). Governmental organizations may use Laws and orders to implement policy. It is impossible to apply instructions to entities that are not part of the government administration, i.e., apart from legal regulations, only soft information tools or incentives may be used there.

A special organization that takes part in implementing policies (and, in fact, in all phases of the policy life cycle) is **implementation units**. The implementation unit is a component of the executive branch of the government. Its primary purpose is to ensure the efficient implementation of public policies. Perhaps the best-known public policies implementation unit is the Office of Management and Budget (USA OMB, 2021), operating (with different names for hundred years since 1921) in the Executive Office of the President of the United States. For public policy to be efficiently implemented, it should be defined consistently. The role of the implementation unit may be to participate in the definition of public policy, in particular, to assess and possibly improve the formal features of the policy definition. Implementation units also control the implementation of public policies. First, the policy should be implemented according to the objectives and indicators set out in its definition. However, modification procedures should be defined due to the complexity of the policies and the environments in which they are implemented. Implementation units react to deviations from indicators in different ways. Usually, the first step is trying to get the policy back on track as planned. This can be done by issuing appropriate instruction and, in more difficult cases, supporting the unit responsible for implementing the policy, such as seconding experienced staff. If it is not possible to return the policy to its original framework, the members of the implementation unit, alone or in cooperation with the implementation unit, formulate requests for change that are passed on to the administrative or political bodies responsible for the policy.

11.8.1 Change in Organization

The change may mean building a road, factory, or airport in infrastructure policies. In defense policies, change is implementing a new type of weapon. In restructuring policies or reforms of the administrative apparatus, change is the definition and establishment of a new organizational structure. In education policies, change is establishing a new school structure and curriculum. In the employment policy, the change means installing a new law and training employees of labor offices in the ways to implement that law. In tax policies, the change means establishing new laws, modifying IT systems, training employees, and publishing information materials. There is no public policy that would not require a change to be introduced or modified. Many instruments implement policies; a complete change is the implementation of changes for each instrument of a given policy (this can take place on different dates). The introduction of each change is project work, hence the importance of projects and project management for public policies. The changes we are talking about here relate to the organization's functioning. The construction of a new road is a change in the organization (of the state) in terms of the communication system.

> Every change to an organization is a project. Since the introduction of change is a central element of any public policy, projects are the most important management elements of its implementation.
>
> The efficiency of project implementation determines the successful implementation of public policies.

In the change process, two main activities should be distinguished: planning and execution (again subject to the possibility of combining and blurring them). Change should be planned – although we remember Gen. Eisenhower's famous statement, "plan is nothing, planning is everything,"[2] which says that before you start acting, you must think it through as much as possible. And the plan is, first and foremost, evidence that you have done some mental work.

The level of blurring of boundaries and the linkage of planning and execution is related to the type of policy. These activities are well separated in the "hardest" policies: infrastructure or military. It is impossible to build an airport without having a detailed construction plan in advance. On the opposite side are social policies. If the goal of social policy is to empower a particular social group, then it would be somewhat contradictory to impose on them ways to achieve that goal. Many researchers point to a blurring or combination of the introducing and exploitation phase of change in social policies (Lipsky, 1971; Sabatier, 1986; deLeon and deLeon, 2002; Section 11.8.4).

A review of issues related to implementing change in an organization can be found in Al-Haddad and Kotnour (2015). However, this work deals mainly with autonomous changes, i.e., the entire process of change, from the idea to the effect, is carried out in one organization. When implementing public policies, the situation is often different: an organization gets to implement a change defined by a policy established by an entity external to the implementing organization (we understand a state as a set of related organizations, not as a single organization). For a literature review on changes in public organizations, see Kuipers et al. (2014). In the life cycle of public policy, its purpose, and hence the purpose of change, is defined outside the process of change – in the process of policy forming. Changes related to implementing a policy established outside a given organization are called exogenous (Koning, 2016).

The factors enabling the implementation of change are knowledge, skills, resources, and staff commitment (Al-Haddad and Kotnour, ibid). However, virtually all authors dealing with change point to leadership as a critical factor in its success (van der Voet, 2014; Kuipers et al., 2014; Griffith-Cooper and King, 2007; Beer and Nohria, 2000). There must be a person in the organization who is deeply convinced of the necessity and benefits of change, has a vision of implementing it, and has excellent communication skills.

Change – as part of the life cycle of public policy – can be considered complete when the policy can be put into operation. In the case of hard policies, it is the moment when cars can start driving on a new road or the moment when a new ship is ready for service. In soft, social policies, where the behavior of target groups is of great importance, the moment of implementation of the change should be considered the moment when a new policy begins to operate, i.e., when a new law or regulation is established, new goals are defined (initially), social workers are trained. In policies with an intermediate level of hardness, for example, those that target people directly, but IT systems essentially influence their implementation, it is possible to pilot the change and full change when lessons are learned from the pilot version.

All issues and techniques related to implementing change in a public organization have their counterparts in the area of project management. When implementing individual policy instruments and the policy as a whole, you must manage stakeholders (Subchapter 3.10), scope of implementation (Subchapter 3.2), communications (Subchapter 3.5), knowledge (Subchapter 3.6), costs (Subchapter 3.4, purchasing (Subchapter 3.9), schedule (Subchapter 3.3), risk (Subchapter 3.8), human resources (Subchapter 3.7), and integration (Subchapter 3.12). Before starting the operation of individual instruments, make sure that the target environment is ready for their proper use. A change can be considered implemented when all outputs of the instrument or policy projects are transferred to the target environment (Subchapter 3.13).

11.8.2 Acting

Acting is the phase in which the policy begins to be applied in the target environment. The new attitude of politicians to a specific issue should start to bear fruit. Taxes are collected according to the new rates. Children in schools learn according to the new curriculum. New planes patrol the airspace. A new model of financing health care is being used.

The elements of policy action are the operation of its instruments, their implementation, and completion. The policy program team also reviews policy to see if the implemented activities are getting closer to their goals.

From the point of view of project management, the operation of a policy is the use of project products (in the language of Public Administration, we refer to the use of the effects of change).

11.8.3 Policy Change

However, environmental conditions may cause the current policy to no longer be effective. Airplanes are becoming obsolete and are not effective against the enemy. New knowledge is created children in schools should learn that. The aging population makes it necessary to increase the financing of health care. The new ruling team may have a different view of energy policy than the previous one.

Public policies often do not have a specific time perspective. Laws usually have a specific point in time from which they are effective, and the vast majority of them have no specific date on which they will expire. Therefore, there is a need to decide on the currently implemented policy. Should it be changed or ended?

Streeck and Thelen (2005) distinguish five types of policy changes:

1. Layering
2. Conversion
3. Displacement
4. Drift
5. Exhaustion

Layering is adding new instruments to existing policy while maintaining its goals. A simple example in the policy of improving agriculture may be the initiation of a project to build fertilizer factories.

Conversion is a change in the goals of the functioning of existing institutions. For example, during the systemic transformation, the office is implementing a project transforming the office that previously issued permits into the registration office for established enterprises. Conversion often takes place due to a change of party in power.

Displacement is re-use, this time for other purposes, of previously used but abandoned instruments. For example, a project to transform the institution of border control formerly performed to restrict citizens' access to a different culture into a tool to fight the COVID-19 pandemic.

Drift is a gradual, sometimes unplanned, instrument adaptation to new operating conditions and requirements. In line with the ruling team's changing preferences, public mass media are starting to prefer the new team's point of view without formally modifying the policy.

Exhaustion – the instrument slowly loses its purpose and ceases to fulfill its tasks, but it is not liquidated. An example is the preservation of the institution of the monarchy in many democratic countries, although it does not perform its primary task for which it was established: state governance.

Policy evaluation (Subchapter 11.10) can lead to policy change.

11.8.4 Top-Down and Bottom-Up Implementation

Linear, sequential implementation of policies has always been used, at least since the power of emperors was absolute, the hierarchy of power was strictly defined, and the subjects had to carry out all the orders of their superiors. Over the years, rulers have had to limit their power, adjusting it to established laws and norms. But the hierarchy was still in force, and the arrangements made had to be carried out as precisely as possible. In modern times, this has meant understanding public policy as defined by the government and executed by society. Such implementation of public policy is called **top-down implementation**. This included the approach of the father of political science, Harold Lasswell (1902–1978). He defined a phased, sequential cycle of public policy (similar to that described in this chapter) from intelligence to its application and appraisal (Lasswell, 1951).

Michael Lipsky (1971) turned attention to the role of street-level bureaucracy – the staff who are directly involved in final policy implementation. He analyzed the work of teachers and judges working directly with the target groups of policies and, on this basis, introduced the concept of "street-level bureaucracy," i.e., public employees working directly with policy's target groups. Lipsky's work has led to the singling of an implementation mode in which the "street-level bureaucracy" has a pivotal role in implementation. This method of implementation is called **bottom-up implementation** (Sabatier, 1986; Pülzl and Treib, 2007; Hill and Hupe, 2014). It consists of determining the implementation manner, based on the objectives defined by policy-makers, by entities involved in its implementation. Lipsky claimed they also have a deciding role in policy formation; they develop the final shape and result during implementation. Supporters of this approach are Sætren and Hupe (2018) and Winter (2012), among others. According to deLeon and deLeon (2002), the policy-making option by which street-level bureaucrats play a central role should be favored when implementing policies.

This is true only for certain types of public policies. In others like infrastructure or military policies, the role of street-level bureaucrats in policy formation is negligent. For infrastructure policies, employees of construction companies should probably be recognized as Lipsky's type of "bureaucrats." But their task is to develop a specific infrastructure element, not discuss its design while working on its implementation. After constructing 15 km of the highway, employees of construction companies do not debate which way the next 5 km should go. Street-level "bureaucrats" are even less important in military policies. It can be said that their role is wholly excluded. In the Manhattan project, the goal was to build nuclear bombs dropped on Japan. The pilots of bombers who flew to drop a bomb on Hiroshima should be considered "street-level bureaucrats." Could you imagine that these soldiers would have any freedom in carrying out their tasks? Any digression would probably have been called rebellion.

Berman (1978) distinguished two implementation modes: **programmed implementation** (top-down) and **adaptive implementation** (bottom-up). The programmed implementation is a kind of top-down implementation. The adaptive (bottom-up) implementation assumes that the original plans may be modified during implementation, which may improve its effects. Berman assumed the active participation of policy implementers in the adaptive process, whose knowledge should be considered in implementation.

If there are two main ways of implementation, it would be advisable to define the criteria that determine the choice of one of these methods. For Berman (1978), the choice of the implementation mode should be determined by the scope of changes (small – programmed implementation), the certainty of definition and means of implementation (large – programmed implementation), the existence of conflicts as to the goals and means of the policy (no conflicts – programmed implementation), interconnectedness in the implementation environment (strong – programmed implementation) and stability of the environment (high – programmed implementation). If the environment does not meet most of these criteria, the adaptive implementation should be used. Sabatier (1986) adds to the top-down factors the existence of the dominant policy program, interest in the average implementation score, legal structuring of the implementation process, engaged and qualified officials, and support from interest groups. Pülzl and Treib (2007) extend the set of criteria to include the elitist model of democracy in a given country.

Projects apply to both approaches to policy implementation. The sequential (waterfall) project life cycle is more appropriate for top-down implementations, while the agile approach is better for the bottom-up one.

11.9 Closing Policy Implementation

The termination of the policy implementation is the cessation of the implementation of its constituent elements, i.e., the policy program. The most natural and desirable way to end a policy is to meet the agreed targets. However, not all public policies have and can have clearly defined goals leading to their conclusion. Since public policy, primarily social, is often an emanation of certain political views, its ending may occur due to a change in the ruling team. In Poland, in the area of housing, the Housing for the Young policy was terminated due to the electoral failure of the Civic Coalition. It was replaced by the Apartment Plus policy, implemented by the Law and Justice team without an objective comparative evaluation of both these policies. Project-based policies, such as infrastructure policies, are relatively easy to assess. The rail transport accessibility policy ends when all cities with a population above a certain threshold (e.g., 20,000 people) are accessible via rail transport. More often than social policies, such policies may be continued regardless of changes in the ruling teams.

11.10 Evaluation

Evaluation is a comprehensive assessment of the effects of policy implementation, i.e., its policy programs. Evaluation is often described as a separate phase of policy implementation; however, it can also be done while the policy program is being implemented in our model. The primary purpose of the evaluation is to check whether the policy has been successful or is on the way to success. And its most important element is the reference to the expected goals. The evaluation task is also to assess how the policy program is implemented (Cochran and Malone, 2005). Suppose an evaluation is carried out during the implementation of the policy program. In that case, it may have two different goals: deciding to discontinue the policy or providing a basis for its modification. Evaluation can also be done at the beginning and end of the policy program. At the beginning of the policy program, the evaluation concerns pilot implementations and can be the basis for a policy program change (Jann and Wegrich, 2007). According to the Sunset Law regarding specific US policies (Gale, 2008), the implementation may be extended if the evaluation proves its continued usefulness.

The evaluation may concern the outcome of the policy program, which is driven by its outputs. The outcome is an observable change in the policy area, its social consequence (Anderson, 2015). A product is what is produced by the policy program, such as goods or a change in the way processes are carried out. The product of the policy program may be an atomic bomb, and the outcome – victory in the war. The product of the Public Administration restructuring policy program may be changed or new organizations, organizational units, and operational processes, and the effect – faster, cheaper, with a reduced number of mistakes, the performance of public services.

The success of the policy program, and thus of the policy it implements, can be viewed in three dimensions (McConnell, 2010): process, program, and politics. Policy succeeds in the process dimension if the processes are carried out as planned, deliver the planned outputs, and there is general agreement on how the process should be carried out. For example, in health policy, an adequate number of vaccinations have been performed, and a planned number of health centers have been established. The policy's success in terms of the program means that it has brought about the expected social effects. For example, health policy has reduced the incidence of a specific disease and increased life expectancy. Political success brings political benefits to leaders and ruling coalitions. The area of intervention is perceived as controlled by the ruling team; trust in it increases in the context of managing the policy area. The opposite of success determines policy failures by non-execution and non-delivery of outputs; failure to bring about positive effects and loss of confidence in ruling leaders and coalitions. Intermediate situations are possible in each of these dimensions. Therefore, complete success means success in each of these three dimensions. McConnell's model shows the complexity of the concept of success and failure in public policies.

The difference between the intended and achieved policy goals is **the implementation deficit** (or gap) (Hill and Hupe, 2014; Bredgaard et al., 2003). So, a failure is a significant implementation deficit. The failures of policies can be caused not only by mismanagement (Pressman and Wildavsky, 1973) but also by political factors: divergent interests of various groups or ideological differences. Failures of public policies may also be caused by other factors beyond the control of the implementing team, such as a financial crisis or a natural disaster (Van Nispen, 2011). The failure of public policy may also be due to its poor initial forming (Jann and Wegrich, 2007).

Evaluation is rarely an objective process that uses only formal models and rules. People with their values, experiences, and views are involved in the evaluation process, reflected in the evaluation process and outcome. Therefore, people representing different views and values must participate in the evaluation process (Bachtler and Wren, 2006). To try to objectify the evaluation process, it is advisable to define its indicators. They can be divided into four groups (Grzeszczyk, 2012), regarding efficiency, regarding effectiveness, regarding the durability of effects, and regarding relevance. Relevance is understood as addressing the problems actually faced by the policy's target group.

Example. The goal of the policy was to improve the demographic situation. For this purpose, money transfers were made for each child born. After four years, it turned out that the number of births did not increase, but the lowest level of poverty among children was eliminated.

Additional criteria may relate to people, teams, and policy-making organizations, such as empowerment and gaining knowledge and experience. The knowledge about the evaluation process itself can also be improved during this process.

11.11 Types of Public Policies

From the managerial point of view, the most important classification of policies is their division into project-based and operations-based policies (discussed in Subchapter 11.4). Policies can also be classified in many other ways. Each of them contributes to a better understanding of this concept.

11.11.1 Acting and Non-acting

There are areas where the government (and society) is satisfied with state action. In these areas, the state policy may be to do nothing, that is, to use only the instruments of doing nothing. A second reason for not carrying out an intervention may be a lack of resources. Governments, too, can do nothing in a specific area if there is no political agreement to pursue any policy. A typical example of a conflict that prevents action is controversy: more state or less state. In the face of the economic crisis, the clash between these two options may make any action impossible. An intervention is a policy of carrying out activities that modify (or introduce) specific solutions in a given area.

11.11.2 Substantial and Procedural Policies

The classifications from the instrument level can be transferred to the policy level, for example, the division into substantive and procedural policies. The substantial policy is based on the distribution of certain products and a procedural one – on changing or establishing a mode of action to a certain extent. But this classification is not a simple summation of instrument types. For example, the substantial policies may consist of instruments that are not only substantial. The poverty reduction policy in a given region is substantial. Still, it may consist of substantial instruments (e.g., food distribution) and procedural instruments (e.g., the implementation of the process of growing appropriate plants). Defense policy is undoubtedly an example of a policy relating to the mode of operation, i.e., procedural. Still, it must use substantial instruments relating to implementing new types of weapons.

11.11.3 Level of Stakeholder Engagement

Policies can directly or indirectly affect members of society, i.e., their non-governmental stakeholders. For example, education policies affect people directly because residents (especially children) are directly involved in the learning process. The same is true for poverty reduction policy – residents of poor regions or countries are included in the process both as recipients of goods and as people learning to produce these goods. Infrastructure policies, such as the transport policy, impact citizens less intensively in a different way. While citizens are road users, their influence can be important in consultations on the structure of road connections, but not in the process of building them. Foreign policy has an even less direct impact on citizens: people observe its effects, e.g., through the possibility of traveling to other countries, but they are rarely included in discussions about types of relations with other countries and are not included in diplomatic efforts. Citizens have an even smaller share in forming defense or intelligence policies – the latter excludes any knowledge or involvement of people other than professionals in their definition and implementation.

Stakeholder engagement may be assessed not only quantitatively but also qualitatively. Public policy stakeholders can be positive, indifferent, and negative. A positive stakeholder would like the policy to solve a specific problem. Usually, travelers are positive stakeholders in transport infrastructure policies. A negative stakeholder is interested in the failure of public policy. A hostile state is a negative stakeholder in the foreign policy of another state. The producer of competitive business good is a negative stakeholder in producing a similar good by another entity. In practice, most stakeholders of most public policies are indifferent – every resident of a country is a stakeholder in any public policy in this country. But usually, apart from the immediate target groups, hardly anyone is interested in natural resources policy, administrative policy, or penitentiary policy.

Target policy groups are a particular type of stakeholder. Policies can be classified according to the entity's level that benefits them (Anderson, 2015). Benefits may be targeted at individuals, communities, or society as a whole. An example of the policy of the first of these categories is the income redistribution policy. Examples of policies aimed at society as a whole are defense policy or environmental policy. Target groups can also be positive or negative stakeholders. Negative public policy stakeholders happen in cultural policies. For example, the policy of eradicating religion met with a strong adverse reaction from many Eastern Bloc peoples after World War II.

The level of the possibility of involving stakeholders in the policy process is one of the most important criteria for their classification.

11.11.4 Policies of State Development Phases

As mentioned above (Chapter 7), countries undergo three development phases: ensuring security, infrastructure development, and ensuring welfare. Public policies have different goals at different phases of a state's development. Therefore, they can be divided into:

1. Basic (military, defense, and order ensuring), most useful in the first phase
2. Infrastructural, implemented mainly in the infrastructure development phase
3. Social, implemented in the welfare phase

This classification is essential from the point of view of the involvement of the authorities and society in shaping and implementing public policies.

The government is fully responsible for basic policies. The instruments for implementing the policies of this category are not only the armed forces but also the law, in particular criminal law. The government has to define policies, implement them, and be responsible for their effectiveness. In this phase, society is primarily the subject of policies. If basic policies are also implemented in the following development phases, the influence of society on them is manifested in the electoral process; it does not deal directly with policy formation – on the contrary, very often, the work on defense policy instruments is done in strict confidence. Society is forced both to implement policies and exploit them – that is, to live in a given country. The exploitation of policy instruments of this type may be the daily, peaceful functioning of relevant services or the implementation of war projects where the decision-making of the performers is limited as much as possible. Both project-based policies (military, defense) and operation-based policies (normalizing domestic life) are implemented in this phase.

The government defines infrastructure policies with possible public participation. The government has a decisive role, but consultations are conducted, for example, on the route of expressways or the location of classic or nuclear power plants. The government approves infrastructure decisions. The change being the base for a policy (Section 11.8.1) in the case of infrastructure policies is building some facility. Although private companies often carry out the construction itself, the government is entirely responsible for these activities. Here, the division into the phases of the life cycle is the sharpest. The use of an infrastructure instrument as a part of an appropriate policy must first be decided and then built to be operational. Project-based policies are the most common in this phase.

The most blurry and complex – and therefore the most rewarding subject of research by social policy specialists – are social policies. In them, society is involved at every stage of their life cycle. The vast majority of the literature on the subject treats it. Society, particularly in the governance model (Subchapter 12.3), is involved in the formation and the change necessary to implement policies. Both the government and society exploit the results of social policies (e.g., education, payment, and collection of cash benefits). These types of policies most often include iterative and incremental processes.

To ensure welfare, first and foremost, operation-based "soft" public policies are needed (e.g., participation in cultural life, reduction of poverty, improving the level of education). Project-based policies are supportive here – for example, building schools or cultural facilities. The contemporary science of public policies has just been created in highly developed welfare countries – and hence probably does not acknowledge the role of project-based hard policies that are characteristic of the countries' infrastructure development phase.

Table 11.3 compares the main features of the different types of policies.

11.11.5 Policies in the Electoral Cycle

Public policies are not static; what is more, their variability results not only from the transition between the above-described phases of the public policy life cycle. Policies also change due to the phase of the electoral cycle. Van der Waldt (2016) distinguishes four hierarchical levels of policies, occurring in different electoral phases.

Table 11.3 Implementation of Policies of Different Types

Policy type	Formation	Change	The dominant type of policy
Basic (military, defense, order)	Government	Government	Project-based/ operation-based
Infrastructural	Government/ society	Government	Project-based
Social	Government/ society	Government/ society	Operation-based/ project-based

First, there is the politics level. Public policy projects are on the agenda of political parties. At this level, completeness is not required – in the sense of covering the entire area of operation of the Public Administration. Political parties focus on selected issues that are most important from the point of view of their value systems and are essential to society. For example, the neoliberal governments in the United States and the United Kingdom have promoted efficiency and reduced bureaucracy in the administration. In times of epidemic, many parties emphasize issues related to supporting enterprises and citizens in the face of the threat of economic downturn and unemployment.

The party that wins the elections must turn its political program into a plan of action for the whole state. Many action areas remain the same – no country could withstand a simultaneous change in all action areas. After forming the new government, the areas of activity can be divided into those that will implement the existing policies and those where changes will be made. This is the governmental policy level. At this level, laws are prepared and passed.

The next level of policies is the executive level – important above all for new or changed policies. At this level, politicians work with civil service managers to identify ways of policy implementation. Projects necessary to implement the policies are identified. It is determined whether new institutions will be created, new processes will be needed, or it is enough to modify the existing ones.

The lowest policy level is the operational/administrative level. Concrete actions to make the policy work. Implementation of projects. Allocation of resources, determination of working methods, implementation of projects – both operational and investment. Identification and implementation of policy implementation control methods.

11.11.6 Covering the Areas of State Activity with Policies

At every moment of its existence, the government implements a set of policies that cover all of its areas of interest. Policy-related activities appear in the work environment of the entire government, that is, in the domain of other policies. From the point of view of relations with the existing policies, they can be divided into four groups (ordered from those requiring the most of new activities, to those requiring the least):

1. Covering a new area with the interest of the state (state expansion)
2. Modification of the policy in the area that was subject to state influence
3. Withdrawal of a state from a specific area (devolution, collapse of the state)
4. Continuation of the current policy

Expansion is often understood geographically – as gaining new areas by the state. Today, there is probably only one country expanding geographically – Russia, which occupied the Crimean Peninsula. But expansion may also have an institutional meaning – introducing mechanisms controlled or implemented by the state in new areas. A well-known example of the state's expansion is enacting the Patient Protection and Affordable Care Act under President Obama (USA Congress, 2010b), extending the scope of compulsory health insurance in the United States. Another example, also from the United States, is passing the PgMIAA law governing the management of (both: policy and project) programs (USA Congress, 2015).

An example of an area where all states operate is tax collection. The modification may consist in introducing or withdrawing a specific type of tax, changing the rules for paying taxes (e.g., introducing tax exemptions), or changing tax rates. An example of a state withdrawal may be the abandonment of the required licenses to practice certain professions. The continuation of implementing the existing policies is probably quantitatively the most common approach to implementing public policies. No state could withstand the simultaneous modification of most or many ways of carrying out its tasks.

This classification is complemented by the classification of Masci (2007) and Anderson (2015), who distinguish two types of policies: regulatory and deregulatory. Regulatory policy is included in items 1 or 2 of the previous classification, and deregulatory policy – in items 2 or 3. Regulatory policies' goal is to increase the number of regulations in force in a given area, and deregulation policies – to reduce them. In post-communist countries, deregulatory policies, for example, those related to the market, dominated in the first phase of state reform. One of the essential components of this policy was the change in the mode of establishing companies: from the approval of the state apparatus to the mode of business registration. Regulatory policies appear as a result of the expansion of the state's activities or a change in the state's attitude to freedoms. As a result of the emergence of radio technology, it has become necessary to regulate some issues related to it, for example, the allocation of frequency bands. It was forced by the technical necessity of not overlapping the broadcasts of different broadcasters. Attempts to regulate Internet communication are examples of a regulatory policy related to the expansion of the state. For a long time, the Internet managed to cope without regulations, so their introduction is state-driven, not a technical necessity.

11.11.7 Unambiguity and Conflicts as Policy Classifying Factors

An important and frequently used model is the classification defined by Matland (1995), which takes the level of conflict caused by a policy and the level of unambiguity of its definition as the main factors that distinguish policy (Table 11.4). The value of each of these attributes can be specified as "low" or "high."

A policy with a low level of ambiguity and a low level of conflict is one everyone supports, and you know what needs to be done. An example is a policy of ensuring fast communication between major cities in the country. The activities implementing this policy are projects to build sections of motorways and build sections of high-speed rail. You know what needs to be done, and political forces do not argue about it. Such an implementation is called an administrative implementation.

A policy with a high level of ambiguity and a low level of conflict is one that everyone supports, but the ways to achieve the goal are not well known at first. An example is a policy of gaining a significant position in global sport. The goal can be defined as achieving the first place in the top ten of the Olympic Games within ten years. But the state may not know how to train

Table 11.4   Ambiguity and Conflict as the Basis for the Classification of Policies

		Conflict	
		Low	*High*
Ambiguity	Low	Administrative implementation	Political implementation
	High	Experimental implementation	Symbolic implementation

Source: Matland (1995).

athletes to get them high enough. Therefore, projects are implemented to implement various training methods, and then those that lead to victories are selected. Such an implementation is called an experimental implementation.

An example of a policy with a low level of ambiguity and a high level of conflict may be reducing traditional coal-fired energy for the benefit of renewable energy sources. The most important activities are pretty well defined: mine closure projects and projects to build solar, wind, and nuclear power plants. However, such a policy causes conflicts with workers and trade unions in the coal mining industry. Thus, political action is necessary to persuade and legislate or even use force if necessary. Such implementation is called political implementation.

The situation is most difficult when the policy can cause many conflicts and the criteria for achieving the goal are not well known. An example is a project to reform the education system in a developed country (Zahariadis and Exadaktylos, 2016). Conflicts arise between the political and executive levels, leading to only symbolic, superficial (e.g., changing the names, not the tasks of organizational units) and not a real improvement of education. In this type of implementation, visible actions are exposed, not those that are effective (Flores and de Boer, 2015). It is called symbolic implementation.

11.12 Management and Public Policies

In the area of public policies, policy-making and management are pretty distinct. Policy-making is generally about setting goals. And management includes projects and other activities aimed at producing desired services or target states (Asquer and Mele, 2018), including efforts to co-ordinate the use of resources. The word "management" is intended, among other things, to suggest that these are activities analogous to those performed in the private sector (Hill and Hupe, 2014). Public management must consider the role of the market and the performance of public sector organizations (Bouckaert and Halligan, 2008).

For example, in policy to improve road safety, Policy-making means identifying a rate by which accidents are to be reduced in terms of kilometers traveled and checking whether this rate is met. And the task of public management is the implementation of projects for the construction of collision-free intersections or traffic lights. Benefits are managed at policy, and its program-level and outputs are managed at the project level.

Politicians are responsible for making politics, while public managers for public management. They operate within the powers vested in them on the political level; politicians, on the other hand, should not directly interfere with the activities of public managers. Public managers can achieve the objectives of public policies in various ways (Hodgetts and Luthans, 1997), which creates space for decision-making. But according to Asquer and Mele (2018), public managers are also irrelevant in making policies which is consistent with the separation of the political and administrative layers. Johansson (2009) considers them intermediaries between politicians and public organizations that implement policies. This excludes management functions within public organizations. But in turn, Bovaird and Löffler (2009) extend public management also to organizations of other sectors that provide public services. For this book, I assume that a public manager is anyone involved in project management or other activities involved in implementing policies. The direction of responsibility of public managers is important (Khal, 2011). Their possible accountability to society would remove some responsibility from politicians. But politicians, when hiring managers, define working conditions and thus take full responsibility for their actions. Consequently, managers are accountable to politicians, and only politicians – to society.

Accounting for public managers by society would help politicians to avoid accountability. The public agency, not the private company, is responsible for delaying the road construction project by a private company contracted by the public agency (of course, the private company should be accounted for by the agency, but it should be treated as an internal issue of the agency).

DeLeon and Steelman (2001) mention communication, information management, conflict resolution, mediation and facilitation, an appropriate level of ethics, and values as necessary competencies for public managers. And according to Frederickson et al. (2012), managers should fulfill not only technical but also leadership functions, particularly the ability to motivate teams to take desired actions. The same authors emphasize using human resources in the definition of public management. In public management, Löffler (2009) distinguishes communication with stakeholders and operational management, directed at the implementation team, the so-called back office (Radu et al., 2008). And Jreisat (2011) emphasizes data, best practices, and work optimization. On the other hand, Patapas and Smalskys, in the area of public management, pay attention primarily to the mobilization and efficient use of resources and emerging opportunities (Patapas and Smalskys, 2014). Managers should consider their managerial flexibility, i.e., the ability to choose from among the perceived options (OECD, 2005). In management, managers should consider cultural factors (Negandhi and Prasad, 1971). For example, some cultures prefer hierarchical decision-making (rather than decentralized), cultures that prefer secure decision-making (rather than tolerate risk), or that like action-competition (rather than collaboration) (Hodgetts and Luthans, 1997).

This exchange of views on management shows how early is the knowledge of management in the area of Public Administration. Project management has a well-coded canon of knowledge, where there are recently only two main approaches on the global market: created by PMI with the top product PMBOK® Guide (PMI, 2017a) and created by a British government agency, Axelos, with the main product Prince2® (UK OGC, 2017). The mainstream of project management research has not dealt with discussions on a set of areas to be managed for several decades. This may be why researchers in Public Administration deal with particular areas of management in a way that gives the impression of being unsystematic. Perhaps structured knowledge in the area of project management would assist researchers in Public Administration.

Notes

1. Organizations of level lower than government may also have their local policies, but they are out of the scope of this book. However, some here presented considerations may be also applied to them.
2. However, it should be remembered that General Eisenhower operated in a military area that is extremely vulnerable to enemy actions, the main purpose of which is to prevent the implementation of our plan. In non-military projects, there is usually no enemy, so Eisenhower's statement should not be carried uncritically into the peace sphere.

Chapter 12

Phases of Understanding Public Administration

The tasks of the state are performed by the public administration, i.e., "government in action" (Wilson, 1887). The point at which these activities became the subject of scientific interest is believed to be the publication of the above-cited article entitled "The Study of Administration" in the Political Science Quarterly by the future US President Woodrow Wilson. Since the publication of this article, science and knowledge about public administration have been developing intensively, going through successive phases. Each of them made a significant contribution to the understanding of the subject of their analyses and was the basis for improving public administration activities (at least those interested in improving their functioning for the benefit of the governed societies).

Public Administration (capitalized) is a science that studies public administration (Hill and Hupe, 2014). It is primarily focused on studying the phenomena that actually occur there. Public Administration uses the scientific foundations of law, organization, and management (Bauer, 2018).

The following sections discuss the main currents and phases of understanding public administration.

12.1 Classical Public Administration

Wilson's paper began the phase of what is now called Classical Public Administration (CPA) (Easton, 1965; Osborne, 2006; Frederickson et al., 2012). During this period, Maximilian Weber built his model of the functioning of the state (Weber, 1947, 1952, 1994) based on hierarchy, procedures, and qualifications. Hierarchy is one of the essential concepts in the field of public administration, in particular in the Weberian model of the state. Hierarchy is the ordering of the components of a structure according to the dominance relationship between them. In the past, in autocratic, absolutist, and centralized states, subordination was complete: the ruler could decide about the organizations and people living in the area where he ruled to support him. In modern states, the scope and manner of hierarchical subordination to higher units are strictly defined by the constitution and the law. A common type of relationship between organizations from

DOI: 10.1201/9781003321606-14

different levels of the state structure is setting the goals of the subordinate unit, determining the budget, and monitoring and enforcing the achievement of goals. Each official has one superior whose orders he is obliged to follow and to whom he reports on the results of his work. Power is depersonalized; actions result from the role held, not from the characteristics of the person who performs that role or from the relationship between the applicant and the official. Officials act as performers of specific roles. Officials and their teams are specialized. Each of them has a strictly defined scope of work and authorizations, resulting from the provisions of law and the definition of the role. Clear, unambiguous procedures govern the working methods of officials. They must not perform activities that are not prescribed by law, and those performed should be performed to the best of their knowledge. The law and operating rules should minimize the discretion of officials. The results of work and decisions made must also be documented. Officials are recruited based on the compliance of their competencies with the requirements of the posts. Promotion is based on qualifications, not on acquaintances and personal relationships (in the United States, this principle was introduced in 1883 through the Pendleton Act, established as a result of the assassination of President Garfield by a man who did not receive the expected office after Garfield took office). Officials are impartial in making decisions. All these rules should guarantee the proper functioning of the administration.

The state described by Weber is called a **bureaucratic state**. It is schematically presented in Figure 12.1.

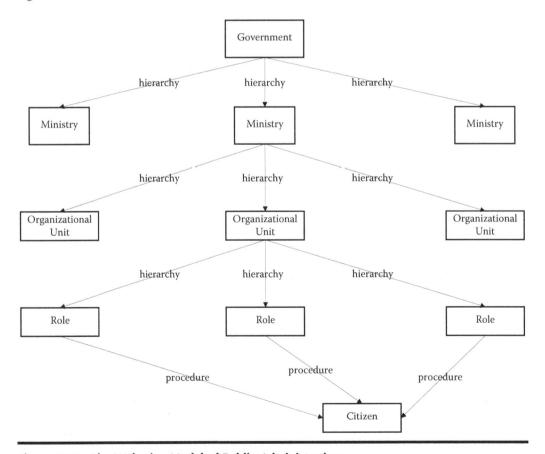

Figure 12.1   The Weberian Model of Public Administration.

Scientific management methods are used to evaluate the activities of public administration (Taylor, 1911/2010), according to which the numbers of public services provided are determined and compared with the requirements.

According to the classic approach to the functioning of the administration, if a medical service is needed, the patient reports to a state hospital with specific operating procedures and performs the appropriate procedure. After the service has been completed, it is possible to analyze whether it has been performed following the applicable procedures.

In the classical period, there was a division into political institutions forming policies and public administration based on the public service, which professionally performed the orders of politicians (Wilson, 1887; Goodnow, 1900). Work has been done to increase the efficiency of public administration. The POSDCORB model was defined, according to which the primary management functions are planning, organizing, allocating people, directing, coordinating, reporting, and budgeting (Gulick, 1937).

Later in the classical phase, the division into politics and administration was questioned. The relationship between knowledge about administration and other fields of expertise, such as psychology, sociology, or economics, was considered (Simon, 1947; Dahl, 1947).

The main sentence characterizing CPA can be taken: there must be order.

When CPA was the primary model of state functioning, the public sector started implementing projects. The first work to be considered a project was the production of atomic bombs for the US army to bomb Japan in 1945 (Gosling, 2010). This project was nicknamed "Manhattan." In the following years, project management in the US administration, particularly the Department of Defense, was used more and more often. Still, it was not included in the scientific models of government functioning.

12.2 New Public Management

The Weberian approach was criticized. Probably the essential elements of this criticism relate to the administration's attitude to procedures rather than their effects, the omission of informal structures, and the threats to democracy associated with an excessive expansion of administration (Merton, 1963; Blau and Meyer, 1987; Fry, 1989; Wilson, 1989).

Margaret Thatcher is associated with the beginnings of a new way of thinking about public administration, later called the New Public Management (NPM). The main idea of this school was to increase the efficiency of the state by introducing into it management methods used in the private sector (Hood, 1991, 1995; Osborne, 2006; Ferlie, 1996; Frederickson et al., 2012). In line with the NPM approach, the market mechanisms, competition, and contracts are used to perform public administration tasks. NPM promotes efficiency, effectiveness, innovation, benefits, and quality (Lane, 1994). Instead of emphasizing procedures (which still have to be followed), the result of the action becomes the most important. Recipients of public services are treated as clients in the private sector. It is permissible to adapt public services to the needs of recipients. The use of resources is optimized, for example, by implementing efficiency standards and controlling the results of service provision, both quantitatively and qualitatively. Audits of the activities of organizational units are carried out. Cost control is introduced for elements of organizational structures, and relatively small structures are distinguished so that the supervision over finances can be as accurate as possible. The creation of detailed analytical accounts also serves for greater cost control. Competition between organizational units providing similar services is being introduced. Separate agencies are created to provide specialized services. Some public services may be provided under contracts by companies from other than public sectors. Through agentification and contracting out, the state apparatus is significantly reduced. The function of superiors in public administration is management, not administration. They are referred to as managers, not administrators. The model of state functioning according to NPM is shown in Figure 12.2. The concept of managerialism (Enteman, 1993; Locke and Spender, 2011; Pollitt, 1990) is born, denoting the vital role of managers in the implementation of public policy.[1] Public sector workers receive a variable wage based on their performance.

NPM enriched, and did not replace, the traditional Weberian model of a bureaucratic, hierarchical state (Pollitt, 2007; Bouckaert et al., 2011).

The NPM view of medical service provision is different from that of CPA. The state considers whether it will be cheaper to maintain a medical facility or pay for services in private companies, possibly many in competition. The patient wonders which facility provides the services he needs best and reports to this one. It is then assessed whether the service has been performed well, cheaply, and effectively.

While traditional public administration functioned in many, if not most, countries of the world, NPM was implemented in relatively few administrations – the United Kingdom, the United States, Australia, New Zealand, and other Anglo-American countries (Torres and Pina, 2004).

The main sentence characterizing NPM: act as a market company.

Before NPM was defined and implemented, project management infiltrated from the public to the private sector, becoming one of the primary management modalities. Due to their effectiveness and efficiency, the projects were introduced for application to the NPM core toolkit and

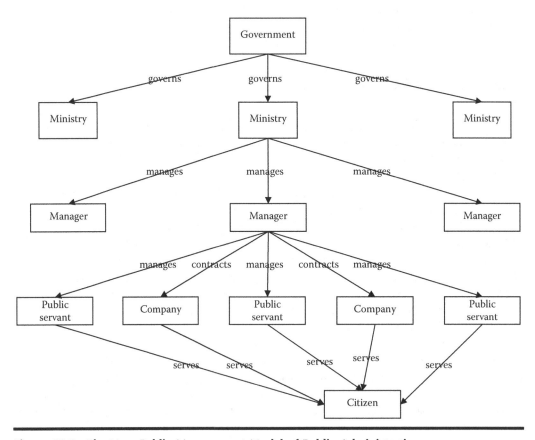

Figure 12.2 The New Public Management Model of Public Administration.

are used successfully in all areas of public administration. New models of state functioning do not negate project management, although they do not pay much attention to it.

12.3 New Public Governance

As knowledge about public administration developed, it was noticed that the mechanisms of hierarchical authority, even supplemented with efficiency issues introduced by NPM, still do not meet the needs of societies and do not support the effective implementation of government tasks (Ramesh and Howlett, 2017). The two main currents of thinking about the state that have arisen out of criticism of the NPM are New Public Governance and the Neo-Weberian State.

The NPM did not notice that the state consists of many related entities, not necessarily hierarchically dependent on each other (Liddle, 2018). The results of the state functioning are also influenced by factors other than hierarchical power. The first such factor is the market and the economy. There is a market in every country. Economic operators produce and trade goods and services. Demand and supply shape prices. Purchase and sale transactions are performed. Market participation opportunities determine the level of welfare of a country's citizens. All countries have economic entities that produce goods that are useful to society. The state regulates economic phenomena in free-market countries only to a limited extent. On the other hand, economic phenomena have tremendous and various influences on its functioning. They define what regulations are needed. Due to the amount of taxes collected, they determine the level of possible government actions. Economic entities form their associations. The market is largely beyond the control of the government and must be considered one of the main factors determining how a state functions.

Citizens also influence the functioning of the state, and in particular by their organizations and associations. For example, the National Rifle Association in the USA is committed to upholding the right of citizens to own guns. Business Botswana is an association of private entrepreneurs in Botswana and representing them. The Cikananga Foundation is an organization dedicated to the conservation of wildlife in Indonesia. In every democratic state, citizens' associations are regulated by law but are not subject to authority. In Poland, in 1980, the central axis of the conflict was to force the communist leaders to establish trade unions independent of the government (and in fact of the communist party and the Soviet Union). Organizations independent of the government, if they have the appropriate number of members and the power of influence, can influence the functioning of the state, including law-making. The role of the National Rifle Association in the United States in protecting the right to own a gun is well known. Medical self-governments can be the body that evaluates doctors and determines their right to practice. Free media are a specific type of entity influencing the functioning of the state. When legislating in a particular area, the state may consider the existence and operation of such associations.

The state functions well if it enables good functioning and cooperation of all entities operating on its territory. The three main types of forces interacting with each other are the hierarchical state, citizens and their associations, and the market (Löffler, 2009; Ramesh and Howlett, 2017). These observations became the basis for a new understanding of the functioning of public administration: New Public Governance (NPG; Osborne, 2006). In line with the NPG approach, public services are delivered through networks of cooperating entities. With this approach, one of the main tasks of the state is to ensure the functioning of social and market entities and to ensure painless interaction between them. One of the main areas of state activity is stakeholder management – each social and market entity is a state stakeholder. Considering these entities, the state functioning pattern becomes more complicated, as shown in Figure 12.3.

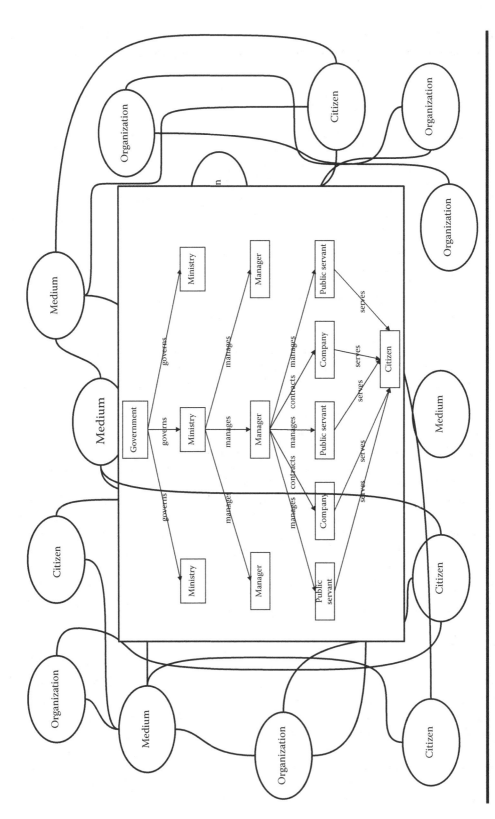

Figure 12.3   The New Public Governance Model of Public Administration.

Health services can be provided by privately owned hospitals financed directly by patients, private insurance companies, or the state budget. There may be many entities providing similar medical services on the market. They use personnel educated in private or state universities. Healthcare facilities contract investments, as well as some services, from the private sector. Patients can join associations of victims of the health service. All of this is described and assessed by independent media. The state's role is to ensure, usually by enacting laws and other regulations, that such a network of related but independent entities functions efficiently.

Unlike NPM, this approach promotes cooperation, not competition, between entities providing public services (Torfing and Triantafillou, 2013).

The main sentence characterizing NPG: efficient functioning of a network of independent entities.

The NPG meets with criticism that may contribute to the emergence of new schools of thinking about the state. For example, Howlett (2020) alleges that the NPG is little interested in providing services directly by the state, hierarchical management structures, and minimally in the policy-making process.

CPA was interested in the procedural hierarchical functioning of the state. The NPM supplemented the knowledge developed in the Classic Public Administration with issues related to the efficiency of service provision. NPG adds to this knowledge issues related to the management of networks and stakeholders.

12.4 Neo-Weberian State

NPG was created out of criticism of the failure to take into account the role of the citizens and their communities in the functioning of the state. The second line of criticism of the NPM was more fundamental. Its main principles were criticized, mainly relying on efficiency, transferring business principles to public administration, and generally not distinguishing between the state and business companies. Drechsler (2005) criticizes *en block* all NPM principles, including project management, but he does not refer to any work here. It seems to be based solely on his attitude. The result was creating a new approach to public administration, called the Neo-Weberian State (NWS).

In 1999, the OECD defined the principles of public administration (OECD, 1999), which significantly referred to the concept of the Weberian state. These principles were then described by Pollitt and Bouckaert (2004). They are usually quoted in full and consist of parts derived directly from the Weberian state and new elements.

The NWS has Weberian elements:

- The state is the leading facilitator of solving social problems.
- Representative democracy legitimizes elements within the state apparatus.
- Administrative law normalizes relations between the state and the citizen.
- Public services have distinguished status, culture, and terms and conditions.

The NWS has new elements:

- Emphasis on satisfying citizens' needs and wishes instead of emphasizing bureaucratic rules.
- Enriching the role of representative democracy through consultation and direct consideration of citizens' opinions.

■ In government resource management, be more focused on delivering results than following procedures.
■ Professionalization of the public service. Bureaucrats become experts not only in the field of law but also managers oriented toward meeting the needs of citizens/users.

The main sentence characterizing the NWS: order is the basis of efficiency.

A sick person comes to a clinic. The lady at the desk knows which surgery to send him to. The patient is taken over by a specific doctor who guides him throughout the disease. If a sanatorium is needed, the patient's preference is indicated. After completing the treatment, the patient fills in a questionnaire regarding the quality of treatment.

Despite criticism of the NPM, new state elements are borrowed from this model of state functioning.

It is natural to ask: Why can the functioning and the NWS model not consider the rules concerning the market and citizens that underlie the NPG?

It is difficult to assume that NPG or NWS, a certain school of thought about public administration, is the last such school. What will be next? Maybe syncretic, connecting all three major schools of thought so far? Maybe an emphasis on risk management? Or maybe project management will be recognized as the main factor determining the success of public administration? This way, we would have the Projectified State. This book goes in this direction. When you look at what is going in the project management area, another plausible concept is the Agile State.

Note

1. Managerialism is the opposite of politicalism (Childe, 1964), that is, making decisions based solely on political preferences. Each style of public decision-making is a mixture of politicalism and managerialism.

Chapter 13

Quality of Government

13.1 Introduction

An excellent private company is one that, acting within limits of the law, provides profits to its owners. An excellent public organization is one that efficiently and flawlessly performs its tasks, and above all, the provision of public services. It is natural to ask what kind of government is good? Scientists (e.g., Quality of Government Institute, operating at Gothenburg University) and practitioners (World Bank; Kaufmann et al., 2007, 2008, 2009) for whom information on the quality of government is the basis for supporting the development and channeling aid to specific countries, have been trying to answer this question for some time.

Virtually every publication concerning public administration or public policies and other related fields is related to the issues of state quality. Also, the general idea behind the writing of this book on public administration, public policies, and project management was the conviction that it would allow governments to function better. Publications from the descriptive current of social research[1] try to understand the functioning of public administration, and from the prescriptive – to define the methods of its effective and efficient operation. This means that a full assessment of the quality of government should take into account a huge number of factors – or at least those appearing in the results of prescriptive research – for which there is consensus among researchers and scientific communities. Then the definition of good government would be good government implements all generally recognized prescriptive recommendations in the area of its operation. The role of descriptive publications would be to show the starting point from which to start transforming public administration into the target state, described by the prescriptive trend. But such a definition would have no operational value because it would not be possible to assess any government due to the enormous amount of work required. So, when you think about the quality of government, you need to simplify it.

The second factor that hinders an unequivocal answer to the question about the quality of government is the diversity of operating environments and needs of different countries (Andrews, 2010). When improving governance, it is necessary to consider culture, especially management culture (Hodgetts and Luthans, 1997), organizational culture (Linder and Peters, 1989), and political culture (Elazar, 1984). Many factors influence management methods from many areas, including law, economy, sociology, politics, and education (Negandhi and Prasad, 1971). Tradition is also important (Jann and Wegrich, 2007).

DOI: 10.1201/9781003321606-15

The size of the problem and the multiplicity of factors influencing the ways and quality of governing show the complexity of analyzing and assessing government quality. However, an increasing number of researchers are facing these challenges. They deal with issues such as the definition of the concept itself, the benefits to citizens of good governing, the ways and principles of good government, and the indirect and long-term effects of good government.

13.2 What Is a Good Government?

The very definition of the concept of good government causes many problems (Charron et al., 2014). The government is accountable to its citizens directly and indirectly by doing or not doing activities in every area of its jurisdiction. The government may regulate (some) prices of goods and services or leave pricing to the market forces. However, in both cases, asking the government about the availability of goods, and perhaps poverty or well-being of citizens, will be justified. If the government does not interfere with the area of religion, it is also a specific type of attitude toward this area of citizens' beliefs and practices. Retracting from activities in a particular area does not relieve the government of its responsibility for that area (although these issues are viewed differently by governments with left-wing preferences and differently by liberal governments). So, the answer to the question about the quality of government may apply many, if not all, aspects of the functioning of the state and society.

Defining good government in this way, we refer to its actions indirectly. Overall, the quality of a government depends on the quality of its institutions (Andrews, 2010). Hence, a good government consists of institutions well performing their public services for citizens, their groups or associations, and commercial organizations. To be able to refer to improving the way government functions, it is advisable to provide a more operational definition. Andrews (ibid) formulated his definition differently, according to which a good government is the one that solves problems faced by the government and society, in particular those resulting from demographic challenges (in developing countries – large population growth, and in developed countries – aging of the society). The definition of Andrews should also include the efficient performance of daily activities, the skillful prevention of problems, and the effective use of opportunities facing the state as a whole. Governments should not solve problems, but most of all should prevent them from arising. If the ICT infrastructure is underdeveloped, it means that the government has not taken care of its development (directly or indirectly) at the right moment. The low level of qualifications of civil servants is a result of not having previously provided their appropriate level of education and qualifications. Finding natural resources in the country should not be treated as a problem; this is an opportunity that the state under government administration should exploit. Public management should be effective (Maldonado, 2010; World Bank, 2002; Andrews, 2010; Bauhr and Nasiritousi, 2012). According to La Porta et al. (1999), a good government fosters capitalist development, while the free market will satisfy the needs of society.

The most general and acceptable definition was given by Agnafors (2013), who says that good government is the one that performs sustainable actions concerning socially desirable goods concerning the resources it has. According to this definition, a government that creates a very efficient police force to stay in power against the public's preferences is not good because it does not meet the social need for the police force. It also cannot be considered a good government that creates good conditions for foreign companies to exploit local natural resources and allows for exporting these goods and the transfer of profits abroad because the effects of such exploitation will not be observed after its completion, i.e., they will not be sustainable.

13.3 Functioning of Good Government

For citizens to experience the effects of good governing, the government should exercise its political, economic, and administrative powers in a specific way (SIDA, 2002).

13.3.1 Politics

In the political sphere, there should be institutions that guarantee the election and replacement of the authorities (World Bank, 2002). The crucial mechanism is free elections organized by the government. Decentralization of power and democratic governance at its lower levels should be pursued (USAID, 2019). Therefore, problems should be solved at the lowest possible levels. Good government, in general, should be characterized by a stable democracy (Halleröd et al., 2014; Rothstein and Teorell, 2008). We can speak of a stable democracy if the governing party is changed several times as a result of elections. The political quality of government needs to strengthen the legislative process (USAID, 2019). A legislative procedure should be defined and followed, in which, for example, it should be specified who can submit legislative proposals, what are the roles, tasks, and deadlines for the next steps – e.g., giving opinions and approving proposals, and finally, the act of approving and implementing the law itself. The legislative process is one of the elements of the public policy process. In the less measurable area, the state should be characterized by a democratic culture (SIDA, 2002). The organizational culture (and the state is an organization) are values, beliefs, convictions, assumptions, interpretations of beliefs, emotions, and methods of action considered obligatory in it (Schein, 1992; Kotnour and Landaeta, 2002). The state can influence culture through examples given by its representatives or media. In a well-governed country, relations between the military and public administration should also be regulated appropriately (USAID, 2019) – the military should not influence the ways of governing the country. It should be a tool to ensure the country's security.

13.3.2 Efficiency and Effectiveness

Government must be effective (SIDA, 2002) and efficient (Maldonado, 2010; La Porta et al, 1999; IMF, 1997; Charron et al., ibid; Agnafors, 2013; Bauhr and Nasiritousi, 2012). Effectiveness means achieving set goals, for example, defined in public policies. To achieve the objectives of public policies, it is necessary to efficiently implement their policy programs, project programs, and projects included in them. A government is effective in that if it sets the unemployment rate at 3% in three years, it actually causes that to be the level of unemployment. The effectiveness of the government is negatively affected by political pressures – the administration should be operationally independent of them (Kaufmann et al., 2008). Operational independence does not mean that legitimate political structures do not influence the administrative institutions but that once the goals are defined, the political sphere does not interfere with the ways in which actions are implemented. This is the structure of power already seen by the future US president, Woodrow Wilson, as an academic teacher (Wilson, 1887). Effectiveness should be manifested in implementing the main task of the government – the implementation of public policies. In other words, the government must be efficient in implementing them (Kaufmann et al., 2008; Holmberg et al., 2008; USAID, 2019; La Porta et al., ibid) and must be credibly involved in implementing them (Kaufmann et al., 2008). If government officials do not demonstrate support for a particular policy, it is hard to expect any good results. Governments should have defined procedures for their operation that vary according to the needs of governments and societies (Andrews, 2010). Therefore, practices and procedures for project implementation must be specified in particular. The efficient implementation of public policies is influenced by the quality of the public service (Kaufmann et al., 2008) and the entire public administration (SIGMA, 2014). In turn, the low quality of government institutions hampers development (Rothstein and Teorell, 2008).

Governments should have a defined development framework (Maldonado, 2010) containing directions and further goals to be achieved (this topic will be discussed in more detail in Chapter 14). Objectives should be defined using indicators (Agnafors, 2013). It is not enough to say that "the level of education will be raised." It is necessary to indicate by what percentage, for example, the number of people graduating from universities will increase and by what date this goal will be achieved (SMART model; see Section 3.2.1). Development is fostered by participation in economic globalization (Al Mamun et al., 2017). The opposite of globalization is autarky that cuts off the state from contacts, especially economic ones, with the outside world. The elements of globalization are the exchange of knowledge between countries at different levels of development (including knowledge about project management methods to which this book is devoted), the flow of capital, enabling appropriate actions in countries with a lower level of wealth, and finally the flow of people (here, a negative phenomenon is the brain drain, i.e., the outflow of educated people) from less developed countries to countries with higher wages).

13.3.3 Solutions in the Most Important Areas

Good government should not only meet general conditions. Researchers indicate areas significant for the quality of government. First, the government must be able to implement projects (Teorell et al., 2017). Good governments should have an efficient system for the recruitment and career of civil servants (Rothstein and Tannenberg, 2015) (again, a reference to the Weberian model of the state). A good government must efficiently collect taxes (Rothstein and Tannenberg, ibid); without a financial base, no organization can function, and taxes are the basis of every government's finances. Good government must have an efficient public finance management system (Andrews, 2010; Al Mamun et al., 2017). The government should have developed IT systems (Al Mamun et al., ibid); information and communication technology is a fundamental factor in the success of virtually all organizations in our time. An important component of good government is audit chambers (Rothstein and Tannenberg, ibid), which can audit financial management and the rationality of public management.

13.4 Rules of Good Government

Good government should have not only the appropriate processes but also some general principles. Firstly, it should be impartial, i.e., the method of implementation of its activities should be determined only by factors described in procedures, instructions, and recommendations, and not by personal issues, for example, relationships with people whose matters are taken care of (Rothstein and Tannenberg, 2015; Charron et al., 2014; Rothstein and Teorell, 2008; Kaufmann et al., 2009; Bauhr and Nasiritousi, 2012). Corruption is a factor that is contrary to impartiality. Thus, one of the measures to ensure impartiality is combating corruption (Bauhr and Nasiritousi, 2012; Charron et al., ibid; Kaufmann et al., 2009; USAID, 2019; IMF, 1997). Anti-corruption is also favored by government by informing citizens about its actions (Maldonado, 2010), openness to society (SIDA, 2002), or transparency in management (Al Mamun et al., 2017; Maldonado,

2010; USAID, 2019) – these features have also its significance. It can be achieved by using IT resources (public organizations' websites, discussion forums) and publishing the operating procedures of public organizations. Openness means that public organizations should have specific procedures for interacting with citizens. The government generally should be fair (SIDA, 2002). The sources of good government operation must be based on the law (SIDA, 2002; IMF, 1997; Bauhr and Nasiritousi, 2012; Andrews, 2010). In the sphere of market activity, the main rule is that any activity that is not prohibited is allowed. In the area of public administration, the main rule is different: no actions are allowed that are not explicitly indicated by the applicable laws or regulations. So, every action of public administration must result from a specific document. Good governing is fostered by the responsibility of government representatives (World Bank, 2002; IMF, 1997; SIDA, 2002; USAID, 2019), their accountability, i.e., bearing organizational consequences for the actions performed (Al Mamun et al., 2017; Maldonado, 2010). La Porta et al. (1999) believe that good government should be adequately large to be able to meet all the requirements I wrote about here.

13.5 Results of Good Governing

The ultimate goals of government action should relate to citizens. Governments seek to create a welfare state (Hill and Hupe, 2014). Better is the government where people have more goods at their disposal. The bad quality of government is one of the main impediments to improving the standard of living (Bauhr and Nasiritousi, 2012). But prosperity is not the most general goal; more important than it is citizens' feeling of happiness. Well-being is one of its components that may be subjectively differently perceived by each citizen. Still, it is assumed that the components of happiness are the quality of public services, life expectancy, or low child poverty (Rothstein and Tannenberg, 2015). The result of the functioning of a good state understood in this way is mentioned, inter alia, by Rothstein and Teorell (2008) and Charron et al. (2014). Happiness research was initiated in Bhutan and is currently conducted in many countries (Helliwell et al., 2020). In welfare states, the state can play an important direct role in achieving such a situation, and in liberal states, it affects the well-being of citizens mainly by giving them freedom of action.

The main factor determining the standard of living is the per capita income (Halleröd et al., 2014; Agnafors, 2013), measuring the level of the economy. Since the starting level of per capita income (and other parameters determining the state of the economy) is received by the current government when it comes to power, it is not responsible for the absolute level of income. Therefore, the level of economic growth is considered a more objective indicator of the quality of government (Agnafors, 2013; SIGMA, 2014; Al Mamun et al., 2017). As a consequence of economic growth, the level of material welfare is increasing (Charron et al., 2014). Charron et al. (2014) also argue that income inequalities in the country should be minimized. However, it should be remembered that income inequality has a mobilizing effect on development. In a country where everyone would be guaranteed the same income, there would be no incentive for new activities. But this remark does not apply to the level of affluence of children – the elimination of their poverty is considered by Rothstein and Tannenberg (2015) and Charron et al. (ibid) to be the result of good government. The good government provides citizens with high-quality public services in areas for which it is responsible, for example, health care, education, and the creation and maintenance of infrastructure (La Porta et al., ibid; Kaufmann et al., 2008). The effects of government action on citizens are sustainable (Charron et al., ibid). Teaching farmers how to cultivate plants efficiently can be considered a sustainable activity, and the direct supply of foodstuffs is perishable (although in many cases necessary). From the group of economic factors related to individuals, an important indicator of government quality is the level of unemployment (Agnafors, 2013), which good governments minimize. From the group of non-economic factors, the effect of good government perceived directly by citizens is the expected life duration in health (Rothstein and Tannenberg, 2015; Helliwell et al., 2020). It is not the life expectancy itself that is important, but the period in which a person can function efficiently in society. Health factors also include generally good health care outcomes (Charron et al., ibid) and low child mortality (Rothstein and Tannenberg, 2015; Rosling et al., 2018). Among the factors relating primarily to young people, an indicator of good government is good, universal, and free education (Rothstein and Tannenberg, 2015; Charron et al., ibid).

But the government can act for citizens not only in the sphere of measurable services. Activities in the area of politics (La Porta et al., ibid) and democracy are also important. In a well-governed state, citizens are guaranteed political freedoms (La Porta et al., ibid) and participation in governing (Maldonado, 2010; USAID, 2019). In particular, there is the freedom to form any kind of social association and political party. Citizens have the possibility of passive and active participation in elections. Their voice is heard in a well-governed country (Andrews, 2010). Usually, the essential factor in ensuring that citizens' opinions are listened to is the free media. The government is representative of society and reflects its pluralism (USAID, 2019). There are liberal laws (Halleröd et al., 2014), and gender equality is guaranteed in the public sphere (Rothstein and Tannenberg, 2015). Good governance requires specific actions and behaviors not only on the government's side but also on the side of citizens who respect the laws and institutions of the state (World Bank, 2002).

Improving the quality of government may also have an indirect effect on social behavior. In countries with higher-quality governments, particularly those with higher-quality public organizations, there is more significant support for democracy (Boräng et al, 2017). Perhaps this is another confirmation of the hierarchy of human needs (Maslow, 1962), which shows that people must first have their basic existential needs met and then reflect on less tangible values. In well-governed countries, leaders have a stronger mandate to implement reforms (Charron and Lapuente, 2010) – people know that power can work well. In such countries, internal armed conflicts are less likely (Charron et al., ibid; Rothstein and Teorell, 2008), e.g., because there are more goods available, that could be their object. In poorly governed, poorer countries, a kind of a closed circle is created: the emphasis on improving the quality of state governance is weaker than in more affluent countries (Charron and Lapuente, ibid). There may be a development trap that intensifies the implementation of inefficient administrative processes in this situation. It would be more beneficial to completely reorganize them – for which there are no resources (Andrews et al., 2017; Repenning and Sterman, 2002). In well-governed countries, citizens tend to favor medium- and long-term investment over current gains (Charron and Lapuente, ibid), probably because they are guaranteed a relatively high level of income and living.

13.6 Projects for Good Governments

In fact, in every statement about good government, including those mentioned above, we can find explicit or implicit references to project and project management. This is explicitly mentioned, for example, by Teorell et al. (2017). Take some examples. The elections that are at the heart of good democratic governance are projects. Each implementation of a new or only changed public policy requires implementing changes, i.e., the implementation of projects. The definition of Andrews (2017) referring to problem-solving unequivocally, although only implicitly, indicates projects as the fundamental organizational solution determining the quality of government. The definition of Agnafors (2013) includes the efficient implementation of properly selected projects which are the primary way of creating goods and services, along with the operational activity. Each investment deciding on enabling the implementation of new services or improving the quality of services provided so far is a project or a project program. For instance, information systems mentioned by Al Mamun et al. (2017) that are a prerequisite for good governance can only be implemented by projects. The same is true for any investment activity. Any state effort to achieve a higher quality of government must include – or be all of – projects.

> Projects are the essential management tool for improving and maintaining the quality of governments and governing.
> And this is one of the main reasons why this book was written.

Note

1. Three trends can be distinguished in social research: descriptive, prescriptive, and normative (Bell et al., 1988). The descriptive trend includes those studies that try to detect relationships between the phenomena studied and do not value them. Their goal is simply to understand phenomena. The prescriptive trend includes research the purpose of which is to define an appropriate, correct way of performing activities. The normative trend includes activities that are used to create standards, i.e., models that are the basis for the assessment of implemented activities.

Chapter 14

State Capabilities

14.1 What Is Capability?

A person who grows plants should know what phases of life the plants go through, what they require, and should be able to perform the activities that cause them to grow. It should also have adequate resources: seeds, land on which the plant will be planted, machinery and equipment needed for cultivation, perhaps funds. If he/she has the knowledge, skills, and resources, he can grow a plant and at the right moment will obtain the results of cultivation – fruits. Likewise, any organization, such as a hospital, should have the knowledge, skills, procedures, and medications needed to achieve its goals to get the expected results. Qualified medical and administrative personnel should work at the hospital. If the actions are not made repetitive, and each time they are organized ad hoc, then there is no capability (Winter, 2003) but rather re-inventing the wheel. The capability is a kind of potential that can be activated as a result of the occurrence of certain circumstances – for example, a command given by superiors, changes in the environment, or as a result of a citizen applying for a specific service – and there is a high probability that the action will bring the intended effect. Baser and Morgan (2008) define the concept of capability very generally, for whom capability is the ability to do something. If the knowledge and practices generated in the course of activities are documented and can be reused in the future, we are talking about creating the organization's capability. The concept of capability is strongly related at the organizational level with the capability to implement processes (SEI, 2010), including the ability to implement projects and the knowledge acquired at the individual and organizational levels (Gasik, 2011). Capabilities may be area-related, related to a specific function or specialty – for example, treating sick people, teaching at schools, and running the education system. There may also be horizontal capabilities – for example, project management which applies to all areas of any organization's activity. Another concept than capability is capacity, which is the measure of doable goods.[1]

All this applies to a particular type of organization: the state. Good governance of the state and the performance of any other activity by individuals and their organizations require appropriate skills (Bester, 2015). Administrative capability tells how well the state can perform its tasks (Bäck and Hadenius, 2008). A state may or may not have defense capabilities or capabilities in any of its areas of activity. For the state to be of high quality, in a sense I discussed in the previous chapter, it should have high capabilities in every area of its activity. State capabilities are divided into three

DOI: 10.1201/9781003321606-16

main components (Dimitrova et al., 2020): extract and monitor (mainly taxes and information and other resources necessary for the functioning of the state), the functioning of the administration apparatus (e.g., coordination between units, employment based on qualifications), and the provision of goods and services. In the language of process analysis, it can be said that the state must have the capabilities related to inputs, executing processes, and their results (outputs).

State capabilities are more precisely defined by Barkley (2011). The government should identify and anticipate the needs of society and then respond to them through effective and coordinated actions, including projects and project programs. A good result of an effort does not have to come from having the capability at the start of the action. The capability can develop during the activity. The United States did not know how to build an atomic bomb when President Roosevelt established the Uranium Advisory Committee in 1939, but the Manhattan Project ended in dropping nuclear bombs on Japan in 1945 (although they failed to use them against Germany in Europe, Gosling, 2010). For a state's capability to create real value, decision-makers must have the will to mobilize the capability and the resources necessary to implement it (Baser and Morgan, 2008). Many less developed countries are unable to combat starvation effectively, although they do have the required capabilities because they do not have adequate funds. The state may be able to build bridges, but it is unable to do it due to the limited availability of qualified personnel and funds.

For the state, the most important ability is the ability to implement policies (organizational policy capability, OPC; Gleeson et al., 2011; EPRS, 2017). Rogers and Weller (2014) simply call this state capability. For an organization to be able to make policies, it must have analytical, managerial, and political capabilities (Ramesh and Howlett, 2017). Policy Analytical Capability is an appropriate organization of work, educated analysts, an open culture focused on communicating analyzes results, and budget availability, and proper information (Oliphant and Howlett, 2010). Policy forming and implementation also require capabilities of personnel development, stakeholder relations, government cooperation, linking analysis results with implementation, policy monitoring, and evaluation, as well as an appropriate organizational culture (Gleeson et al., 2011). The government must be able to solve problems that arise during the implementation of policies (Ramesh and Howlett, 2017). These capabilities need to be backed up by political capabilities that ensure continued support for the policies under implementation. Policy-making capabilities must be increasingly complex due to the increasing complexity of public concerns (Tiernan, 2012).

Countries with greater policy-making capability may choose from a larger set of potential policies (Chuaire et al., 2014). Countries with a lower level of policy capability increase the effects of their actions extensively by engaging more resources. Countries with a higher level of policy capability can achieve similar effects more intelligently by implementing new types of policies (Chuaire et al., ibid). According to the definition of public policy, which says that the set of public policies covers all areas of the state's functioning, the policy capability applies to the entire state – but it is only one of the views on capability development. In addition to policy capability, it may relate to two other types of activities: technical (e.g., agricultural, transport, educational) and functional (e.g., project management, planning, strategic management) (Bester, 2015).

14.2 Capability Development

To better and better perform their duties toward citizens, states should acquire new capabilities or improve the existing ones. The purpose of acquiring capabilities is development, not progress (Riggs, 1970). Progress is the acceptance that the community has a primary goal to achieve and only carrying out activities that bring it closer to that goal. For example, the goal of acquiring new capabilities in the countries of the former Soviet bloc was to build a communist state. On the other hand, development increases the state's capabilities in response to emerging challenges, opportunities, problems, or other circumstances. Development was a way of holding back the expansion of communism (Loveman, 1976). Suppose the level of water supply in the country is too low at a given moment. In that case, measures should be taken to develop the supply capability – which does not require placing this objective in a broader context of any "progress." Capability development is critical to the world's growth and development (Farazmand, 2009). The implementation of the development process does not exclude the existence of aggregate goals, for example, increasing communication capability, which consists of growing road, rail, water, and air transport capabilities. Development goals may also be horizontal, involving many or all of the country's organizations – for example, developing project management capabilities (as discussed in this book) or strategic planning and management (UNDP, 2009). In relation to a single country, we should talk about many development processes, not about one process encompassing all areas of state activity. The state should have its capability development strategy (UNDP, 2009) – this does not mean striving to achieve a single goal, but arranging the implementation of development processes so that effective planning is possible. The development strategy and its implementation may be supervised by the national capability development facility (UNDP, ibid).

The effects of development work may be changed status (e.g., in terms of knowledge), changed processes, and new products: tangible and intangible, e.g., new policies formulated and operating (Otoo et al., 2009). Development activity should have a specific goal, but knowing the desired state is only a part of the success of state capability building (Levy, 2004). For a state to have a specific capability, it must meet certain conditions on three levels: environment, organization, and individual (UNDP, 2009; Bester, 2015). At the environmental level, for a capability to develop, conditions must be met regarding public policies, laws, power relations, social norms that generate mandates, modes of action, and civic involvement in achieving the capability. At the level of the organization responsible for the development, there must be appropriate routines in place regarding own resources (including the most important: people) and relationships with other organizations involved in the action. At the individual level, people should have the experience and skills necessary to carry out the activities foreseen for them. At every level, from the state and society as a whole, through individual organizations to individuals involved in the development, the essential role is played by the knowledge necessary to create or develop capabilities (United Nations, 2006; Gasik, 2011).

There are two concepts related to acquiring capabilities: capabilities building and capabilities development. Building presupposes that the specified capability was not owned by a given entity and is constructed entirely from scratch. And capability development (which has attracted a lot of interest since the 1990s, EPRS, 2017) is an attempt to improve the previously possessed capability. Capability development is carried out at all organizational levels: the state as a whole, institutions, and individuals. Development is often implemented with a significant share of foreign forces and resources. Still, it should be endogenous, i.e., driven by the needs of entities subject to development and not by external organizations providing development activities (Lopes and Theisohn, 2003). The owners of the development process should be the inhabitants of developing areas; they should play the role of a driving force in development processes (EPRS, 2017). Conducting aid projects by employees of external aid organizations usually ends with the lack of involvement of local entities. A consequence of the endogeneity of development is the need to consider the characteristics of countries undergoing development and a shared understanding of development activities between organizations supporting development (e.g., the UN) and developing countries (Bester, 2015).

Capability development processes are not always successful. There are various reasons for this situation. Stakeholders engage too little in development projects (World Bank, 2012). Policy instruments that are effective in one country are not effective in another (Andrews et al., 2017). Organizational arrangements do not work as intended (World Bank, ibid). The effects of development work are not sustainable. They disappear after the end of foreign support (EPRS, 2017). Development activity benefits only its immediate beneficiaries and is not perpetuated at the level of the organization and the state as a whole (McEvoy et al., 2016). In some countries, there is the earlier mentioned capability trap (Repenning and Sterman, 2002; Andrews et al., 2017). However, a set of recommendations may contribute to overcoming the problems of capability development. Several recommendations have been developed to overcome these and many other problems in building and improving state capabilities.

In development projects, individuals, organizations, or society as a whole create or develop individual capabilities (United Nations, 2006; EPRS, 2017; Bester, 2015). In line with these recommendations by UNDP (2009) and the World Bank (Otoo et al., 2012), a capability development project should begin with an introductory activity in which a process leader is identified who is presented with an overall outline of the project. Then, with the participation of representatives of the local community, the needs are identified and assessed, and the beneficiaries of development appear. The current situation and its weaknesses are assessed, and development goals emerge. Knowing the current and target status allows creating an action plan that is then implemented. During the implementation of the development project, which will enable you to see the area of operation in great detail, usually, there are modifications to the target state and the action achieving it (adaptive implementation; see Section 11.8.4). After some time of using the new capability, an evaluation is carried out to check whether the planned development effects have been achieved.

Each development process is a project in which project implementation capabilities should be used. This leads to the concept of meta-capability.

14.3 Projects as Meta-capabilities

Projects are the basis for any development activities (Biggs and Smith, 2003). But project management processes and techniques are helpful not only for managing single projects and project programs. One role of project management in the government requires particular emphasis: enabling other changes. A meta-capability is the capability to acquire other capabilities, and a meta-policy is a policy about making policies. To date, some authors have dealt with the meta-level in the area of public policy with each understanding this concept differently (Kiel et al., 2012; Barzelay, 2003; Peters, 2010; Elazar, 1984; Baser and Morgan, 2008), but they did not adequately highlight the role of projects at this meta-level. Central government reforms include capability building and learning (Alari and Thomas, 2016). The capability to manage projects and project programs is one of the factors affecting the ability to implement any reforms and changes in governments. Projects cause organizational development. Public administration reform is a fundamental horizontal reform because it sets the framework for implementing other policies (SIGMA, 2014). Public administration reforms as a whole and government reforms take the form of many changes, i.e., projects aggregated into project programs. The acquisition of state capability is carried out through the implementation of projects.

Projects are the essential capability required to acquire other capabilities, i.e., they are meta-capabilities.

The lack of this meta-capability can be a significant impediment to the development of state capabilities. Hence, one of the first major horizontal reforms in any administration should be a (meta-)project to improve its approach to public project implementation.

Note

1. The terms "capability" and "capacity" are used in different ways by different authors; for the purposes of this book, I have adopted the above definitions.

PUBLIC PROJECTS ARE SPECIFIC

Organization and project management researchers paid much attention to the differences between public organizations and organizations in other sectors. One of the ways to improve the management of public projects could be indicating differences between public projects and projects of other sectors and then, based on those differences, defining improved managerial methods adequate to public projects.

This part contains a systematic attempt to organize the concepts related to differences between public-sector and other-sector projects. It may have both practical (developing better ways of public project management) and theoretical (systemizing the knowledge about differences between public and other-sector projects) implications.

In Chapter 15, we start from looking at approaches to the differences between public sector projects and other projects found in project management literature. Chapter 16 contains an overview of the models of general organization differences and an attempt to extend them to the project level. After demonstrating that neither of these models is adequate for the project level, I define a model of differences most adequate for this level of management. The next two chapters are devoted to describing the results of a survey carried out in a group of specialists engaged in public projects. Chapter 17 deals with the differences in complexity between public sector projects and projects in other sectors. The last of this part (Chapter 18) describes success factors specific to public projects.

DOI: 10.1201/9781003321606-17

Chapter 15

Approaches to Differences between Research on Public and Other Sector Projects

The number of publications on project management in the public sector has been growing rapidly in recent years (Moutinho and Rabechini, 2020). When proceeding with the analysis of public projects, it is first good to note how the researchers have addressed the inter-sectoral differences at the project level up to now. Did they think it is worth noting the differences between projects from different sectors? Are public projects managed the same way as other sectors' projects, or do significant differences exist? The project management research community is not coherent in answering this question. When analyzing the literature on project management, one can observe explicitly or implicitly presented different approaches to these differences. Four main research approaches can be observed:

1. The Denying Differences Stream
2. The Public Projects' Research Stream
3. The Averaging Stream and
4. The Stream of Differences Analysis

I will describe them in the following subchapters.

15.1 The Denying Differences Stream

The first group of researchers analyzes public projects. Still, they formulate conclusions in a general manner, not stating that they are valid only for public projects or suggesting that they are valid for any projects (Bhuiyan and Thomson 1999; Boersma et al., 2007; Chou and Yang, 2012; Couillard et al., 2009; Dey, 2000; Chapman, 2016; Duffield and Whitty, 2016; Dvir et al., 2003; Fu and Ou, 2013; Hossain and Kuti, 2008; Olsson and Magnussen, 2007; Stummer and Zuchi, 2010; Sutterfield et al., 2007). This can conclude that they believe that public projects, at

DOI: 10.1201/9781003321606-18

least in the analyzed areas, do not differ from private projects. A similar approach is shown by researchers who expressly believe that the results of their research carried out in the public sector can be generalized to all projects (Aubry et al., 2011; Padhi and Mohapatra, 2010; Rose and Manley, 2010). This group also includes researchers who do not refer to the sector where the projects are implemented (Savelsberg et al., 2016; Boh, 2007; Damm and Schindler, 2002).

15.2 The Public Projects' Research Stream

The second group of researchers conducts studies on public projects. Some openly formulate the thesis that public sector projects are different from other projects (Tabish and Jha, 2011; Rose, 2006). Others explicitly or implicitly state that the results of their research are valid only for public projects (Faridian, 2015; Kim and Lee, 2009; Kwak and Smith, 2009; Adler et al., 2016; Arrain, 2005; Chang et al., 2013; Chih and Zwikael, 2015; Hung and Chou, 2013; Kuula et al., 2013; Procca, 2008; Skulmoski and Hartman 2010).

15.3 The Averaging Stream

The third group consists of those researchers who, when characterizing the sample of projects studied, report that both public and private projects have been studied and then formulate conclusions regarding the entire population of projects (Helm and Remington, 2005; Martinsuo et al., 2006; Mengel et al., 2009; Reich, 2007; Eden et al., 2005; Male et al., 2007; Schmid and Adams, 2008). From this approach, it can be concluded that these researchers believe that there are differences between public and private projects, but the inclusion of representatives of both these groups in the studied sample results in averaging the values, allowing for drawing valid conclusions for the entire population of projects.

15.4 The Stream of Differences Analysis

The fourth group of researchers deals with the projects from different sectors trying to check whether there are any intersectoral differences. Some of these studies show differences (Dilts and Pence, 2006; Coster and Van Wijk, 2015; Hvidman and Andersen, 2014; Rwelamila and Purushottam, 2012), some do not (Hobbs and Aubry, 2008; Ramos et al., 2016).

The existence of – implicitly or explicitly defined – four different approaches to the issue of differences between public sector projects and projects from other sectors shows how unstructured knowledge is about public projects, particularly about the differences between public sector projects and projects from other sectors. There is no general model in the literature that would clearly articulate the differences between the public sector and other sector projects. The further development of knowledge about public projects and their application may be hampered by the lack of a systematic framework for these differences. I will look at these differences in more detail in the following chapters.

Chapter 16

Differences between Public Projects and Projects of Other Sectors[1]

16.1 Projects as Organizations

Gasik (2018) analyzed the differences between the public sector projects and projects from other sectors. In projects, a structured group of people works toward achieving a common goal, so projects are organizations (Subchapter 8.1). Projects are defined as a sub-type of temporary organizations (Lundin and Soderholm, 1995; UK OGC, 2017; Turner et al., 2010; PMI, 2017a, and many others). The taxonomy of organizations with permanence versus temporariness as the main classifying criterion is schematically presented in Figure 16.1.

Project management can be defined in two dimensions: the type of sector and the type of management. We focus only on dividing the public sector and other sectors in the sector dimension. In the management dimension, we have project management and operational management. Hence, public project management is defined by project management, public-sector organizations management, and sector independent organization management. These relationships are shown in Figure 16.2.

Public project management involves the processes, techniques, and functions:

- Specific to the public sector
- Sector independent for areas of
- General management (applicable to projects)
- Project management

Hence, the discussion about cross-sectorial project management differences should include:

- The discussion about the nature of the differences between organizations from different sectors
- The discussion about the nature of differences specific for the project level

DOI: 10.1201/9781003321606-19

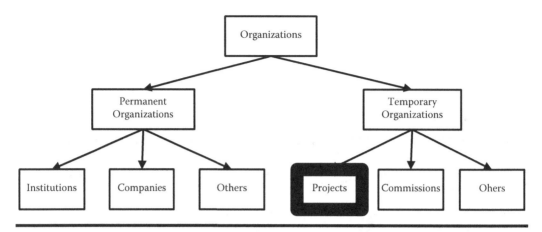

Figure 16.1 Types of Organizations.

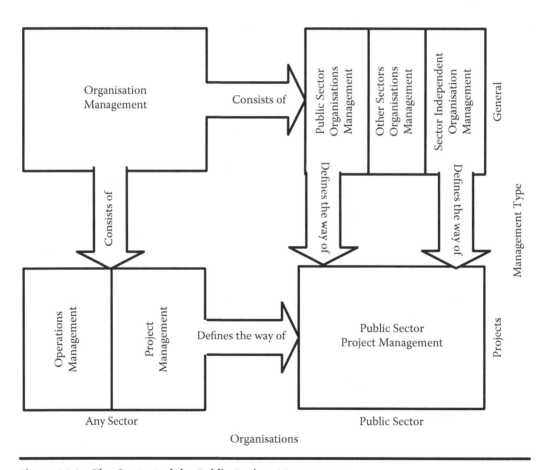

Figure 16.2 The Context of the Public Project Management.

16.2 Constituting Differences between Public Projects and Projects of Other Sectors

The research on organizations has built three main models of the differences between public-sector organizations and organizations from other sectors (Scott and Falcone, 1998):

1. The generic model
2. The core model
3. The dimensional model

I will call the differences between public projects and projects from other sectors included in this meta-model the **constituting differences**.

According to the generic model, there are no fundamental differences between public and business organizations (Lau et al., 1980; Murray, 1975; Scott and Falcone, 1998). This approach would mean that there are no differences between public projects and projects of other sectors in the area of projects. The general principles of building a schedule or formulating requirements are the same for public sector projects and other types of projects. Representatives of the first group of researchers mentioned earlier in the Denying Differences stream (Subchapter 15.1) might be considered representatives of the generic model at the project level.

According to the core model, there are substantial differences between the public and organizations from other sectors (Bozeman and Bretschneider, 1994). The differences stem mainly from public organizations' formal status, which entails fundamental differences in the processes implemented in organizations of different sectors. In projects, such an approach may manifest itself in procurement management or stakeholder management in which fundamental differences result from legal regulations, for example, regarding public procurement. Work on the core model was initiated by Sayre (1953), claiming that "private and public organizations are alike in all unimportant respects." In the area of project knowledge, the second Public Project Research (Subchapter 15.2) and third Averaging (Subchapter 15.3) group of researchers described above can be considered as representatives of the core model.

The dimensional model is the one in which the "publicness" of an organization can be viewed in several dimensions (Perry and Rainey, 1988). These dimensions are:

■ Ownership
■ Funding
■ Mode of public control (market or polyarchy – control from many different subjects)

A continuum from fully private to fully public may be observed for each. Therefore, organizations can be more or less public. The most public are state-owned organizations, financed by the state, and subject to social control solely by the state. This group includes all government agencies, i.e., ministries and central offices. Most private organizations are privately owned, financed, and controlled by market forces. But, in line with the dimensional approach, there may be organizations public in ownership and funding but subject to market control – an example being public, state-owned, and funded universities that have to compete with private universities. There are privately owned, state-funded organizations subject to social (not necessarily financial) control, such as private companies responsible for maintaining public roads. On the other hand, an example of an organization owned and financed by private entities but subject to state control is, for example, energy supply companies.

Dimensions of organizations' publicness are also important to projects. Projects can be wholly public if they are state-owned, financed by the state, and subject to polyarchical control by various actors. Examples include military projects or public organization restructuring projects. Most private projects are owned and funded by private companies subject to market control, such as building a private factory. Among them, there are, for example, philanthropic projects – financed and owned by private owners (e.g., creation of cultural institutions), controlled by both the owner and members of society, future users. Another example is public-private partnership projects that are subject to control by various entities (e.g., users of facilities under construction, chambers of financial supervision, ministry of infrastructure), owned and financed by the private sector (or jointly). To evaluate project audiences in line with the Perry and Rainey model, in our analysis, we replaced the three dimensions they provided with project management areas. Representatives of the dimensional model at the project level are members of the fourth group (Subchapter 15.4), who try to determine the differences in individual management areas, i.e., they assume that differences may occur in some areas and not in others.

In the practice of government involvement in implementing projects, a fourth feature can be included in the group that constitutes their level of publicness. It is the involvement of the government in their implementation. Governments in some countries actively support privately-owned projects (see Chapter 21). The decisive criterion for involvement is the project's significance for developing the state's economy. Such involvement consumes public funds. If the state gets anyhow involved in implementing the project, it takes some responsibility for its success. If a project implemented by a private entity but supported by the government fails, it makes sense to ask the government why did it happen? Government involvement in management can range from full support to nothing. It is the fourth factor, next to ownership, funding, and mode of public control, which determines the publicness of projects. Hence, the enriched set of dimensions of differences between public and other sector projects finally includes:

- Ownership
- Funding
- Mode of public control
- Participation in management

One might consider transferring the way of thinking about models of organizational differences (core, generic, and dimensional) to thinking about differences in projects and managing them. To verify the validity of such a meta-model, I will analyze the differences in managing public and other sector projects. This will be the basis for building an appropriate model of the differences between public sector projects and other projects.

16.3 Managerial Differences

Besides the fact that public projects, such as organizations, may differ in their constituent features, they may also differ in how they are managed.

Project management knowledge is a well-coded area compared to general organization management. Project management knowledge may be divided into several areas. The PMBOK® Guide (PMI, 2017a), followed by ISO 21500 (ISO, 2012) divides this knowledge into ten areas (specified below) that become the dimensions for potential managerial differences. The analysis of the differences between public sector projects and other projects can be based on this classification. As a particular, additional area of analysis, due to its fundamental importance for public sector projects, I have also included the area of governance, which is considered a separate area of project management.

The division into knowledge areas, defined by PMI for project management, has been extended upwards to management in organizations of any type. In any organization, just like in projects, you can talk about cost management, quality management, human resource management, or other areas of project management.

When considering the possibility of applying the differences identified at the general level of the organization to projects, it is necessary to refer to the features that characterize the projects, i.e., distinguish them from among the overall organization. There are two such features (they must occur simultaneously): time limitation and delivery of a specific product closing its activity. Mere time constraint is not enough for an organization to be considered a project (Figure 16.2). Standing committees in elected bodies (parliaments) are limited in time. Still, they cannot be considered projects as their purpose is not to produce a specific product (after which their work would be terminated). So, suppose the difference defined at the general level of the organization does not result from the timeframe or non-production of a specific product ending its activity. In that case, it also applies to projects.

It is possible to learn that public projects differ from projects of other sectors in some management areas while they do not differ in others. It moves the issue of project intersectoral differences from the "constituting" dimension to the managerial dimension of integration management, quality management, etc. So, for the area of management, one can also construct a meta-model similar to that built by Perry and Rainey (1988) for the area of constituting differences. The existing models for the managerial differences between public-sector organizations and other organizations, extended to the differences between public-sector and other-sector organizations, create a hypothetical meta-model shown schematically in Figure 16.3.

A question has to be asked: are any of these models appropriate for projects and their management? In the following sections, I describe the managerial differences between public and other sector projects to answer this question.

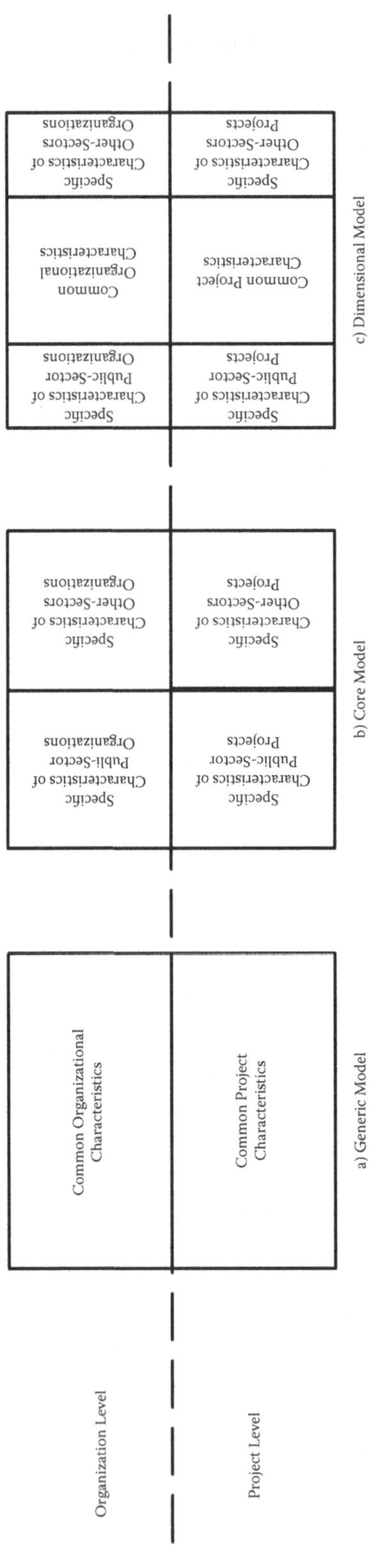

Figure 16.3   Possible Models of the Managerial Differences between Projects of Different Sectors.

16.3.1 Governance

Project governance and project management in the public sector are more precisely separated than in the private sector – someone else makes the fundamental decisions, and someone else manages (Williams et al., 2012). This results from a division between the political sphere and administration in a traditionally understood state. A public organization is part of a larger whole – the state – and therefore has limited ability to decide on implementing major projects independently. They result from government policies and their programs. The influence of the higher management sphere is visible in many aspects of public projects. In public sector projects, unlike in the private sector, higher-level non-sponsoring organizations, in addition to the power to order projects (e.g., legal regulations regarding employment conditions and processes may force modifications of IT applications in all sectors of administration), also have the power to veto them – even in areas formally left to government agencies to decide (Dilts and Pence, 2006; Baldry, 1998). Public projects have a greater number of external centers of power and influence involved in the governance process, which, in addition, often operate independently of each other (Rainey, 2014). Public projects are exposed to political influences far more significant than private projects (Kwak et al., 2014; Kassel, 2010; Pūlmanis, 2015; O'Leary and Williams, 2008; Dilts and Pence, 2006). Due to many influential stakeholders in public projects (Pūlmanis, 2015), including entities directly involved in their implementation, it is essential to establish and apply formal governance structures (Barkley, 2011; Kwak et al., 2014), removing possible conflicts regarding decision-making.

On the other hand, in the dynamic dimension, public projects are simpler because there are usually fewer decision gates in the public sector than in private-sector projects (Williams et al., 2012).

16.3.2 Integration Management

Integration management involves integrative processes, such as identifying, unifying, and co-ordinating activities that are crucial for achieving an organization's success (Subchapters 2.4 and 3.12).

16.3.2.1 Project Goals

Public sector values differ from private sector values (Subchapter 8.7), which fundamentally affects how their objectives are defined. Hence, public projects differ from private ones in terms of ways of defining and achieving their objectives. The goal of Public Administration units is to implement public strategies and policies. Some projects are implemented as components of policies as a direct result of government decisions, and all must consider public policy goals defined at a higher organizational level. Project objectives may be unstable due to changes in public policies, resulting from changes in ruling groups, political leaders, and responsible officials (Pūlmanis, 2015). This significantly impacts the selection and set of projects implemented by public organizations. All the criteria for determining the objectives of public projects described below complement the goals of public policies defined at the political level. This approach strongly distinguishes public sector projects from private companies' projects which have complete autonomy in defining their goals and selecting projects to implement them.

The value, and hence the goal of a public sector project, can be viewed as meeting social needs, following an approach that emphasizes public values (Moore, 1995). The main stakeholders who can define the success of a public sector project in various ways are the communities for which the projects are implemented (meeting public requirements) and their contractors (achieving a fair planned profit, Jackson, 2004). The purpose of public projects is also to support the development of the state's economy.

Public projects may have many different goals resulting from a variety of public-sector organizations and projects engaged (Mihăescu and Țapardel, 2013). Different institutions may have different requirements (Rainey, 2014). Additional difficulties in identifying public sector project objectives result from their implementation in organizations that themselves have difficulty identifying performance and mission measures (Wirick, 2009). This diversity increases the need to clearly define project objectives to avoid allegations of their non-fulfillment or improper implementation (Cats-Baril and Thompson, 1995). Particularly important is the definition of negative requirements indicating what will definitely NOT be implemented in the project (MoSCoW model; see Section 3.2.2). It is also important in public sector projects to demonstrate early that goals and benefits are realistic and achievable and define a methodology for assessing these benefits (Kwak et al., 2014).

The goal of a project is to meet its requirements. Requirements are different from constraints (Subchapter 3.2). There is a tendency in the public sector to equate compliance with requirements and compliance with constraints, in particular regarding project implementation procedures. As many procedures, regulations, and guidelines drive the public sector, it is also necessary to emphasize the importance of achieving goals – and not the very compliance with the procedures. In this way, we come to the concept of red tapes, which can affect all management areas, very important in public sector projects. There is a more significant impact of red tape on the functioning of managers in the public sector than in the private one (Bozeman et al., 1992; Baldwin, 1990). In other words, public organizations are more formalized. The rules of bureaucracy change according to the level of political influence. The more an organization is influenced, the more bureaucracy it has (Bozeman and Bretschneider, 1994).

16.3.2.2 Initializing

The relevant analyses often are missing when initiating public sector projects (Pūlmanis, 2015). Politicians may change decisions concerning the commencing of public projects developed based on appropriate analysis. Power and politics play an essential role in selecting public projects (Spittler and McCracken, 1996; O'Leary and Williams, 2008). Governments sometimes decide which public sector projects to select without considering community views, which, in turn, reduces support for the project (Mouly and Sankaran, 2007). When initiating projects, open discussions should be held between stakeholders (government, regional administrations, provinces, city boards) to check their support (Shiferaw, 2013).

The assessment of the expected project benefits should consider, as far as possible, the benefits for all stakeholders, including project opponents (Klakegg, 2009; Ng et al., 2012). This does not apply to hostile stakeholders, for example, the criminal community for public safety improvement projects.

16.3.2.3 Planning

Planning influences the achievement of success in implementing projects in each sector, including the public projects (Toor and Ogunlana, 2010). But in the public sector, project managers have less planning autonomy as goals are set by statute and government regulation (Fottler, 1981). Private sector projects are more precisely organized (Coster and Van Wijk, 2015). The planning of public sector projects is more geared toward external dependencies, while in the private sector – toward internal (Bretschneider, 1990). For example, in the public sector, it is necessary to plan purchases more precisely (due to legal regulations) and the management of external stakeholders (due to their number and possibilities of influence). Public sector project planning should move backwards from effects to products to activities (Barkley, 2011). Pūlmanis (2015a) mentions quality assurance, an effective cash flow plan, effective use of resources, an accurate definition of project objectives, ensuring good problem solving, and precise determination of project progress as the benefits of public sector project planning.

The main reasons for poor project planning in the public sector are delusion, deception, and bad luck (Flyvbjerg et al., 2009). Illusion is making decisions based on optimistic assumptions, not on facts. Deception is a deliberate overestimation of revenues, underestimating costs by the main actors of projects: politicians, planners, project leaders, to increase the likelihood of obtaining financing. Such phenomena may result in initiating an unfavorable project and not initiating favorable projects. Both illusions and deceptions can be overcome with an outside audit.

16.3.2.4 Execution and Control

In the public sector, project execution and control procedures are more often determined by external entities than the project implementing organization (Williams et al., 2012). Public sector projects must be coordinated within these organizations with the operational activities (Kwak et al., 2014). There are often conflicts between using the same people (and other resources) in projects and operations. Participation in projects usually has a lower priority for employees, whose primary focus is to carry out their daily duties. Therefore, it is beneficial to develop procedures for engaging line workers in projects in advance.

The constraints resulting from the project-control process are more significant in the public sector. In the public sector, oversight mechanisms overlap, for example, from regulatory bodies, audit offices, financial chambers, legislative bodies, and elected officials (Wirick, 2009).

The organization of the control process is easier in the private sector because, in this sector, there is one main criterion of effectiveness – profit (Fottler, 1981). In public sector projects, it is necessary to balance the community's requirement for supervision and control and the hostile atmosphere that can arise in contractors from a sense of distrust among private contractors resulting from this control (Kassel, 2010).

16.3.3 Scope Management

Scope management involves processes required to ensure that organization performs all of them and only the work needed to achieve its success (Subchapter 3.2).

In non-project, permanent organizations, the scope is defined historically and/or by their statutes. Permanent organizations are usually established to pursue the same goals over long periods. Hence, scope management is more important for projects than for general organizations. There was little reason to analyze the differences between public and private organizations in the scope management area. One of the differences in this area identified at the organization level is greater production stability in the public sector than in the private one (Meier and O'Toole, 2011). A difference identified at the organization level which may be relevant to projects is the greater complexity of work in the public sector (Boyne, 1998).

In projects, the scope is determined by the project charter and the Work Breakdown Structure (PMI, 2017a). Each project has specific, defined initially needs and requirements. These requirements, due to the larger number of stakeholders of public projects (Subchapter 3.10), may be varied and processes of their definition are more complex than in the private sector (Mihăescu and Ţapardel,

2013; Kwak et al., 2014; Pūlmanis, 2015; Wirick, 2009). Therefore, these processes should be more precisely defined. Specific for the project level is the difference of scope sizes: the public projects are usually larger (Kwak et al., 2014), and the scope is more diverse (Shen et al., 2004; Wirick, 2009). The public sector deals with this problem by defining project programs consisting of smaller, manageable projects and sub-projects (Kwak et al., 2014). The project scope and the level of ambition of stakeholders in public sector projects significantly change over time (Pūlmanis, 2015).

The main difference between public sector projects and other sector projects in the area of scope management is the greater complexity of the requirements definition process and the larger size of public sector projects.

16.3.4 Schedule Management

Schedule management involves processes needed for the timely completion of work (Subchapter 3.3).

Schedule – which should have a well-defined end – is more specific to projects than other types of organizations. If an organization builds a non-cyclical schedule, it usually means that it is a project and not a temporarily unlimited process. That is probably why researchers are less interested in schedule management and its inter-sectoral differences in organizations other than projects.

Public projects are characterized by the long life cycle of their products (Kwak et al., 2014). The period of use of buildings, bridges, or roads is usually at least many decades. This is inconsistent with the projects' planning cycle, which may be shorter than in the private projects due to electoral cycles (Wirick, 2009; Van der Waldt, 2011) and the annual funding cycle (Kwak et al., 2014) rather than the nature of the projects (Tabish and Jha, 2011). And this is also contrary to the generally longer duration of public projects than of private ones (Hobbs and Aubry, 2008). However, in some types of public sector projects, e.g., implementing ERP systems, the cycle is shorter than in private sector projects (do Céu Alves and Amaral Matos, 2013). Perhaps these contradictions cause that public projects have the highest rate of schedule overruns (Zwikael, 2009). Factors that positively affect project delays (i.e., reduce them) in the public sector are, among other things, coordination of the participation of stakeholders, the way of implementing changes by the owner during the project, and careful preparation of schedules and changes (Hwang et al., 2013).

The limited schedule is a specific feature of projects. Public sector projects have the highest over-schedule rate. The life cycle of public sector projects is inconsistent with the electoral cycle – which is not the case in the private sector.

16.3.5 Cost Management

Cost management involves processes needed to complete work within the approved budget (Subchapter 3.4).

Possession of an organization by the government significantly affects the processes of budgeting and accounting (Rainey and Bozeman, 2000) both on the input (budget) and output (service charges) side. The two main characteristics of the structure of the budget planning process of public projects are multi-source financing and the need to consider the budget cycle (which does not have to be the case in the private sector). Public sector projects use public resources (Pūlmanis, 2015; Mihǎescu and Ṭapardel, 2013), resources of international agencies (Moe and Pathranaraku, 2006), or multiple entities, each of which may have more than one source of funding (Torres and Pina, 2004) and private projects are financed with own resources or loans (Moe and Pathranaraku, ibid). Public projects are sometimes forced to lower the budget in order to avoid spending on unnecessary requirements

(Spittler and McCracken, 1996). The reason for lowering the budget and overstating the benefits is the desire to raise the funding more easily, which Flyvbjerg and his colleagues repeatedly pointed out (Flyvbjerg et al., 2002; Flyvbjerg et al., 2009). The cost calculation process is different in the public sector than in the private sector in decision-making. Private-sector managers are more likely to make decisions based on analysis and are less likely to support them when they result from negotiation. By contrast, public sector managers are more likely to support budgetary decisions based on negotiation (Nutt, 2006). In other words, budget decisions in the public sector are less likely to be based on hard data from reliable analyzes.

Because there are many sources of financing and many directions of public projects subordinating, their budgets are often approved by bodies external to them (Bretschneider, 1990). Usually, there are no direct financial relationships between the service provider and the customer (Spicker, 2009) – governments pay them – therefore, there is no natural mechanism of pressure on project costs. The way to overcome these problems may be to perform independent audits and allow for price competition by simultaneously providing similar services by public and private-sector organizations and companies. The essential financial decisions related to public organizations' revenues and incomes are also made outside of them. The profit expectations in public sector projects are lower than those of other sectors (Holt et al., 1995). This affects the cost of services. Public organizations tend to have higher, not optimal, unit prices than in the profit-oriented private sector (Spicker, 2009). At the same time, in public services, cost efficiency cannot be achieved by reducing production, so it must be achieved by lowering unit costs or eliminating losses (e.g., hospitals cannot stand empty) (Spicker, 2009). Since public services rarely consider customer preferences, the power of demand is minimized, the efficiency of resource allocation is lower in the public sector than in the private sector (Rainey, 2014). Public customers also less frequently delay their payments, suggesting that they are less concerned with project finances than private customers (Bageis and Fortune, 2009). In public sector projects, financing is often done on the "spend it or lose it" basis within the budget for a given period which is not conducive to cost optimization (Van der Waldt, 2011). Generally, all these facts prove that the public sector is less interested in the costs of projects than the private sector (Hasty et al., 2012).

Most public sector projects experience cost overruns (Flyvbjerg et al., 2009; Pūlmanis, 2015), although external consultants often support them in improving economic aspects (Kwak et al., 2014). The reason for cost overruns is often inadequate reporting of their implementation (Pūlmanis, 2015).

The main differences in cost management between public sector projects and projects in other sectors are that public sector projects are not profit-oriented. There are no direct financial relationships between contractors and clients. Public sector projects are less financially effective and often exceed their budget. Public sector financial management needs to fit into the budget cycle.

16.3.6 Quality Management

Quality management involves processes responsible for satisfying the needs for which the work was undertaken (Subchapter 3.11).

In public-sector organizations, there often is no competition for providing services (e.g., penitentiary services, traffic regulation, granting of permits) which is not conducive to improving their quality (Boyne, 1998; Fottler, 1981; Meier and O'Toole, 2011). Still, in many developed countries, some public services, e.g., those related to health or education, operate in a competitive environment – both public and private organizations can provide them. In this situation, a competitive pressure that may lead to improvements in quality is present.

Another factor of poor quality of public projects is the lower quality of the staff of public organizations (Mouly and Sankaran, 2007). Regardless of that (or because of that), Kwak et al. (2014) suggest developing especially high-quality-management processes for public projects. Another rather natural way to improve the quality of the staff is by increasing the salaries of project team members and particularly their managers. The hiring of Richard Granger in 2002 as the director for the implementation of the IT system for the National Health System with the highest salary among civil servants in the United Kingdom at that time (UK EHI, 2002) contributed to a significant improvement of the implementation process of that system (O'Dowd and Cross, 2007). Still another way of quality management in the public sector may be the introduction of customer surveys assessing the quality of services provided by projects.

16.3.7 Human Resource Management

Human resources (HR) management involves processes for ensuring that work will be performed by a collaborating team of qualified people (Subchapter 3.7).

Researchers pay a lot of attention to issues related to the personnel of public organizations. Possession of an organization by the government significantly affects the processes of HR management (Rainey and Bozeman, 2000) as well as the behavior of employees. In the public sector, managers are not owners of organizations. Public project teams advocate public interest (Rose, 2006).

Numerous studies report lesser work involvement of public-sector workers than those of other sectors (Goulet and Frank, 2002; Buchanan, 1975; Rainey et al., 1986; Boyne, 2002, 1998; Subramanian and Kruthika, 2012; Rainey, 2014). This is explained by the imprecise, fuzzy goals of public organizations that do not allow direct verification of the effects of work (Buchanan, 1974). To ensure more involvement, the work goals should be defined more precisely, which will allow for a more precise assessment of their achievement. Disengagement is in a sense sanctioned by public sector managers who have lower expectations of subordinate involvement than in the private sector (Buchanan, ibid). The lower involvement is accompanied by more minor external satisfaction from working in the public sector (Wang et al., 2012).

Workers in different sectors differ in their vulnerability to different types of motivation. Public-sector employees are less vulnerable to external stimuli (Buelens and Van den Broeck, 2007; Jałocha et al., 2014). They attach less weight to financial incentives (Rainey and Bozeman, 2000; Rainey et al., 1986; Rainey, 2014). Public-sector workers are motivated by achievements, and private-sector workers by power (Andersen, 2010). In the public sector, the factors most motivating staff to work are stable, secure future, a chance to learn something new, and the opportunity to use special abilities (Jurkiewicz et al., 1998), the importance of public services, participating in the implementation of public policies, sacrifice for others, responsibility and integrity (Rainey and Bozeman, 2000). In the private sector, the prime motivating factors are high salary, the chance of being a leader, and promotion (Jurkiewicz et al., ibid). Hence, motivation systems in the public sector should be aimed at internal rather than external incentives. When recruiting team members for public projects, their significance for the community of beneficiaries should be emphasized and material incentives to a lower extent.

The sectors differ in terms of the managerial methods used. Executives in the public sector are less willing to delegate power (Rainey, 2014). Because of many relationships between individuals and organizational units in the public sector, the need to define clear leadership and accountability rules is more significant than in the private sector. Public sector managers usually do not have the appropriate authority to efficiently manage their personnel (Rainey and Bozeman, 2000) and have fewer opportunities to use financial incentives (Fottler, 1981). Managers in private organizations have a greater variety of internal organizational activities, greater autonomy of their application, and better options for using the environment (Hvidman and Andersen, 2014). In the public sector, the responsibility of project managers is blurred (Pakistan Shah at el., 2011), e.g., because they must obtain permits from their supervisors to act. For example, they usually do not have the authority to obtain the resources necessary to implement projects (Van der Valdt, 2011). Management is more often done by persuasion than by coercion (Buchanan, 1975). Managers in public sector projects need to interact with, negotiate, and participate in resolving disputes in their teams and in the external environment for which the projects are implemented (Torres and Pina, 2004). Due to more frequent management volatility in the public sector, which may result in volatile support for the project. Due to the constraints resulting from regulations, the need to convince employees to implement project products in their environment in public projects is greater (Cats-Baril and Thompson, 1995). Because of all these reasons, leaders' charisma in public organizations is more important than in the private sector (Fottler, 1981). The choice of a project leader having both technical, political, and interpersonal skills is particularly important (Cats-Baril and Thompson, 1995).

Controlled by committees, the public sector attracts the project management elite less than the private sector (Peled, 2000). The lack of competitiveness of public organizations on the labor market, compared with private companies, results in a lower quality of staff in public projects than in private ones (Mouly and Sankaran, 2007). These features of the public sector make it difficult, for example, suggested by Gomes and colleagues (2012), to appoint people from the private sector to be managers of public sector projects. The unattractiveness of the public sector may also cause that public project managers concentrate on one project. In contrast, in the private sector, they are interested in many projects implemented in one company (Coster and Van Wijk, 2015).

Dependencies other than to project managers relate to project team members, although the influence of team members on project success is less in the public sector than in the private sector (Nagadevara, 2012). Public sector projects require more human resources to control compliance with constraints imposed by public sector oversight of public sector projects (Wirick, 2009), i.e., public sector project teams must be larger than corresponding private projects. Wirick (ibid) states that the long-term nature of employment in the public sector, resulting from the protection system of civil servants and employment, results in limited opportunities for creating and modifying project teams. This phenomenon also entails the presence of strong group norms – loyalty to a small group is more important than loyalty to the entire organization (Wirick, ibid).

16.3.8 Stakeholder Management

Stakeholder management is the processes of interaction with the organization's stakeholders, including their engagement in the organization's decisions and assuring their satisfaction, if possible.

The external environment of organizations is an important sector differentiating factor (Bretschneider, 1990). Because of their social role, public-sector organizations are subject to greater involvement of external authorities and interest groups (Rainey, 2014). They are subject to more external influences than private companies (Torres and Pina, 2004). Public organizations are subject to ongoing external evaluation (Fottler, 1981).

The external environment influences organizations in various ways, facilitating or hindering their functioning. Formal constraints arising from supervision by legislators, the hierarchy of executive agencies, regulatory agencies, and the courts apply to public-sector organizations. These constraints apply to salaries, promotions, and disciplinary action in public-sector organizations (Rainey, 2014). Public organizations are strongly influenced by a particular type of stakeholders –

politicians involved in the process of their governance (Rainey, 2014; Spicker, 2009). In addition to formal, public organizations have to deal with a large variety and intensity of informal political influences, lobbying, public interest, the influence of various interest groups, clients, or constituting institutions. There are greater challenges in balancing external political relations (Rainey, 2014). Political impact on public organizations is more significant than on private ones due to, inter alia, the need to raise funds and mandates for action in a non-market way. The number of external interventions and interruptions imposed by external interest groups and political factors is expected to be greater in public organizations than in private companies (Rainey, ibid). But on the other hand, the set of options to block the influence of the environment is greater in the public sector than in the private one. For example, red tape mechanisms may be used for this purpose. The utilization of the domain is more effective in the private sector because private managers have more options for acting (Meier and O'Toole, 2011). For example, they can shape purchasing policy more flexibly. But the external environment of public organizations is more stable (Meier and O'Toole, ibid), among other things, because both services (e.g., health care, education) and their recipients are more stable.

The researchers working at the project level mostly confirm results identified at the general organizational level. Public projects have more stakeholders than private ones (Mihăescu and Ţapardel, 2013; Kwak et al., 2014; Pūlmanis, 2015; Kassel, 2010; O'Leary and Williams, 2008; Dilts and Pence, 2006). In practice, the mere identification of the stakeholders of a public project may be difficult. Public projects are more exposed to external factors than private ones (Gomes et al., 2012). The most important public project stakeholders are taxpayers and the communities for which these projects are performed (Wirick, 2009), in which environment they are implemented (Shiferaw, 2013), to which they are responsible and accountable (Mihăescu and Ţapardel, 2013; Pūlmanis, 2015). They must be taken into account when deciding on implementing public sector projects (Shiferaw, 2013). Other essential project stakeholders include the legislators whose requirements must be met (Kassel, 2010), whose requirements should be subordinated to the interests of the groups mentioned above. But it may happen that due to legal regulations, the customer's perspective is less important than the regulatory perspective (Krukowski, 2015). Stakeholders in public sector projects are often business communities (Kwak et al., 2014), such as sector-specific or professional associations. In public infrastructure projects, environmentalists change the approach to management, for example as a result of their research (Smith, D., 2015). For public projects, a specific type of external stakeholders is other public agencies with which inter-agency agreements are established (Kwak et al., 2014). Public projects operate under media scrutiny (Wirick, 2009). The oversight mechanisms of the public sector acting on many levels and having possibly conflicting interests further increase the number of project stakeholders (Wirick, 2009).

This diversity of stakeholders means that public sector project managers have to consider a much greater level of interdependence between organizations involved in their implementation than private-sector project managers (Bretschneider, 1990).

The way public sector projects are implemented and especially their criticism often affects the government's image in the eyes of society. Hence, public projects must also take into account the interests of politicians (Kwak et al., 2014) belonging to various political parties (Kassel, 2010). It happens that they are not always familiar with project management (Pūlmanis, 2015). The political environment, like opposition, may be hostile to projects (Kassel, 2010). Public projects are vulnerable to political changes (Kwak et al., 2014). Elected politicians and executives in the public sector have enough power to start, kill, or change projects (Dilts and Pence, 2006).

Continuous stakeholder engagement is a critical success factor for a public project (PMI, 2014). In the public sector, external stakeholder management needs to be planned more thoroughly than in the private sector (Bretschneider, 1990); the ways of gaining their support should be considered from the very beginning of the project (Gomes et al. 2012).

In line with the approach of Ng and colleagues (2012), communities should be involved in drawing up the project strategy itself and designing the way they will participate in the project and then in the process of planning the project and making decisions about its implementation (Jänicke et al., 2001). In the planning phase, communities should be involved in the preparation, modification, and finalization of the overall project plan. For the success of public participation, it is important to believe that it really influences the project, that not everything has been agreed upon in advance (El-Gohary et al., 2006). Participation in the social and environmental impact assessment should be public.

Community involvement should be maintained throughout all project duration. The optimal features of the team representing the community are full knowledge of the needs, requirements, and interests of stakeholders, knowledge related to the manufactured product, and good inter-action with the project developer, government, and other stakeholders (Peled and Dvir, 2012). The benefits of including representatives of the community for which the project is implemented are, among others, psychological involvement with other stakeholders, better definition of re-quirements (Markus and Mao, 2004), having information about the progress of work, better product quality, and use – rather than rejection in the future (Gallivan and Keil, 2003).

Support for a public project should also be built within the organization implementing it (PMI, 2018a). Public sector projects are more challenging to implement because they require the cooperation and effectiveness of the organization beyond the project team (Wirick, 2009). Public sector projects must be coordinated within these organizations with the operational activities carried out (Kwak et al., 2014). There are often conflicts between the use of the same people (and other resources) in projects and operational activities. Also, the number of internal customers may be greater in public projects than in private ones (Hobbs and Aubry, 2008).

The number and importance of stakeholders mean that stakeholder management is one of the most critical areas of public project management and should use proven techniques, such as stakeholder analysis or stakeholder circle (Bourne and Walker, 2008). It is desirable to involve the most important stakeholders in project management bodies. It is essential to involve stakeholders in the process of project definition and evaluation of its effects (Peled and Dvir, 2012).

16.3.9 Communications Management

Communications management involves processes responsible for providing all organization members and its environment with needed information (Subchapter 3.5).

Communications management can be considered as part of the stakeholder management area.[2] Communication includes relations with all the stakeholders mentioned in the previous section. Several differences between public projects and other-sector projects in the commu-nications area stem from a variety of public projects' stakeholders.

Public organizations are more transparent than private ones. They transmit more information about their processes and decisions to their environments (Meier and O'Toole, 2011). Information about public projects must be accessible to many stakeholders, especially the public, and cannot be kept secret, like in the private sector (Rosacker and Rosacker, 2010). Unlike in the private sector, information on the implementation of public projects must be available to external stakeholders, based on FoI-type regulations (e.g., USA Congress, 1966) or project implementa-tion regulations (e. g., Argentina Congreso de la Nacion, 1994). Due to a large number of stakeholders in public projects, good communication, both with internal and external stakeholders, is a critical success factor for these projects. Disseminating information about public projects promotes stakeholder engagement. Publishing the main project documents – its charter, plan, periodic reports allow obtaining knowledge from stakeholders that can be used in the further project implementation. Publishing reliable documents helps increase the trust of stakeholders in the purposefulness and correctness of project implementation. This way, uncertainty is reduced, and corruption is also limited (Van der Waldt, 2011).

The higher level of red tape which characterizes the public sector is associated with less effective communications. The red tapes strictly define the scope and procedures of communicating; in particular, they limit the scope of informal information (Ning, 2014). Studies conducted in the public sector have shown that clarity of objectives is positively correlated with the effectiveness of communications, both internally and externally. Hence, the impact of red tape on communications may be overcome by the clarity of objectives and appropriate organizational culture (Pandey and Garnett, 2006).

The project's vertical (top-down or bottom-up) communication is easier in public organizations than in private companies, but external communication is more difficult in public sector projects.

Public sector projects are more transparent, and they need to convey more information to the outside world. The lack of clarity of public sector projects' objectives adversely affects the communication effectiveness of projects in this sector.

16.3.10 Risk Management

Risk management is responsible for decreasing the probability and impact of threats and increasing the probability and impact of opportunities (Subchapter 3.8).

Public projects are exposed to different risks than projects in other sectors, e.g., political risks, risks related to the participation of public stakeholders, and risks stemming from public regulations. Governance, management, and contracts are considered the main sources of risks for large public projects (Patanakul, 2014). Public projects are inherently risky due to longer planning horizons and the more complex environment (Pūlmanis, 2015). Risk can manifest at any phase of major public sector projects (Baldry, 1998). Therefore, it is essential in the public sector to develop contingency plans and formal risk monitoring processes and then use them (Kwak et al., 2014). Meanwhile, inadequate risk information is the norm in public sector projects (Pūlmanis, 2015). In public-sector organizations, there is more emphasis than in the private sector on risk avoidance (Fottler, 1981). Public employees are more cautious (Rainey, 2014). On the other hand, it is specific to the public sector risk management process that it disregards early warning signals used in the private sector (Williams et al., 2012). Penalties for non-compliance with regulatory restrictions and possible criticism from the political opposition are conducive to a negative risk attitude factor in public projects (Wirick, 2009). The level of control is more robust in public-sector organizations, and research has shown that managers in organizations with higher levels of internal control are less willing to take risks than in those where the level of control is lower (Bozeman and Kingsley, 1998). As noted earlier, the objectives of public organizations are less clearly defined than in the private sector, and managers whose goals are vaguely defined are less likely to take risks than those who have well-defined goals (Bozeman and Kingsley, ibid). A high level of control and vagueness of goals are two next factors implying a negative attitude to risk in the public sector.

All of this can lead to hiding and not reporting project risks. The way to overcome this problem should be independent reviews of project plans, performed in order, among other things, to identify risks.

16.3.11 Procurement Management

Procurement management involves buying or acquiring goods and services outside the performing organization (Subchapter 3.9).

There is a two-way relationship between contracts and project management. Contracts play a special role in public sector projects, and they can also be implemented using project management techniques. As public sector projects rely heavily on purchasing, it is imperative to ensure efficient collaboration with the procurement staff and an efficient purchasing process (Kwak et al., 2014).

The primary characteristic distinguishing procurement in public-sector organizations from procurement in private-sector organizations is its formalization. Excessive formalization of public sector purchases seems to be their inherent feature. Public purchases are made based on legal regulations, while in the private sector, they can be made based on arbitrary decisions of the owners. The government has a significant legal impact on purchasing processes in public-sector organizations (Rainey and Bozeman, 2000). Hence, the flexibility of the procurement process is minor in public projects than in other projects (Drew and Skitmore, 1997; Shen et al., 2004).

Public projects' bid evaluation criteria differ from private projects (Bretschneider, 1990). The price criterion often plays a decisive role in public projects, not least because the input parameters in public sector projects are easier to control than the output parameters (Fottler, 1981).

Due to the negative risk attitude in public organizations (Fottler, 1981) "off-the-shelf" solutions are, where possible, suggested for public projects rather than the highly risky development of new products (Kwak et al., 2014).

The parties involved in public projects are more constrained in implementing relational contracting because regulations prohibit specific behavior by public officials, making it difficult to form relationships (Ling et al., 2013). Public-sector regulations often prohibit taking into account the history of previously completed contracts in the process of bid evaluation which could favor relational contracting, which is natural in the private sector (Rahman and Kumaraswamy, 2004).

The way to overcome some problems with procurement is to create, at the level of the project-implementing organization, a well-trained, specialized purchasing team, supporting project procurement processes. At the regulatory level, non-price criteria should be allowed or enforced, and the history of cooperation with the supplier should be considered.

The procurement process is more complex in public than private projects (Rahman and Kumaraswamy, ibid). Procurement management in public sector projects is more formalized than in other projects. This makes it difficult to implement relationship-based contracts. Supplier selection criteria in the public sector relate to price more often than in other sectors. Off-the-shelf solutions are recommended for use in public sector projects to minimize risk.

There are differences between public sector projects and projects from other sectors in each analyzed management area. Some of these differences were identified during the general research of the organization, and some – during the analysis of projects. Public sector projects inherit the differences between public organizations and organizations in other sectors.

16.4 The Layered Model of Differences

After discussing the differences in individual management areas, I can return to the general model of differences between public sector projects and projects from other sectors.

The presence of – minor or significant – differences in each management area could suggest the adequacy of the dimensional model of differences. In some management areas, the sector's impact is more significant (e.g., stakeholder management, procurement management, human resource management). In others, it is less (e.g., quality management, schedule management, or scope management). Hence, the dimensional model seems to be appropriate. But at the same time, many project management techniques and processes in each management area are common to public sector projects and other types of projects. There are no known arguments for not using the same processes, for example, risk management (risk identification, assessment, defining responses, etc.) or quality management (planning, quality assurance, quality control) in all sectors. So, in project management, each area has both shared features different from public sector projects and other projects. For example, the basic activities (activity identification, relationship definition, critical path-building algorithm, etc.) are the same for different sectors for schedule management. Usually, these are technical and managerial features. On the other hand, the business characteristics discussed in the previous sections, above the basic layer (taking into account the annual and electoral cycle, frequent delays ...), are different. Consider procurement management. In the public sector, it is necessary to consider different legal conditions of purchasing than in other sectors – which does not change the overall structure of the procedure (identification of purchases, the definition of the procedure, etc.). In stakeholder management, the structure of the process in public projects is also analogous to that in other sectors (stakeholders identification, their analysis, determination of the manner of their involvement, etc.). Still, due to organizational and social conditions, the content of these processes is different. Each area (dimension) of management has its common (managerial and technical) layer and an extra (business) layer. I will call this model of differences the **layered model** of project inter-sectoral differences (Figure 16.4).

The layered model allows further analysis of the adequacy of adopting general differences between public organizations at the project level. The existing literature analysis also shows *terrae incognita* of research on these differences. Are the differences in quality management or scope management really less important, or has insufficient research been performed there? The other

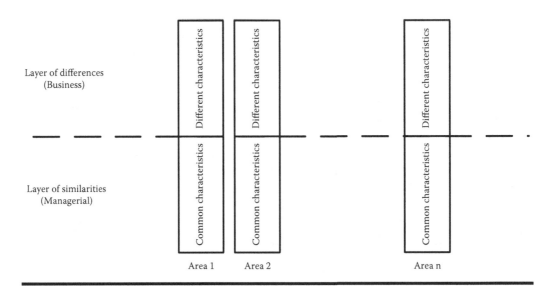

Figure 16.4   The Layered Model of Project Inter-Sectoral Differences.

proposition is that, at least in research in those management areas where the differences are more significant, the project sector should be regarded as an explanatory variable of studied phenomena. This variable's influence – or lack of influence – could be the basis for further elaboration of a model for differences (if impact is identified) or general project management models (if the sector's impact would not be significant).

Up to now, for all organizations, three models of inter-sectoral differences have been defined (Subchapter 16.2): the generic model, the core model, and the dimensional model. You can consider extending this set of models defined for any type of organization with a fourth inter-sectoral difference model defined here for projects: the layered one. At the business level, where there are differences between sectors, you can apply earlier defined difference models, particularly the dimensional or the core model. In some managerial areas, there are differences at the business level, while in others, they are smaller or do not exist.

Notes

1. An earlier version of this chapter was published by the author as a paper entitled *A Framework for Analyzing Differences between Public and Other Sector Projects* in "Public Governance" quarterly – the journal of Malopolska School of Public Administration, Cracow University of Economics, 2018, 3 (45): 73–88.
2. Until 2008, PMBOK® Guide (PMI, 2008a) treated stakeholder management and communications management as one area of knowledge.

Chapter 17

Differences of Complexity

17.1 What Is Complexity?

The term "complex" means "composed of something, composed of several (many) parts, elements; having a complicated structure" (Szymczak, 1989, vol. 3). Complexity is a property of objects composed of many related elements (Baccarini, 1996). A complex project has an ever-changing, unpredictable political, economic, and social environment with many interconnectednesses. Some stakeholders can force radical changes which do not consider existing decisions or schedules (Chapman, 2016). Complexity concerns the stakeholder relationship network (Whitty and Maylor, 2009). An important factor influencing their complexity in public projects is the variety and variability of project goals (Section 16.3.2) (Van der Waldt, 2011). A project is complex if uncertainty plays a significant role (Bosch-Rekveldt et al., 2011). Brady and Davies (2014) define two dimensions of complexity: structural and dynamic. Structural complexity is the arrangement of components and subsystems into the overall system architecture. Dynamic complexity is the changing relationship between the parts of the system and the environment during the project implementation. Geraldi et al. (2011) extend the concept of complexity to five dimensions: structural, uncertainty, dynamics, implementation path, and socio-political. A structural dimension is a large number of related elements. Uncertainty is the discrepancy between the target state and the information held and the level of ambiguity in that information. Dynamics means changes to the project. The path is the time at which the project is to be done in relation to the abstract optimal time. Socio-political complexity is the importance of the project, stakeholder support, convergence/alignment with opinions, interests, requirements, transparency, and the presence of hidden stakeholder agendas. In turn, according to Bosch-Rekveldt et al. (ibid), the complexity of large projects can be characterized in three dimensions: technical, organizational, and environmental. Environmental complexity involves stakeholder perspective, local conditions, and market conditions. Baccarini (ibid) proposes two dimensions of complexity: differentiation and dependence, distinguishing between organizational and technological complexity. Differentiation refers to the number of components (tasks, specialists, lower-level components, etc.).

17.2 Survey on Project Complexity

The complexity of public sector projects is an important issue, both from a purely cognitive and practical point of view. Finding that these projects are more complex than those of other sectors might impact the project management processes of this sector in all management areas. For example, if similar public sector projects are more complex than other projects, longer duration, higher costs, and greater risk burden would be expected.

The above-discussed qualitative characteristics of the managerial differences between public projects and projects in other sectors were supplemented by surveys on the relative complexity of public sector projects. The research was part of a larger survey.[1]

Five hundred twelve people responded to the survey. The respondents came from 61 countries from all continents. People working for various types of organizations: public, private, academic, and others (e.g., freelancers) participated in the survey. The respondents worked in various areas of activity. The areas of IT/Communication and Construction/Infrastructure were the most numerous in the study. People performing various roles in the project environment participated in the survey. They play the managerial role (majority of respondents – 237 persons), team members, people from the project implementation environment (subject experts, product users, PMO members, etc.). People with various levels of experience in project implementation participated in the research – both in the implementation of public sector projects and projects from other sectors. The respondents were also classified from the point of view of their project management certificates. They may be treated as project management experts as 440 had some project management certificates.

17.3 Survey Results

Questions with three possible answers were used to test whether public sector projects are more complex than private projects:

   0. Private projects are more complex than public ones
   1. There are no differences, or it's hard to tell
   2. Public projects are more complex than private ones

This question was asked for each management area. The obtained results are shown in Table 17.1.

As a whole, public sector project management is considered more complex than project management in other sectors. Respondents also believe that managing public sector projects in each area is more complex than managing private projects. Three groups can be distinguished among the areas of management:

1. The most significant difference group is those areas with a relative complexity between 1.77 and 1.59. This group includes stakeholder management, procurement management, and communications management. The dominant and median of responses for all these areas was 2.
2. The group of mean differences is those areas whose relative complexity is between 1.44 and 1.35. This group includes human resource management, scope management, integration management, cost management, schedule management, and risk management. The dominant for these areas was 2, but the median is 1 or 2.

Table 17.1 The Relative Complexity of Public Sector Project Management Areas

	As a whole	Stakeholder management	Procurement management	Communications management	Human resource management	Scope management	Integration management	Cost management	Schedule management	Risk management	Quality management
Mean	1,58	1,77	1,74	1,59	1,44	1,41	1,41	1,36	1,35	1,35	1,15
Median	2	2	2	2	2	1	1	2	1	1	1
Dominant	2	2	2	2	2	2	2	2	2	2	1
Standard deviation	,588	,511	,534	,587	,666	,614	,607	,734	,667	,706	,696

3. A group of minor differences. It includes only the area of quality management. The ratio of the relative complexity of public sector projects is 1.15.

The group of areas that are relatively the most complex is stakeholder management, procurement management, and communications management. These three areas relate to the relationship, in whole or in large part, with the public project's environment. Their high relative complexity indicates which areas of management project managers and other persons responsible for project management in public organizations should pay particular attention to. At the organizational level, it may be creating appropriate management practices, and at the project level – building the right team and acquiring the necessary knowledge.

Factor analysis was performed to identify the main factors influencing the complexity of public sector project management. The input variables described the relative complexity of public sector project management areas. Convergence was obtained in three iterations after the factor analysis with the Varimax method (variance maximization). Two factors of the relative complexity of public sector projects have been identified. The results of the factor analysis are shown in Table 17.2. Management areas are sorted according to the value of their loads. Management areas included in individual factors (with the highest appropriate loads) are marked in gray.

The first factor includes:

1. Quality management
2. Schedule management
3. Cost management
4. Risk management
5. Scope management
6. Integration management
7. Human resource management

Table 17.2 Complexity Factors of Public Sector Project Management

	Factors	
Management area	1	2
Quality management	0.74	0.10
Schedule management	0.73	0.04
Cost management	0.67	0.12
Risk management	0.62	0.33
Scope management	0.58	0.25
Integration management	0.56	0.42
Human resource management	0.53	0.15
Communications management	0.24	0.65
Procurement management	0.19	0.63
Stakeholder management	0.02	0.79

The second factor includes:

1. Communications management
2. Procurement management
3. Stakeholder management

The first factor is the areas requiring less project management involvement in contact with the project environment. Work Breakdown Structure can be built only using a computer. Building a schedule or budget does not require intensive contact with the external environment. Quality control is mainly the relationship between controller and product. In the components of the first factor, people outside the project team are less important – although, of course, both to gather relevant information, enabling the creation of a schedule, budget or scope, and to execute the schedule, budget, or scope of the project, as well as other major artifacts – interaction with people both inside and outside the project team is necessary.

The second factor is the areas that need constant, intensive contact with the project's external environment. Managing stakeholders or contracts requires continuous communication with external entities.

There is an interesting convergence between the assessment of relative complexity and the factor analysis results. The more complex areas: stakeholder management, procurement management, and communications management, simultaneously constitute the second factor. It includes those areas of management that are implemented in relation to the project's external environment. Stakeholder management (irrespective of the formal definition of this area) is considered to involve external stakeholders. Although the project team is one of the main stakeholders of the projects, activities involving team members and the team as a whole are usually performed in a specific area – human resource management. Procurement management also applies to a component of the external environment: suppliers and subcontractors, while stakeholders other than contractors can be influenced mainly through communication; the project team has no direct means of issuing orders and enforcing them.

The first factor may be called **internal management,** and the second – the **external management**.

The main reasons for the differences between public projects and projects from other sectors are formal regulations governing public organizations (including projects) and greater exposure of public projects to external influences.

17.3.1 The External Management

The relatively most complex areas are stakeholder management, procurement management, and communications management. At the same time, these three areas constitute a variable collectively referred to as "external project management."

The diversity and number of stakeholders, their ability to influence public projects, and their importance for multiple stakeholders (Subchapter 3.10 and Section 16.3.8) make stakeholder management the most difficult area to manage in public sector projects. My survey confirmed this. Hence, the public sector should develop project stakeholder management tools and methods specific to this sector. Project managers and other members of project teams, as well as decision-makers responsible for the shape of project management in public organizations, should be educated differently than in private companies. It is necessary to consider mechanisms other than those based on classic contractor-client relationships to ensure the parties' satisfaction from the implementation of public sector projects. For example, they need to be aware that the product's low price need not be a factor of the satisfaction of the recipient of the public product since recipients of public benefits are not intended to pay the contractor directly. Reducing the complexity of stakeholder management in public sector projects is usually impossible by reducing the number of stakeholders.

The main feature influencing the complexity of procurement management, the second most complex area, is its formalization (Section 16.3.11). If the project needs goods from outside the project organization, the buyer cannot simply go to the store or the manufacturer and buy them. Procedures are always necessary, sometimes very complex. At the same time, most public projects are implemented in whole or in large part through purchases. The method of implementing public sector projects, and thus also procurement, is usually determined outside the purchasing organization. Therefore neither the organization where the project is implemented nor the project team can influence these processes. Consequently, it is necessary to educate public organizations that implement purchasing projects by the imposed bureaucratic constraints. An additional consequence is the creation of similar teams in private companies dealing with supplies for the public sector.

The complexity of communications in public projects is a derivative of the number of their stakeholders (Section 16.3.9). The complexity is further exacerbated by the requirement of transparency, i.e., providing stakeholders with the main information about the project implementation. Transparency is one of the fundamental characteristics of public organizations, so project managers must pay particular attention to its form and content. The relatively high complexity of communications can be understood as the importance of its precise formulation for success in stakeholder management and thus indirectly for the entire success of the project.

Managing public sector projects is complicated by their external environment.

The high relative complexity of the three management areas described above indicates which areas of management project managers and other staff responsible in public organizations for project management should pay special attention to. At the organizational level, it may be creating appropriate management practices, and at the project level – building the right team and acquiring the necessary knowledge.

17.3.2 The Internal Management

The area of human resource management (Section 16.3.7) is the most complex of the internal areas of project management. This is probably because, similarly to external areas, people are the main management subject. The difficulty of managing the personnel results from the lower quality of the personnel and, at the same time, the greater sustainability of its employment which results in difficulties in shaping the project team. In the public sector, managers have fewer opportunities to apply financial incentives, and team members are less engaged in their work than in the private sector.

Contradictory to the higher complexity of public projects is the lower quality of the staff working there. It would be natural to pay more for harder work. Here, the public sector offers a lower salary for more complex work. If this trend continues, the effects of implementing public projects may further deteriorate.

Integration management, the next most complex management area, is a specific, rather abstract area. While when we talk about managing cost, human resource, stakeholder, or other "specific" areas, we know exactly what we are managing. And "integration" is not visible; it is a kind of more abstract area. Perhaps this is why integration management has not been assessed as particularly complex compared to other sectors. And yet, virtually all elements of integration management in the public sector are performed differently than in other sectors (Section 16.3.2): the nature and definition of project objectives, the level of red tapes, ways of initiating, implementing, and executing projects are significantly different.

The complexity differences between scope management in the public sector and other sector projects are moderate (Section 16.3.3). The main reason for the greater complexity is the larger size of public sector projects. It is also more difficult to agree on the requirements of a project when it has more stakeholders.

The non-profit nature of public sector projects and the lack of financial relations between the contractor and client of public projects probably caused that their relative complexity was not rated too high (Section 16.3.5). The frequent cost overruns and fast payments in the public sector suggest that cost is not a particularly important area in public sector projects. A complicating factor in cost management is funding public projects by multiple entities that may have different financial requirements.

Managing the schedule in public projects is relatively little different from managing in other sectors (Section 16.3.4). There is no reason why building and maintaining a schedule for public sector projects should be remarkably different from other projects. The critical path has no version for public projects. But, on the other hand, frequent overruns suggest more significant scheduling problems for public projects than for projects in other sectors.

Quality management was considered the least relatively complex area of management (Section 16.3.6). It is believed that there is less pressure on quality in the public sector than in other sectors.

Note

1. The results of this survey were in detail described in Gasik (2016b).

Chapter 18

Why Public Projects Succeed

18.1 What Is a Public Project Success?

There is a wealth of literature on the causes of failures and successes of public sector projects. Virtually every publication on project management increases the chance of success or reduces the risk of failure. Discussions on the successes and failures of public projects should begin with determining what we consider to be their success.

Determining the level of success of a public project is more difficult than in the private sector as there is no single success factor – profit. The possible multitude of objectives for public projects (Subchapter 8.7) also makes it difficult to assess the achievement of success clearly. In addition to the multitude of goals, there is also difficulty defining and measuring the value of individual success factors, which may have different measures. How to measure the level of provision of cultural goods, social development, honesty, or moral behavior?

The success (and failure) of public projects, according to Volden (2018), are assessed in six dimensions. The project must be relevant; its products should be needed. An example of irrelevance may be the construction of a high-speed railway or a motorway between towns where there is no intensive movement of people. The project must be effective, i.e., its effects should meet the expectations and requirements. The project, also public, must be efficient, i.e., it must deliver the planned products within the scheduled time and budget. The project must also be cost-efficient, i.e., the costs of producing the products cannot be too high. Project effects must be sustainable, i.e.; they should be observable sufficiently long after the project is completed. Achieving effects other than planned may also affect the assessment of the project's success. For example, as a result of the project implementation, excellent project management processes that could be implemented in other organizations were developed, which was not set as a goal before the project. The seventh dimension of success is compliance with public sector values, e.g., honesty, impartiality. Even if it meets the requirements of the first six areas, a public project with corruption processes will not be considered successful.

The effectiveness of a public project is a social concept. What is a failure for some may be a success for others (Goldfinch, 2007). The most obvious example is elections: for the winning party, they are a success, and for the losing party, they are a failure. It is similar with relevance: road builders consider the construction of a motorway as a necessary project, while environmentalists – as entirely unnecessary.

DOI: 10.1201/9781003321606-21

The successes and failures of public projects can affect the entire community (Dilts and Pence, 2006) and influence the development of nations (Chih and Zwikael, 2015). For example, poor implementation of health care reform will affect society as a whole. This phenomenon does not exist in the private sector, where failures usually affect only one company, the owner. Private companies typically do not cover the entirety of society (probably apart from projects of IT giants). A category of success, and therefore failures, specific to public projects, falling within the area of relevance, is contributing to the development of the country (Damoah and Akwei, 2016), e.g., a failure because the project is not related to the strategic goals of the state and society (Shiferaw and Klakegg, 2012).

In the procedural dimension, public projects can suffer various types of failures: shelved; planned, not implemented; stalled; implemented, but to a lesser extent, malfunction or non-performance of project products (Ikejemba et al., 2017; UK POST, 2003).

18.2 What Influences the Success and Failure of Public Projects?

Determining the success of projects has been the subject of interest of many researchers. For example, Kuula and colleagues (2013) explored ways to identify success factors for an outsourced visa-issuing project by the Finnish Ministry of Foreign Affairs. Tabish and Jha (2011) conducted research in India to determine what influences public construction projects' success. Jacobson and Choi (2008) compared success factors in the public sector and PPP projects. Public sector project success was also investigated by Prasad et al. (2013), Ika et al. (2012), Brown (2012), Gomes et al. (2012), Shah et al. (2011), Yalegama et al. (2016), Songer and Molenaar (1997), and many others. This research allowed for identifying factors influencing the success of public projects.

From the point of view of examining the successes and failures of public sector projects, two groups can be distinguished: factors specific to the public sector that do not occur (or occur to a lesser degree in a different context) in other sectors and factors that also happen in other sectors (e.g., governance, planning, personnel qualification, risk management, specification of requirements). In this book, I only deal with the first group of factors – a general discussion of all the factors for success and failure of projects would far exceed the planned size of this book.

18.2.1 Politics

The most important group of factors influencing the success of a project, specific to public projects, is politics. The relevance of the public project is its connection with public policy (Achterstraat, 2013). Political support is one of the main success factors of public projects (Jacobson and Choi, 2008; Anthopoulos et al., 2016; Damoah and Akwei, 2016). It is essential for the success of a public project to attract as many political allies as possible (Brown, 2012). The most important projects should be supported by prominent political leaders (Anthopoulos et al., ibid). The support may come from the expectation that the implementation of the project will affect the political position of a politician or a political party (Flyvbjerg et al., 2009). Such an expectation is an example of an implicit political agenda that negatively affects the project's chance of success (Ikejemba et al., 2017). A factor in the success of public projects is political inclusion, i.e., gaining the broadest possible political base to support the project (Ikejemba et al., ibid). A tool to broaden the policy base may be developing a third-party verified business case, that is, demonstrating that the project is actually objectively needed. In turn, the lack of political will to implement the project may be the reason for its failure (Batool and Abbas, 2017). Suppose the political will is present among some political forces, and other parties are against the project. In that case, we are dealing with a conflict of interest that negatively affects the project (Achterstraat, 2013). From the typically public factors of failure, Damoah and Kumi (2017) cite party politics, political influence, and changes in the government. Political power is often manifested by influencing the staffing of projects and demanding changes to project plans, most often about the scope of work. This causes misdirection and a waste of project resources (Barkley, 2011). The attitude of the ruling politicians to the project may change after the change of the ruling team. The green party may stop the project of building a coal mine, initiated by a more conservative party. The level of political support may not be overestimated. Even if the same political option remains in power, support for the project may decline (Anthopoulos et al., 2016).

18.2.2 Stakeholders

In addition to politicians, public projects have a much larger set of stakeholders than projects from other sectors (Section 16.3.8). Many researchers indicate the area of stakeholder management, their involvement, and various aspects of working with them as success factors for public sector projects (Pakistan Shah et al., 2011; Ikejemba et al., 2017). The most important, specific group of stakeholders in public projects are the societies and communities for which the projects are implemented (Yalegama et al., 2016). Project successes are strongly supported by the involvement of these stakeholders in the project implementation (Yalegama et al., 2016). It may happen that communities will oppose development projects implemented for them (Damoah and Kumi, 2017) which requires appropriate negotiation skills. In these and other situations, stakeholders should be respected (Jacobson and Choi, 2008), a partnership between the project team and other project stakeholders is important (Tabish and Jha, 2011). The measurable benefits of the project should be indicated (Yalegama et al., 2016), and participatory quality assurance (Achterstraat, 2013) should be applied, i.e., involving stakeholders in the evaluation of the project and its future products.

Other major public projects stakeholder groups are project contractors and financiers (domestic, i.e., taxpayers or external). Failure of public projects may be caused by the non-involvement of relevant stakeholders (Shiferaw and Klakegg, 2012). The success of a public sector project is adversely affected by the volatility of stakeholders, in particular policymakers (Prasad et al., 2013). Stakeholders may have different, conflicting interests, so it is essential to be able to cooperate and build support among unfavorable entities (Achterstraat, 2013). Support should be built both outside and inside the organization implementing the project (PMI, 2014).

There are information and communication issues related to stakeholder management. The success of public sector projects is influenced by information (Prasad et al., 2013) and open communication (Kuula et al., 2013; PMI, ibid), both internally and with external stakeholders (PMI, ibid). The area of success factors related to communication is knowledge management. Proper, effective training, acquired knowledge, knowledge transfer, and the creation and maintenance of knowledge about best practices (Ika et al., 2012; PMI, 2014) affect the chances of success for public sector projects.

18.2.3 The Specificity of Public Organizations

Public organizations are driven by procedures that often take the form of red tapes. This makes flexible project management difficult. Also, the lack of competition means that they are not focused on continuous adaptation to customers' changing needs (Barkley, 2011). A contracted service, even if demand for it decreases or diminishes, will usually be delivered. Red tapes and the need to follow procedures also interfere with many other situations in public projects. Obtaining approval for a project change can require multi-level approvals and a long time.

18.2.4 Corruption

Corruption is a very often mentioned cause of failure of public projects (Aladwani, 2016; Damoah et al., 2018; Locatelli et al., 2017). The greater the freedom of the officials, the economic rent of decision-makers, or the weakness of the institutions, the greater the exposure to corruption. Bribes may decide to initiate projects. Corruption causes deterioration of governance (Aladwani, 2016) which is one of the main factors influencing the success of projects. Corrupt phenomena may make it impossible to select the best suppliers from the private sector. Non-substantial filling of positions in projects can contribute to the failure of a project.

18.2.5 Early Project Phases

The project awarding process influences the success of public projects (Ikejemba et al., 2017). The reasons for the failure of public projects are the wrong definition of the needs and goals of the project (Shiferaw and Klakegg, 2012). The success of a public project is threatened by difficulties in prioritizing projects related to the multitude of values and stakeholders in the public sector (PMI, 2014).

The reasons for the failure of public sector projects are delusion and deception (Flyvbjerg et al., 2009). Deceptions result from multi-level decision-making structures, and there is a temptation between them to present the project better, consistently with the principal's expectations. In addition, the higher level has less information and cannot verify lower-level details. Politicians expect that the implementation of the projects will improve their political situation.

18.2.6 Human Resources

The specificity of human resource management in projects, resulting from permanent employment and usually lower remuneration than in the private sector, affects the success of a public project. But the impact of team members on the project's success is less significant in the public sector than in the private sector (Nagadevara, 2012). This impact may be enhanced by participatory decision-making (Zhu and Kindarto, 2016), which promotes employee involvement in projects, increasing the chance of public success (Kuula et al., 2013; Songer and Molenaar, 1997). Hierarchical decision-making structures, specific to the public sector, are not conducive to the success of projects.

Project success is well influenced by linking the future personal position of the project management with the project outcomes (Vit, 2011), which is not obvious in the public sector, not least because of the sustainability of employment.

A significant factor in the success of public projects is the introduction of independent people into governance structures (Achterstraat, 2013) and monitoring and control by external entities (Tabish and Jha, 2011). Independent people have no interest in concealing the poor condition of the project. Therefore it is possible to react earlier to emerging threats and, in particularly justified cases, to decide to terminate the project.

18.2.7 Management Methods

The most important public projects success factor is simply the consistent application of documented, formalized project management methods (PMI, 2014). Governance (Pakistan Shah et al., 2011) and especially establishing proper relations between many stakeholders of public projects greatly influence the success of a public project. Barkley (2011) even argues that public sector projects fail not because they lack resources but because they are mismanaged in wealthy Western societies. But PMI (2014) claims that, on the contrary, the success of public projects is primarily threatened by budget constraints. Probably, Barkley's opinion applies to the situation in developed, affluent countries and PMI – to countries with a lower level of wealth. Projects should be guaranteed management autonomy, which is particularly difficult in the hierarchical public sector (Prasad et al., 2013).

The reason for the failure of a public project may be a mismatch between management methods and project needs. For example, exerting pressure to complete a project quickly when in fact, quality is the critical success parameter (Sauser et al., 2009). In the space or military projects, the safety of team members or project users is a critical factor, not time or cost.

Public projects are usually large. The project's success is influenced by its division into smaller, autonomous components, in which the number of internal dependencies necessary to monitor will be reduced (POST, 2003).

18.2.8 External Environment

Researchers also indicate the influence of external factors from outside the organization implementing the project on its success (Gomes et al., 2012; Ika et al., 2012). This environment is determined by law and regulations. Public procurement law may favor or hinder the efficient implementation of public projects (Batool and Abbas, 2017). Other laws relating to public projects are the law relating to public finances, the law relating to the civil service, and regulations relating to the areas in which projects are implemented. Awareness of these regulations influences the success of public projects (Tabish and Jha, 2011).

Delays may also influence the failures of the public project in the mobilization of funds by the government (Batool and Abbas, 2017) or donors.

18.3 Public Projects Success Factors

The above-discussed public project survey also looked at the success factors of public projects. Based on a literature review and a review of practices in public sector project management, 15 factors that may affect the success of public sector projects have been defined (Table 18.1). For each factor, the lowest organizational level was determined at which decision on its implementation may be made: government (G), organization (O), project (P). Due to the hierarchical subordination in Public Administration structures, each higher level may decide on the operation of the lower levels. For example, a project manager may decide to inform stakeholders about projects (F6). Still, such practices may also be introduced by the decision of the organization's management board or by general government regulation. On the other hand, neither the project manager nor the organization's management board can regulate public procurement (F1).

For each of these factors, respondents were asked about the level of its impact on the success of public sector projects. The respondents could provide answers on a scale from 0 – no influence to 4 – decisive influence.

Table 18.1 Influence of Individual Organizational Factors on the Success of Public Sector Projects

Id	Factor	Min. Level	Mean	Median	Domi-nant	Std. Dev
F1	Compliance of the project portfolio with the strategy (country, state, territory)	G	2,82	3,00	4	1,19
F2	Regulations on public procurement	G	2,75	3,00	4	1,28
F3	Methodologies of managing public sector projects	O	2,72	3,00	4	1,25
F4	Acknowledgment of project management as a strategically important capability of public organizations	O	2,70	3,00	4	1,27
F5	Project governance processes	O/G	2,66	3,00	4	1,31
F6	Informing the stakeholders about the projects	P	2,66	3,00	4	1,25
F7	The external environment	G	2,58	3,00	3	1,00
F8	Identification of ways to improve the implementation systems of public projects	O	2,53	3,00	3	1,20
F9	Engaging stakeholders in the implementation of projects	P	2,52	3,00	4	1,24
F10	Managing public project portfolios	O/G	2,51	3,00	4	1,34
F11	Evaluation of project maturity of public organizations	O	2,48	3,00	4	1,33
F12	Defining the rules allowing only verified contractors to participate in the projects	G	2,47	3,00	3	1,26
F13	Regulations on public sector projects	G	2,38	2,00	4	1,34
F14	Support for project implementation by public Project Management Offices	O	2,37	3,00	3	1,31
F15	Setting up rules to allow only qualified project managers to be included in projects	O	2,08	2,00	3	1,37

The influence of individual organizational factors on the success of public sector projects is presented in Table 18.1. The factors are sorted according to the significance of their average impact on the success of the public project ("Mean" column).

Factors influencing the success of public projects can be divided into three groups, taking as dividing points relatively significant differences in the value of their impact on the success.

18.3.1 The Most Important Success Factors

The first group consists of factors with the highest impact ranging between 2.82 and 2.66.

The most important factor influencing the success of public projects was F1 Compliance of the project portfolio with the strategies and the policies of national or territorial governments. Public projects are helpful when they are an element of the overall functioning of the state. Projects defined in isolation from governmental policies and strategies are not considered successful. The government's task is to define the rules to launch suitable projects in subordinate institutions.

Public procurement regulations (F3) significantly influence the success of public projects. Without clearly defined rules for organizing tenders and implementing projects, it is difficult to guarantee the success of a public project. Well-defined regulations guarantee impartiality, prevent corruption, and influence project goals' proper definition and evaluation. This practice also falls under the sole responsibility of the government.

Public projects should be implemented following well-defined methodologies (factor F3). Teams implementing public projects should receive the necessary knowledge in a documented form. Defined methodologies may be a factor that compensates to some extent any gaps in knowledge of entities involved in project implementation. Implementing a methodology is a task of the organizational level not lower than the organization board.

For public projects to succeed, the organization should recognize project management as its strategic capability (factor F4). Usually, such an approach to projects entails specific actions – for example, creating a PMO, implementing training in project management, or defining career paths in this area.

The governance processes were recognized as the fifth most important success factor (F5). However, it should be noted that among the factors directly affecting the processes of implementation, it is the second factor after the methodologies. It can be interpreted that the methodologies cover the entire project implementation process, and among them, the governance processes have the most significant impact. Without well-structured governance processes, it is difficult for projects to achieve success. Some aspects of project governance – governance of projects – may be defined at the organizational level. Still, their full scope should also consider the role of projects in the overall functioning of the government. Therefore, it was recognized that the lowest level to decide on its implementation is both the organization's management board and the government.

According to the respondents, stakeholders should be informed about projects and their implementation (factor F6). Stakeholders of public projects are primarily the communities for which projects are implemented and all taxpayers, indirect project owners. These groups must be informed about the projects implemented for them to avoid creating unacceptable deliverables and guarantee the possibility of reacting in the case of improper project implementation. The project management team by itself may decide to inform project stakeholders about the project and its progress unless higher organizational levels introduce some practices in this area.

18.3.2 The Second Group of Factors

This group includes factors with an impact level between 2.58 and 2.47.

The external environment influences the success of public projects (F7). The procurement regulation is just one part of this environment. Projects are affected by other regulations, for example, regarding public finances, personnel (civil service), personal data protection, or any regulations concerning areas of project implementation, for example, construction or natural resources. Another element of this environment is the market. All these factors seriously affect the success of public projects.

The next success factor is the methods of improving implementation processes (F8). These methods support management methodologies (F3) and governance processes (F5). Organizations can improve their project implementation processes on their own.

Apart from the fact that they should be (passively) informed about the projects, stakeholders should also be actively involved in their implementation (F9). This is achieved by advisory and consulting bodies cooperating with teams implementing projects. The creation and operation of such teams is a tool for direct response to what is happening in projects. Such practices may be left for the discretion of the project management team – unless higher organizational levels make any decision.

Project portfolio management (F10) has a relatively low position. This may be because of the perception that, in the public sector, project portfolios are shaped by public policies. Public organizations are commissioned to implement projects from higher hierarchical bodies in many cases. Public organizations have less autonomy than private companies, and therefore less important is project portfolio management. But for some types of organizations with greater autonomy, the portfolio management decisions may be made at this level. Thus, this factor has been identified as feasible at the level of autonomous organizations or the government level for portfolios managed under policies fully defined at the government level.

The next place in the second group of factors (but still greater than 2 = significant) has the project maturity evaluation of public organizations (F11). This is puzzling as it might seem that project maturity plays a decisive role in the success of a project. I am inclined to hypothesize that the respondents did not come into contact with the formal assessment of project maturity and therefore rated its significance relatively low.

The last factor from the second group is allowing only verified contractors to participate in projects (F12). Suppose the company has demonstrated qualifications and experience in project implementation before implementing a given project. In that case, it can be assumed that it will also contribute to its success in the next project. Restricting suppliers' access to public projects requires legal regulations from the government level.

18.3.3 The Third Group of Factors

The third group includes factors whose influence was between 2.38 and 2.08. This group also includes the factor F15 which influence differs significantly from other factors. However, it should be noted that the influence of all factors in this group is also higher than 2 (significant). Hence it is justified to deal with them in considering the successes of public projects.

Regulations on public sector projects (F13) are important, but not as much as purchasing regulations (F2). Regulations on public projects cannot function in isolation from the regulations on public procurement and, in this sense, can be considered complementary to them.

The most unexpected is the (relatively) low rating of project support by PMO (F14). Perhaps it can be interpreted as a manifestation of project managers' striving for independence and the perception of PMO support as external interference in implementing projects. PMO may be established at a level not lower than organizational.

The lowest position of influence of allowing qualified managers to implement projects (F15) is difficult to interpret the more that most surveyed persons had certifications. Could this be considered a vote of no confidence in the quality of the certificates? This fact and others in this survey require further research.

In the respondents' opinion, all the proposed factors affect the success of public projects above average. The impact of all of them is between 2 (significant) and 3 (essential). Hence, the staff responsible for implementing public sector projects should pay attention to all these factors.

The survey results showed that successful implementation of public projects is possible thanks to the activities of all three organizational levels: project managers, organizations, and the government level. If project managers do not get support from higher hierarchical levels, most success factors will not be realized. Also, the actions of individual organizations not supported by appropriate steps from the government level are not sufficient for the success of public projects. This is another rationale for dealing with the role of governments in achieving public projects success.

PRACTICES OF PUBLIC PROJECT MANAGEMENT

Implementing any public policy (different from the continuation of the current course of action) consists of implementing a change and operating its effects (Subchapter 11.8). In project-based policies, the role of projects is obvious and essential. In operation-based policies, the new policy changes operations and then implements their latest versions. The change must be completed within a certain period for a new course of action to begin: tax collection, poverty reduction, education, or other public policy. You cannot implement a new method of collecting taxes if, among others, changes in IT systems, training for officials, and an information campaign on a new tax collection method will not be carried out. And each of these activities is a project. So, introducing a change, the first stage of implementing any policy is a project. Hence, the project approach is essential in implementing any public policy – not only project-based.

The relation of projects with policy programs, and thus also with policies and strategies, is shown in Figure IV.1.

Hence, to implement their public policies effectively, that is, to have good organizational policy capability (Chapter 14), governments need to implement public projects effectively. To this end, they implement organizational solutions, processes, techniques, document patterns, and other project management practices for which they are accountable.

One of the main goals of this book is to describe and systematize the ways of organizing project management in Public Administration by the governments that govern them.

In this part, I describe governmental practices in the area of public project management. Governments are concerned with public projects. Any public project can be subject to government management,[1] although, in fact, it is not – some governments do not implement public project management practices. Laws, regulations, policies, ordinances, decrees, memoranda, instructions, etc., issued by authorized governmental bodies describe such practices. When any practice is implemented due to an autonomous decision of the board of any public organization, it is not treated as the government practice.

There are several hundred governments in the world, each of which implements projects. Many of them introduce deliberate project management processes. Some governments have implemented

DOI: 10.1201/9781003321606-22

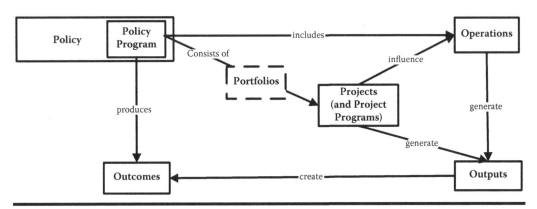

Figure IV.1 Projects and Operations in Policy Implementation (Extension of Figure 11.3).

processes to improve project delivery. Their activities have been the main area of my research interest for several years. Their full review would exceed the capabilities of my one-person team, but I decided to try to gather and systematize as much knowledge in this area as possible. The indicated solutions are just some examples of implementing public project management practices.

I was inspired by the grounded theory approach (Glaser & Strauss, 1967; Strauss & Corbin, 1998). First, I reviewed literature and online material describing practices used by governments for implementing projects practices. Due to the nature of my research – the publicness of projects – much of the material is publicly available over the Internet. Governments inform this way the public and other stakeholders about what is happening in the area of project management and how they expect public projects to go ahead. Then I surveyed public projects. Some of the survey results have been discussed in the previous chapters. This material allowed me to prepare interviews with participants in project implementation processes in several countries. I conducted interviews in the USA, Canada, Argentina, Brazil, and Australia. I reviewed the solutions of central governments and, in federal states, state (provinces, territories, etc.) governments, which have autonomy in the implementation of organizational solutions, in particular, related to project management. People at different levels of commitment have decided to devote time to me: ranging from employees of county offices to a member of the Australian parliament. I also thank them for their help. At the same time, I continued to supplement my knowledge using the resources appearing on the Internet. In organizing the materials, I noticed after some time that practices that would force me not to add new categories appear less and less frequently. I found this to indicate saturation of the research area (Green & Thorogood, 2004; Bowen, 2008). In practical terms, I found the material collected so complete that its presentation could benefit readers, particularly project management practitioners at the government level. As a result of this work, carried out in 72 governments (more than the countries of origin of the survey respondents mentioned earlier), I identified 2,294 practices. I have tried to systematize these practices (Gasik, 2014; Gasik, 2016a; Gasik, 2019). This is what this part of the book is about.

The practices were grouped into two huge territories: Governance and Delivery and two smaller standalone areas: Development and Support. They are described in the four following chapters constituting this part of the book.

Note

1. That is why sometimes I omit the word "public" when I write about projects influenced by governments.

Chapter 19

Governance Territory (Government Level)

Project governance enables an organization to articulate its needs precisely and benefit from project implementation (Victoria DTF, 2012). The benefit is a measurable improvement perceived by stakeholders coherent with the organization's strategy of implementing the project (NSW Government, 2018). Establishing project governance principles is required, for example, by the Scottish Government (Scottish Government, 2013b), the Government of the United Kingdom as a whole (UK HM Treasury, 2007), or the government of the Australian state of Victoria (Victoria DPAC, 2016; Victoria DTF, 2012). The making of public project decisions should be based on reliable data. To provide such data, appropriate institutions and procedures have been established by several governments, e.g., the Independent Evaluation Office in India (India DMEO, 2021), the Quality on Entry procedure in Norway (NTNU, 2013), and the OGC Gateway Process in the United Kingdom (UK OGC, 2007).

19.1 Public Procurement Regulations

Since public projects are very often implemented through purchases, public procurement regulations are a component of the governance of projects defined at the government level. Usually, procurement laws are established earlier than project regulations, so for some time, the procurement regulation is the primary regulation on public projects.

Public purchasing or procurement regulations have been established in many countries, for example, in the United States (USA GSA et al., 2019), Taiwan (Republic of China President, 2011), Brazil (Brazil Presidência da República, 1993), and Poland (Poland Sejm, 2004). The inclusion of external entities in implementing projects takes place based on such regulations. They usually define general rules of conduct in the scope of concluding and executing contracts between the public and private parties, not only in the area of project implementation. The main elements of such regulations include defining the types of contracts depending on the type, significance, and complexity of works, defining the rules for admitting to tenders, defining the rules for awarding contracts, a description of dispute resolution processes and means of appeal.

DOI: 10.1201/9781003321606-23

Public procurement regulations may include special provisions for public sector projects. In the United States, federal-level public procurement must comply with the Federal Acquisition Regulations (FAR, USA GSA et al., 2019). These regulations have been modified many times since their introduction in 1984. Currently, they are about 2,000 pages long. The FAR have a section on projects (i.e., acquisition of major systems) in which they require, inter alia, using Earned Value Analysis (EVA; see Subchapter 3.5) and Integrated Baseline Reviews (IBR). IBR is a project assurance technique (Subchapter 20.5) to check whether a project has a chance to be completed as planned.

19.2 Governance of Projects

In addition to procurement regulations, governments have laws that apply to project implementation. Such regulations may refer to single, specific projects called "special acts". In some countries, legal regulations have been introduced regarding particular types of projects, i.e., sectoral regulations. Regulations with the broadest scope of application apply to the entire set of public projects implemented in a given country.

Schematically, the types of regulations concerning projects are presented in Figure 19.1.

19.2.1 The Project-Level Regulations

Special acts are those that apply to individual projects or project programs. Examples of special acts include the American act on the construction of the Hoover Dam (USA Congress, 1928), the Canadian act on the Hibernia Development Project – the construction of oil production platforms (Canada Parliament, 1990), or the Polish act on the preparation of the European Football Championship (Poland Sejm, 2007). The level of advancement of special acts in the field of management is determined by the level of culture of the administration creating these acts. These laws devote the most attention to budget issues, the principles of establishing entities involved in implementing project programs, the principles of determining the location of facilities and acquiring land, administrative procedures, and special rules for awarding contracts. In the area of management, these laws usually order the creation of a work schedule, order certain entities to exercise control, report, and coordinate the flow of documents.

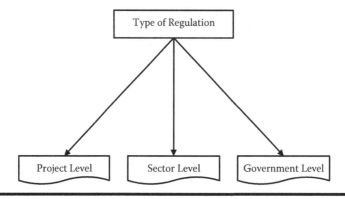

Figure 19.1   Types of Laws Regulating Public Projects.

19.2.2 The Sector-Level Regulations

Sectoral laws apply to certain types of projects such as infrastructure or ICT ones.

An example of a sectorial law for public sector projects is the Public Sector Project Procedures Law for construction projects in Iceland (Iceland Althingi, 2001). This law replaced the corresponding regulation from 1970 (arguably the first regulation for public sector projects globally, Iceland Althingi, 1970). According to this regulation, the project consists of the phases of alternative assessment, planning, construction, and assessment of the completeness of execution.

The state of Colorado has an IT project management law (Colorado GASC, 2012). This Act requires the performance of a feasibility and risk analysis, defines the elements of the project plan, and defines the responsibilities of the IT Office in the field of project management. The Act also requires independent system verification and validation.

There are provisions in the Florida statute regarding IT projects (Florida Legislature, 2016). According to them, Florida has standards for the management and supervision of IT projects.

The advantage of sectoral laws over special laws is that many projects use them. A positive feature about government-wide regulations is the ability to relate to the specifics of certain types of projects – but this advantage of sectoral laws can be offset in governmental regulation by specifying application-specific clauses.

19.2.3 The Government-Level Regulations

The government-level regulations apply to all public projects implemented in a given state. Such laws usually set a minimum size limit for projects, below which these regulations do not need to be applied.

Below I describe some examples of regulation of the level of entire governments.

19.2.3.1 Argentina

In Argentina, public investment law (law no. 24.354, Argentina Congreso de la Nacion, 1994) requires establishing the National Public Investment System, covering all public projects. The law defines the process of initiating public projects. The same law defines the project life cycle, which consists of the pre-investment phase (idea, profile, preliminary Feasibility Study, full Feasibility Study), project implementation, product operation, product shutdown, or recall, and ex-post evaluation (more about the Argentina system in Section 20.9.2). Similar laws operate in other countries of the United Nations Economic Commission for Latin America and the Caribbean (UNECLAC, Spanish: Comision Economica para America Latina y el Caribe, CEPAL), for example, Peru (El Congreso de la Republica del Peru, 2000) or the Dominican Republic (El Congreso Nacional Dominicana, 2006). Countries with similar laws are affiliated with the Network of National Public Investment Systems (ECLAC, 2021).

19.2.3.2 USA, Federal Government

The two principal regulations regarding the project approach to management in the United States are the Government Performance and Result Act (GPRA, USA Congress, 1993), with a modification from 2010 (USA Congress, 2010a), and the Program Management Improvement and Accountability Act (PgMIAA, USA Congress, 2015).

GPRA instructs government agencies to develop strategic plans and sets out how to account for achieving their goals. GPRA considers programs to be the core unit of work for government agencies.

Policy programs are used to achieve strategic goals. Since the programs in GPRA cover the entirety of federal agencies' operation, in our understanding, they are policy programs. Progress in implementing these programs must be reported to the Office of Management and Budget at least annually.

PgMIAA does not limit its area of interest to project programs, i.e., it covers all policy programs (hence, for instance, the PMI Standard for (project) Program Management (PMI, 2017b) may support only a part of US federal programs). It requires, among others, to adopt standards, policies, and guidelines for program and project management, establish a Program Management Policy Council (PMPC), and introduce to each federal organization persons responsible for developing program and project management skills. The Government Accountability Office (GAO) is expected to conduct portfolio reviews for high-risk programs and projects.

19.2.3.3 Canada, Federal Government

National project management principles were developed in Canada by the Treasury Board of Canada Secretariat (TBoCS) as Directive on the Management of Projects and Programs (Canada TBoCS, 2019). The directive requires each agency to establish a project management framework consisting of best practice processes. Projects must undergo a risk assessment. A brief should be prepared for each project, the content of which is specified in the document. Projects approved for implementation must have baselines submitted to the Office of Comptroller General. The document defines the main requirements for the leadership and governance of projects and project programs. Projects and their programs must go through the gateway process. For each government organization, its level of capability in project management is determined (Canada TBoCS, 2013a). This level determines the class (size and type) of projects that the organization can implement on its own.

Government-level regulations regarding project management are created based on the needs of a given country and may contain various legal solutions. Such regulations result from an adequately high level of project management awareness among governments.

19.3 Audit Chambers

National audit chambers oversee public projects, usually working for parliaments. Examples of such organizations are the Polish Supreme Audit Office (www.nik.gov.pl), the British National Audit Office, NAO (www.nao.org.uk), the American Government Accountability Office (GAO, www.gao.gov), and the Australian National Audit Office (ANAO, www.anao.gov.au). Audit chambers may audit projects or programs (of any type, including project programs) at parliament's request or based on their plans. Entire portfolios of projects carried out by public organizations can be audited (USA Congress, 2015). Audit results are presented to owners, parliament, and the public.

Audit chambers have the authority to audit public sector projects and their programs in two areas: financial and operational performance. In the latter area, processes and other arrangements that may affect management (e.g., organizational structures) are usually verified. The necessity to perform audits results from the reluctance of people carrying them out to report problems. As a result of the control, the reasons for the ineffectiveness of project implementation may be determined, and methods of their improvement may be suggested. Audit results can be the basis for applying for budget reduction. Audit chambers may use maturity models in their audits, e.g., ANAO uses the British P3M3®[1] model as a reference (UK OGC, 2010). An audit may be carried out on processes related to the management of public sector projects other than the implementation. Australian ANAO carried out an audit of the manner of implementation of the gate review process by the Ministry of Finance (Australia ANAO, 2012), as a result of which it was found, inter alia, that the intervals between gate reviews are too long, that the implementation of review conclusions is not always checked and that it would be advisable at random also view projects that are excluded from this process.

The US GAO reviewed OMB's implementation of program management resulting from the PgMIAA (USA Congress, 2015). GAO, among others, concluded that OMB's project management standards were too general and that OMB itself had no structures to develop and maintain such standards. OMB has conducted reviews of a few federal agency portfolios and a few risk areas identified by GAO. OMB has not built a system for assessing the agency's level of advancement in project and program management practices (USA GAO, 2019).

Specialized teams may be formed to perform audits of projects – in Texas, the Office of State Auditor to control public sector projects has established a Quality Assurance Team (Texas SAO, 2013). A common practice is to audit how one responds to previously formulated audit recommendations.

Audit chambers shape the ways of implementing public projects by publishing guidelines. GAO has published guidelines for cost management (USA GAO, 2020a), schedule management (USA GAO, 2012a), and technology management in projects (USA GAO, 2016). Together with the Cabinet and Prime Minister Department, Australian ANAO published guidelines for policy program execution and policy implementation (Australia DPMC and ANAO, 2006) and, together with the Cabinet Implementation Unit, guidelines for sponsors (Australia ANAO, 2010a) on how to initiate projects. These guidelines become the reference material for audits. Since the audited entities know these materials, they try to implement the projects in such a way as to avoid objections from the auditing entities.

Audit chambers also assess the overall ability of government organizations to implement projects. In 2016, the British NAO audited the implementation of major projects for the Committee of Public Accounts (UK NAO, 2016). Project owners' accountability, staff ability to implement major projects, project assurance, and too early identification of solutions to problems were identified as the main areas for improvement.

Audit chambers verify the ways of implementing projects, make recommendations for their improvement, and publish recommendations on implementing public projects.

19.4 Standards

According to the ISO definition, a standard is a document established by consensus and approved by a recognized body that provides for widespread and repeated use, principles, guidelines, and characteristics for activities or their results, aimed at the optimal level of ordering in a given context (ISO, 1996). Public organizations use standards as the basis for creating their project management methodologies (Section 20.9.1).

In the United States, the main standard-setting body is the non-governmental American National Standard Institute (ANSI). In the field of project management, ANSI has granted the Project Management Institute (PMI) status as a standard-setting organization. The most important standard developed by PMI is A Guide to Project Management Body of Knowledge, PMBOK® Guide (PMI, 2017a). PMI has also developed an extension of this standard (based upon an earlier version, PMI, 2004) for the government sector (PMI, 2006b). The government standards mentioned in the PgMIAA (USA Congress, 2015) should be based on "nationally

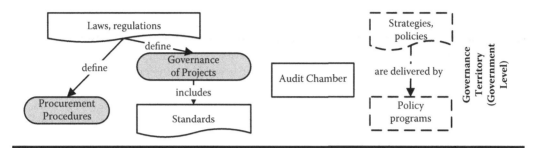

Figure 19.2   The Governance Territory.

accredited standards," which in practice means basing them on documents developed by the PMI. In addition to the PMBOK® Guide, the main PMI's standardization documents concern project programs (PMI, 2017b[2]) and project portfolios (PMI, 2017c).

In the United Kingdom, the standard-setting body is the British Standard Institution (BSI). This institution began standardizing the area of project management in 1968 with the publication of a standard describing a glossary of terms related to the analysis of the project activity network (UK BSI, 1968). In the early 2000s, BSI published a series of project management standards (UK BSI, 2000a, 2000b, 2002, 2006). Recently, the British government developed a standard for portfolio, project program, and project management (UK HM Government, 2018).

Standardization can be done in two ways by authorized bodies or by market exclusion (Bredillet, 2003). The market exclusion occurs when one of the solutions gains a market advantage and becomes the actual standard. Thus, in the United Kingdom, documents established by the Office for Government Commerce (UK OGC) have become standards. UK OGC has developed a set of documents in the area of project management (currently maintained by the government's joint venture Axelos). The most important document in this series is Prince 2® (UK OGC, 2017). The paper describing the methodology is its publication version which plays the role of a standard. The publication version must be adapted to the realities and needs of each organization; its implementation version must be created. Prince2® is required for government projects. It describes project management themes, processes, principles, and the necessary documents. Other relevant documents developed by the UK OGC include the Project Gateway Review Process (UK OGC, 2007), the Prince2® based project, program, and portfolio management maturity model (UK OGC, 2010; more on maturity models in Chapter 24).

The knowledge possessed by audit chambers may help create standards. The British NAO developed recommendations for selected areas and functions of project implementation, which were then taken into account in the development of standards and guidelines by the organizations of the executive branch of the government, for example, concerning assurance (UK NAO, 2010a, 2012) or initiating major projects (UK NAO, 2011).

The structure of the governance territory is presented in Figure 19.2.

Notes

1. P3M3® is a registered trademark of Axelos Limited.
2. The author was a significant contributor to PMBOK® Guide 5th Edition (PMI, 2013a) and Standard for Program Management, 3rd Edition (PMI, 2013b).

Chapter 20

Delivery Territory

This chapter deals with the implementation of projects and project programs. The implementation of projects results from governance arrangements adopted at the government level. Projects are implemented by governmental agencies, but it can also be a project implementing agency established to implement a specific project, project program, or another set of projects.

The territory of delivery consists of seven process areas:

1. Management of portfolio
2. Management of projects
3. Project assurance
4. Engagement of stakeholders
5. Knowledge management
6. Methodologies and systems
7. Actors management

In all aspects of project delivery (as well as in other areas of project management), Project Management Offices play a fundamental role. Therefore, I start this chapter with their brief description.

20.1 Project Management Offices

Project Management Office (PMO) is an organizational unit or team of people whose main task is to ensure projects' success or participate in activities leading to this success. In many countries, organizational units are established to implement such activities for public projects. PMOs vary in size and structure and are placed in different places and at varying levels of organizational structures. The public sector's PMOs deal with all or specific types of public sector projects (e.g., IT or infrastructure) and perform various functions. Still, their overall aim is always to improve the implementation of public sector projects.

The second function of PMO is promoting project management in their administrations, being an advocate and model in the field of project management (Vermont DoII, 2013, Information Technology Advisory Board, PMSC, Missouri OoA, 2013c, Project Management Advisory Workgroup in Montana, PMAW, Montana SITSD, 2021a).

PMOs can work at the Governmental level or the level of individual institutions (or groups of them).

20.1.1 Governmental Project Management Offices

At the government level, project management tasks are performed by specialized organizations or organizational units within the ministries of the executive branch (e.g., finance ministries, ministries of development, prime minister's office). Collectively, I will refer to such entities as the Governmental Project Management Office (G-PMO).

Ministers are appointed to manage units responsible for project management (e.g., Australian Victoria Minister, responsible for Major Projects, Victoria DSDBI, 2013). Victoria's Major Projects are the state's G-PMO. The organizational location of the PMO at the government level particularly emphasizes the role of public sector projects for the functioning and development of the country.

In the United Kingdom, the Infrastructure and Project Authority operates within the Cabinet Office and is also subject to HM Treasury (UK IPA, 2020, 2021). Its main goal is to ensure the efficient implementation of major public projects. This goal is achieved by supporting the development of project management capabilities in government departments, setting standards, providing knowledge to projects, and independent assurance to mitigate project risks. IPA also supports policymakers in defining public policies and works with the market to enable commercial companies to implement projects.

The Ministry of Statistics and Programme Implementation operates in India. Its specialized part, the Programme Implementation Wing (PI Wing, India MoSPI, 2021), monitors the implementation of the Indian Government's major projects using an appropriate software tool. The PI Wing supports projects in removing major problems that threaten them and is committed to improving project implementation processes.

In the United States federal government, the organization responsible for implementing policy programs and projects is the Office of Management and Budget (OMB; USA OMB, 2021), reporting directly to the President of the United States. OMB's primary role is to ensure the smooth implementation of government policies. OMB develops, improves, and implements project management methods in government agencies. OMB oversees the performance of policy programs, starting with their definition. Reports on the implementation of policy programs and projects are sent to OMB. If there are problems, OMB assists departments in solving them.

In Canada, at the federal level, the institution dealing with project management is the Treasury Board of Canada Secretariat (Canada TBoCS, 2021a), dealing with shaping the way projects are implemented and maintaining software supporting project management. Canada TBoCS develops policies and other documents shaping the methods of project management.

Infrastructure projects in New Zealand are handled by the Major Projects team at the Infrastructure Commission (New Zealand Infrastructure Commission, 2021). The team develops obligatory recommendations for major projects. In addition, the team performs reviews and prepares reports on the progress of work on some projects. It is also possible to collaborate in implementing the project throughout its entire life cycle. Also, in New Zealand, the System Assurance team (New Zealand GCDO, 2020) work for all digital projects.

G-PMOs are important in shaping the state's project and project program implementation system. They prepare legal regulations, methods of their implementation and supervise their implementation. In addition, G-PMOs take part in the implementation: from the proposal evaluation phase, through planning and execution, to the final evaluation of the project and its effects.

20.1.2 Local Project Management Offices

Local Project Management Offices (L-PMOs) are located at a level lower than the government. They may be subordinate to governmental PMOs (if any exist). In the US federal government, the role of the government-level PMO is played by the Office of Management and Budget, while most central agencies also have their L-PMO. In some solutions, the entire project implementation is separated from linear structures and transferred to specialized PMOs.

Virtually all functions related to project management in public organizations, particularly those described in this chapter, can be performed by L-PMO.

L-PMOs performing various functions may be located in ministries; for example, the Singapore Center for Public Project Management dealing with infrastructure projects is located in the Ministry of Finance (Singapore CPPM, 2011). There are also, for example, the IT Project Management Office of the California Department of Technology (California CDoT, 2013a) and the PMO of the Vermont Department of Information and Innovation (Vermont AoDS, 2020) at the ministry level. Natural Resources Canada has a Major Project Management Office (MPMO) that performs specific natural resource project functions for all departments and agencies. MPMO deals with the process of obtaining environmental and regulatory permits for the implementation of projects and the improvement of this process and regulation (Canada Government, 2007; Canada MPMO, 2021). Also, in other countries, PMOs are placed at the ministry level (Vermont AoT, 2020; Western Australia DoF, 2020b; Bahrain MoW, 2013; Nevada NDoT, 2013).

Not every PMO employs project management specialists: at the Pennsylvania Department of Transportation's Bureau of Project Delivery, no project manager position was listed among the 187 posts (Pennsylvania DoT, 2013) – this demonstrates the technical understanding of the term "project" within the organization. Most of the PMOs are real organizations with employees. There are also virtual project offices, where people involved in implementing projects, but not their employees, collaborate to collect knowledge, examples of documents, processes, and project management practices.

PMOs can work for the needs of a set of public organizations, several ministries, the so-called Cluster, as is the case in the Government of the Canadian state of Ontario (Liem, 2014). Many PMOs work on single projects. The smallest size of PMO is the so-called Project agents, i.e., individuals, operating in organizational units of state institutions in the area of project management (Prado, 2014).

20.1.3 Other Structures

The implementation of public sector projects involves structures integrated with the entire agency, whose task is to implement projects, i.e., project structures.

The purpose of Action Committees (AC) in Malaysian ministries is, among others, to coordinate and implement development policies and projects, report on project progress, and monitor the impact of projects on the ministry's area of operation. ACs also exist in individual states; they implement projects and policies on their territory (Malaysia Dahlan, 2015).

If the major public works have the nature of project programs, special organizations may be appointed to manage them. For example, the construction of the Australian capital Canberra was managed between 1958 and 1989 by the National Capital Development Commission (NCDC, Australian Government, 1957).

Western Australia's Ministry of Development is essentially an example of a project structure (Western Australia DoSD, 2013a). The components of this ministry are the Resource and

Industry Development Division and the State Initiatives Division. Directly to the Resource and Industry Development Division director, there were subordinated, among other things, the largest projects related to the state's infrastructural development. The State Initiatives Division included, among others, the Project Approval Strategy Department, which was responsible for the process of establishing new initiatives.

In the Philippines, the Undersecretary of State for the Operation of Unified Project Management Offices works in the Department of Public Works and Highways (Philippines DPWH, 2020). His task is to supervise and coordinate the work of Unified Project Management Offices, mainly carried out with funds from foreign support. In 2020, a dozen or so projects were implemented with their PMO reporting to the Undersecretary of State.

Major projects can also be implemented by elements of organizational structures of public organizations. We are talking then about **projects embedded in organizational structures**. This is how, for example, the major infrastructure projects in the Netherlands are implemented, where directorates are appointed in ministries for the largest projects (The Netherlands MvIeW, 2021).

20.2 The Hierarchy of Project Bodies and Roles

Supervision and control as governance of project functions can be performed at the organizational level by several types of bodies (Figure 20.1). The two entities most frequently involved in governance are the Steering Committee and the L-PMOs. The Steering Committee makes the critical decisions for a project or project program. They usually consist of a sponsor who directs his work and is responsible for the project's success, a contractor representative, and a product user representative. Suppose the project or project program is implemented with the participation of many organizations. In that case, its composition should be extended so that the structure of action and decision-making can consider the relationships between these entities and their impact on the implementation of works. The proper definition of powers, responsibilities, and relationships in the Steering Committee is one of the most important factors for success. It is even more important in multi-stakeholder project programs. The Steering Committee is usually primarily concerned with the project products and business outcomes and supports the project manager. Representatives of important stakeholders may be involved in the work of the Steering Committee (NSW Infrastructure, 2020).

Another option of involving public party representatives in implementing projects is creating advisory bodies (USA EPA, n.d; European Union, 2020; USA Congress, 1972). Such bodies may also include field specialists and project management specialists indicated by public representatives. They advise at the project level but can also advise hierarchically higher bodies on selecting projects for implementation. These bodies do not have the deciding influence – this is the role of persons included in the Steering Committee. They have two main types of tasks. First, they observe the implementation of the project. They indicate the places at risk (threats) and places where the project can be better implemented (opportunities), particularly from the point of view of the benefits that the project can provide to stakeholders, including opportunities to initiate projects. Advisory bodies can also, upon request, assist project structures in solving identified problems.

L-PMO is interested in the project from a management point of view. As part of the governance of projects defined at the external (government) level, L-PMO defines project management processes in the organization and supports and enforces their implementation. It can be said that this unit governs project management practices and processes in the organization. In addition

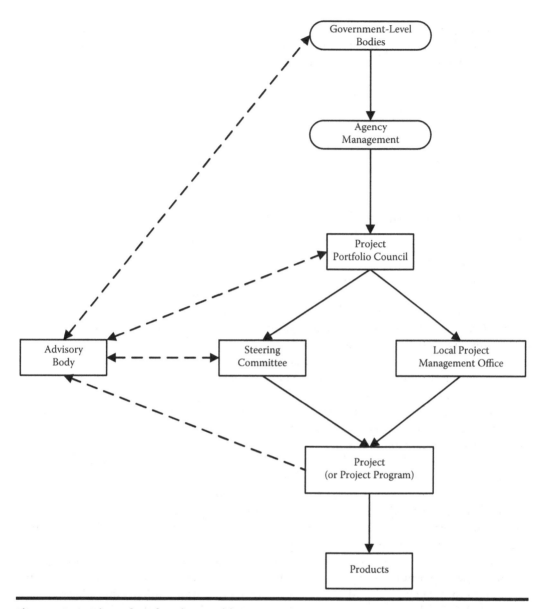

Figure 20.1 Hierarchy of Project Entities.

to the Steering Committee and L-PMO, there is always a portfolio decision-making body: the directorate or the Project Portfolio Council. At the highest organizational level, the leadership of ministries is held accountable for their projects. The most important projects are supervised, from outside the organization implementing them, by governmental institutions (parliament, audit chambers, governmental PMO).

One person should be responsible for each public project (Pakistan Shah et al., 2011). Responsibility for each project product should also be specified (Canada TBoCS, 2019).

The proper definition of the governance process determines the multi-level responsibility and accountability of public projects and project programs.

Oversight powers are an important element of the governance system of public sector projects. As part of this function, the project status may be verified. In particular, essential recommendations or orders (up to and including the project's closing) regarding its implementation may be issued.

External governance structures can be even more elaborate; for example, in US government departments, projects are overseen by their offices specialized in project management and also by the Office of Management and Budget (OMB).

PMOs support the setting up of governance of projects, for example, in California (California CDoT, 2013c) or the Australian state of Victoria (Victoria OPV, 2016a).

20.3 Management of Project Portfolio

The most important meaning of the concept of project portfolio is a collection of projects performed by a specific organization. Portfolio management is decisions about establishing and maintaining the composition of a project portfolio within an organization (Subchapter 5.2).

The only government organization with complete autonomy, covering the selection of implemented policies and projects, is the government that disposes of the state budget. This budget covers all state expenses; hence the government more or less directly approves all public projects in a given country. The question is how directly is it done?

Enacting the state budget is a political decision and depends on the distribution of political power in parliament and/or other governing bodies in a given country. The final, decisive factor for placing any projects, policies, or enacting the budget as a whole is a political decision. The political power may decide to spend billions of dollars for anything they prefer, and only the political opposition will complain about it. Hence, political science knowledge is more valuable than project management knowledge for understanding decisions made at this level (Subchapter 11.7). The fact that a policy or project has a good business case may support but does not guarantee the political decision to include into the budget and implement it. Maybe it would be better if political parties persuaded the electorate that it is worth implementing policies and projects that have a classic business justification, but this cannot be a formal requirement – it would be considered inconsistent with the principles of democracy. The decision whether to develop traditional, nuclear or green energy depends primarily on the balance of political forces in parliament and not necessarily on rational arguments for each of these options. Another political decision may be, for example, to choose the development of health care based on its own universities or by recruiting foreign specialists. If the first option is selected, then the policy for creating own medical universities is defined. In many solutions, the most important projects are simply indicated at the political level as components of policy programs and passed for implementation to appropriate agencies. Such a decision may be the determination of the site for the construction of a nuclear power plant or cities where medical universities will be built.

The creation of public project portfolios is carried out hierarchically. Projects decided at the central level are passed to the relevant public agencies and become part of their project portfolios. In this case, there is a difference between the private and public sectors: private companies decide autonomously about implementing projects in the private sector. In contrast, some projects are accepted for implementation from outside the organization in the public sector. Private companies are subordinate only to owners who do not have higher superiors (although some may have extensive ownership structures). And every public organization (except the government) is subordinated to higher hierarchical organizations.

The criteria for selecting projects for implementation, where agencies can decide autonomously, independently of higher bodies, are also different. The goals of public organizations are different from those of private-sector organizations. Private companies are ultimately accounted for the profit provided to their shareholders. On the other hand, public organizations maximize public values (Subchapter 8.7). The differences in the objectives of the activity result in differences in the criteria for selecting projects for implementation. In private companies, the most important (although not the only) criterion for choosing a project is its impact on the financial result of the organization. On the other hand, public projects can be considered simultaneously in terms of achieving different public values. A road construction project can be analyzed from the point of view of ensuring communication, reducing unemployment, or domestic companies' development. To determine the significance of these factors, it is necessary to build models of achieving public value, which will eventually manifest themselves through the system of assigning weights to criteria or through the structure of project selection.

Table 20.1 shows the main differences between the methods of selecting projects to be implemented in different sectors.

Decisions on the composition of the project portfolio, i.e. whether to include a project or project program or to exclude them, planned or emergency from the portfolio, are made by the Project Portfolio Council, consisting of authorized representatives of the institution's management board and organizational units. They make portfolio decisions resulting from the implemented public policies, level of organization's autonomy, and the budget allocated. The impact of external factors on the portfolio of launched projects may be different. If the government adopts a motorway construction policy and its program, the motorway-building agency has no freedom to select projects to be implemented. On the other hand, the scope of projects carried out by, for example, the audit chamber by the executive government branch has no influence. Between these extremes, there are agencies with different levels of autonomy in selecting projects for implementation.

Public policies sometimes only define goals to be achieved by public agencies. In the area of employment, culture, education, health protection, local transport, or many others, the way of achieving the set goals through the selection of appropriate projects is better performed by an organization of a lower organizational level than the central government. The very definition of specific criteria may also be assigned to the agency level. For example, as part of a central policy for promoting culture, a lower-level institution may define a target number of seats in cinemas, theaters, number of libraries and volumes, etc. Leaving lower-level entities to decide on the selection of projects for implementation is often practiced in the areas of social policies. The level of dependence of such institutions and their presence in the public sector, in general, depends on the constitution of the state and state governance. But all public agencies usually may decide on projects of a maintenance or operational nature, e.g., renovation of a road or public buildings, replacement of equipment, training of staff, or often the implementation of IT systems – many below cited examples come just from this area.

Portfolios can be divided according to the type of impact on the activities of a public organization: strategic portfolio and operating portfolio (Scottish Government, 2013a).

Table 20.1 Differences in the Ways of Initiating Projects

	Goals of operating	*Autonomy*
Private company	Profit	Full
Public organization	Public value	Limited

A strategic portfolio is one consisting of projects implemented to achieve strategic goals. The operational portfolio is a set of such projects that are implemented to enable the daily implementation of the processes of a given organization. For example, housing construction for poor people in an area may be considered a portfolio of operational projects to ensure that homelessness does not increase.

Portfolio management at the national level can be shaped by law (Argentina Congreso de la Nacion, 1994; Peru El Congreso de la Republica, 2000). Special institutions are appointed to develop portfolio decisions at the central level. For example, Infrastructure Australia (n.d.) suggests decisions related to the development of infrastructure in Australia, and in Argentina, there is a National Directorate of Public Investment (Argentina DNIP, n.d.). A national portfolio of major projects may be subject to special management rules (UK HM Treasury and Cabinet Office, 2011; UK IPA, 2020). A portfolio of natural resources projects in Canada is accounted for by the MPMO (Canada MPMO, 2021). A portfolio may be departmental (meeting the needs of a specific organizational unit) or corporate (meeting the institution's needs as a whole, Brazil MPOeG, 2013). There are portfolios of individual administrative units, for example, for counties, cities, or higher administrative units. There is a body empowered to make portfolio decisions at every organizational level. This may be a specially created Project Portfolio Council. Suppose such a body has not been established. In that case, decisions regarding the project portfolio concerning its composition in its autonomous part, including the most important ones, are made by the organization's governing body, i.e., its management or board.

Some public organizations develop project portfolio management methodologies, most often for the IT area. It may include, for example, policies, standards, guidelines, procedures, tools, and templates related to project portfolio management (Vermont DoII, 2010). In Brazil, a methodology for managing project portfolios included in the Information Technology Resource Management System was developed (Brazil MPOeG, 2013). In the Canadian state of Ontario, there is an integrated methodology for managing portfolios, project programs, and projects (Ontario OPS, 2011).

The initiation of the project as well as its supervision and making fundamental decisions, for example, on termination of the activity, although their result may change the portfolio composition, are also part of the project life cycle and are therefore described in Section 20.4.2.

In organizations established to implement a specific project program (such as building a national airport), project portfolio management may play a lesser role because project initiation decisions may be a program management component. When an agency is set up to carry out a single project, this function may not be needed at all.

These principles materialize through planification – the process in which detailed rules and processes for initiating projects within the scope of authority of a given organization (the government as a whole, it's any component or single agency), i.e., for their inclusion into project portfolio, are defined (Chile MINVU, 2009). Planification is carried out in an environment defined by legal regulations concerning public sector projects (Subchapter 19.2). The rules for qualifying projects for implementation may be described in the policy for assessing investment proposals (Arizona ADoA, 2013), in the relevant guidelines (UK NAO, 2011; Canada TBoCS, 2019) or may be the result of the step of determining the principles of prioritizing project initiatives (Argentina MEFP, 2015). This phase involves preparing systems for evaluating project initiatives (New York SOT, 2003) and defining processes that define the business case and related metrics for measuring project success (Vermont DoII, 2010). The PMO can determine initiatives' priorities. Other departments may modify them. Detailed procedures, such as performing feasibility studies (Colorado GASC, 2012), may be defined in the planification phase.

An essential part of planification is to define the criteria for selecting projects. As in other sectors, these criteria relate to the organization's goals.

The criteria for initiating public projects can be divided into multiple groups:

- Social
- Business
- Operational
- Organizational
- Political

Social criteria are those that are related to society and its development. This category includes criteria related to public values. Criteria may be improving the delivery of health services, improving the education system, improving public transport, and criteria relating to all other areas of a country's activity. However, public sector projects sometimes fail to consider social needs, for example, because they are not being analyzed. Planners or decision-makers may also ignore them due to political or personal factors. Social priorities may be specifically defined vaguely to allow for different actions in the future. The priorities and needs of society can also be so complex that planners find it difficult to cover them all (Shiferaw and Klakegg, 2012). Objectives of various types of entities may be contradictory; for example, local governments may have objectives other than government administration.

Business criteria are those that relate to the organization's operating processes and their effectiveness. For example, changes in technology, the need to improve business (NI DoF, 2021), a broadly defined impact on business (Scottish Government, 2013c), the impact of a project on daily processes, adapting to infrastructure priority (Southwell, 2015), or a project's financial return. When initiating public sector projects, a negative business selection may also occur: the public sector implements those projects that are not profitable for private sector companies (Lefman, 2014).

Operational criteria are related to the implementation process of a specific project. Projects that the organization can implement efficiently should be selected, and those for which the organization has no skills to implement should be rejected. The project management approach can be assessed (Williams et al., 2010). The estimated cost of project implementation and budget availability are always essential criteria for project selection (Williams et al., 2010; Scottish Government, 2013c; Iceland Althingi, 2001). A frequently used criterion for selecting an initiative to be implemented is the (low) level of related risks (Canada TBoCS, 2015; Scottish Government, 2013c; Australia DoF, 2016a). There are four levels of tolerable, tactical, evolutionary, and transformational risk in Canada (Canada TBoCS, 2010b).

The organizational criterion is a kind of generalization of the operational criterion: the general ability of an organization to implement projects. For projects financed by aid institutions, this is a general assessment of the ability of countries to implement projects (Siles, 2014).[1] The criterion of the ability of individual organizations to execute a project is used in Australian federal projects (Australia DoF, 2016b). The risk level of the project initiated must align with the agency's level of capability (Canada TBoCS, 2019; see also Subchapter 24.5). If an agency wants to initiate a project that requires a higher capability than it has, it must request it from Canada: Treasury Board of Canada Secretariat (Canada Government, 2020).

Specific types of project initiation criteria are related to politics (Flyvbjerg et al., 2009; Fuertes and Herrera, 2014), i.e., they are related to the interests of power centers, in particular with maintenance of power. These interests do not always comply with social criteria (van Krieken, 2014). Arming or the construction of monuments is not always in line with social needs. For political reasons, projects are sometimes initiated that are not wanted by the public (Townsend, 2014).

Project initiation criteria are given appropriate weights in selecting project initiatives.

20.4 Management of Projects: From Initiation to Evaluation

In this subchapter, I describe the phases that public projects go through. Together, these phases form the project life cycle described here and its non-traditional variants. Since not all projects, depending on their characteristics, have to be subject to the same rules, this chapter also explains the rules for classifying projects.

20.4.1 Project Life Cycle

Project management processes are included in project governance schemes. As a standard, project management processes, in the most general way, can be divided into six groups (Section 2.4.1):

1. Initiation
2. Planning
3. Execution
4. Closing
5. Evaluation

Projects are also subject to activities defined as:

6. Monitoring and control

This is the structure of the description of management processes that I adopted to describe the management practices of public projects in various public organizations. The following sections are devoted to these groups.

20.4.2 Initiation

Choosing suitable projects for implementation is one of the project success critical factors (UK Lord Browne of Madingley, 2013). The final decision to start work on a project or project program rests with the Project Portfolio Council.

Initiation processes are those in which decisions are made about implementing projects. The initiation takes place based on criteria established in the planification phase. The actual process begins with the emergence of the project implementation initiative. Project initiation occurs due to implementing a more or less complex process. The most important element of initiating is the accurate recognition of the current situation in the considered area of project implementation and coming up with a method of operational reference to the existing situation (therefore, this phase is sometimes called the *thinking phase*, New Zealand Infrastructure Commission, 2019).

When assessing the proposal, the result of the institution's project management maturity assessment (Canada TBoCS, 2019) may be used. The assessment result is a decision of whether the project will be implemented in a given year, whether it will be reconsidered in the next year, or it will be rejected. Initiating is part of the portfolio management process (see Subchapter 20.3). As part of the initiation process, final decisions are made on the portfolio's composition.

The decision to initiate a project is usually documented in a project charter, which contains the most important information about the project. There must be a project manager without whom the project cannot be implemented. The governmental PMO may veto the manager appointed by the project implementing agency (UK Lord Browne of Madingley, 2013).

In the state of New York, the project charter includes information on the project background, objective, Critical Success Factors, required resources, constraints, and project authority (NY SOT, 2013). As recommended by the US Federal Acquisition Institute (USA FAI, 2015), the project charter should define the project's purpose and objectives, success measures, high-level requirements, project description and characteristics, milestones, total budget, and authorization for the manager. As recommended by the Alabama Office of IT, the project fiche should describe the scope of the project, expected results, milestones, and the purchasing strategy (Alabama OoIT, 2017). The Canadian Treasury Board of Canada Secretariat requires a fairly extensive project charter. Its most important ingredients are goals, outcomes and objectives, project scope, milestones, deliverables; cost estimates and source of funding; risks, assumptions, and constraints project organization (Canada TBoCS, 2008). The most important role of the project charter is to authorize the start of the project implementation, i.e., to approve the project manager to act.

The final condition for initiating the project is the availability of funds (Iceland Althingi, 2001). In some solutions, after a general decision to qualify a project for implementation, a decision is also needed to start the project on a given date (Naffah Ferreira, 2014).

20.4.2.1 Submitting Initiatives

In most solutions, project initiatives can be submitted by two main groups of entities: communities and public organizations. Local communities are entitled to submit project initiatives (Poland Sejm, 1990; Vermont AoT, 1996). Depending on the applicable regulations, institutional initiatives of public sector projects may be submitted by governmental institutions (e.g., ministries) and local government institutions operating in a given geographical or functional area. In some regulations, other entities may also submit their initiatives, but on the condition that a public organization (e.g., a ministry) is found to represent this project in the appropriate institutional forum (government, parliament) (Iceland Althingi, 2001). For aid and development projects, initiatives may be submitted by countries (operationally: by governments) as a whole (Siles, 2014). Initiatives may come from lower administrative units such as provinces (Argentina MEFP, 2015). They can be reported from institutions responsible for a given area of operation (Argentina MEFP, ibid) – e.g., transport projects by the Vermont Department of Transportation (Vermont AoT, 1996). The authorization to submit initiatives is vested in managers of organizational units or cells (Koenig, 2015).

The main element of the submitted initiative is the business case, which defines the fundamental business determinants of the project. In the United Kingdom, they are grouped into five areas: strategic rationale (compliance with applicable policies), the economic rationale (benefits to society and impact on other stakeholders), the financial rationale (costs and financing method), the commercial rationale (ability to purchase goods and services), and management rationale (organization and management processes) (UK HM Treasury, 2018). Costs should be assessed independently (NSW Infrastructure, 2020). The main elements of a business case usually go to a project charter. A business case can be developed gradually: problem definition, strategic case, and detailed case (NSW Treasury, 2018c).

Cost-Benefit Analysis (CBA) results can be a component of the business case. The CBA of public projects takes into account four groups of stakeholders: consumers, producers, employees, and the government. Not only finances are taken into account but also social, economic, and environmental benefits (NSW Treasury, 2017b). The analysis covers the effects on both households and enterprises. The categories of benefits included in the CBA are savings or avoided costs, government revenues, consumer surplus (lower prices), producer surplus (the difference between the price and the cost of production), labor surplus (the difference between the new income level and the current one) and benefits for the wider community (e.g., lower air pollution level after transport investment). The result of the CBA of public projects is the definition of net social benefit and an indication of the groups that will benefit and those that will bear the costs.

If the business parameters are not known, a Feasibility Study is carried out (Bogucki, 2015) to specify them. The realization of the Feasibility Study may be assisted by PMO staff (Missouri OoA, 2013a; Western Australia DoF, 2020a; Bangladesh DoPW, 2013).

20.4.2.2 Unsolicited Proposals

Some governments (USA Federal Government, USA GSA et al., 2019; New South Wales, NSW Government, 2017) allow submitting unsolicited proposals by private entities. This mechanism enables the implementation of innovative, important projects not provided for by government strategies and policies. They may not be submitted in areas covered by public tenders as this would be unfair competition with other bidders. Unsolicited projects have a specific initiation process.

In NSW, a proposer can start the unsolicited project initiating procedure by meeting with a Treasury representative to determine if the proposal generally complies with the unsolicited project guidelines. To make such a proposal, the proponent uses government-defined forms. After the application is completed, it is preliminarily assessed, and the further procedure for dealing with it is determined; for example, a set of necessary clearances and approvals is determined. A preliminary contract is signed, specifying the parties' tasks in the project initiation process. The government decides whether to qualify the proposal for the next detailed proposal stage.

In this phase, information about the proposal is published on the website. Both parties participate in the detailed proposal development workshop. The implemented reference projects (if any) and the decision-making criteria are defined. A detailed proposal is prepared. The government provides qualified persons to evaluate the proposal. An assessment is made against all criteria. The evaluation team prepares a recommendation to the government regarding the further analysis of the proposal, i.e., the transition to the negotiation phase of the final offer.

In this phase, a negotiation plan is developed. A schedule for negotiating all points of a possible contract is determined, and negotiations are held. As a result, the proposer submits a binding offer, including, in particular, all technical, commercial, and legal conditions. Based on substantive recommendations, the government decides on implementing the project. In the case of a favorable decision, a contract is signed, and the project is implemented.

20.4.2.3 Modes of Initiating Public Sector Projects

The process of initiating public sector projects can be characterized by the general mode, the number, and the content of checkpoints leading to the decision to implement the project.

The two main modes for initiating public sector projects are portfolio and individual modes. There is also a mixed project initiation mode, i.e., combining the portfolio mode with the individual mode.

20.4.2.4 The Portfolio Mode

The portfolio mode is one in which, for specific units of time – usually individual years – within one process, a decision is made on the portfolio of projects and/or project programs carried out in that unit of time (Figure 20.2).

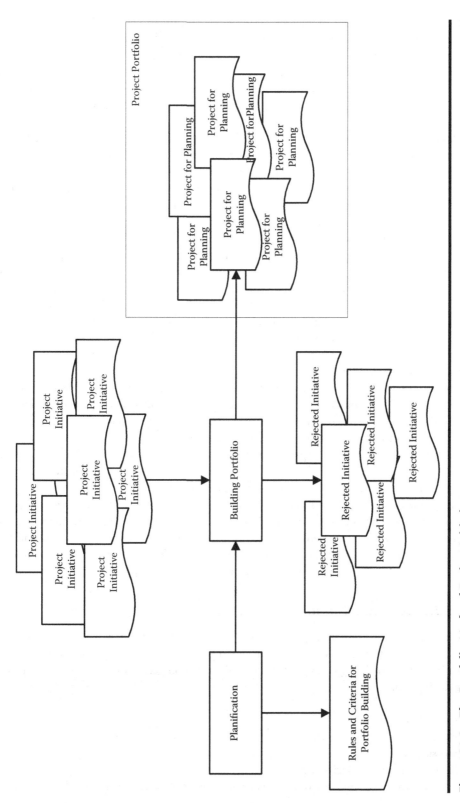

Figure 20.2   The Portfolio Mode of Project Initiation.

The portfolio mode is used by many governments, e.g., for projects in Argentina (Argentina MEFP, 2015) and the Brazilian state of Bahia (Gaino, 2014). This procedure defines the criteria for selecting projects from among the initiatives submitted up to a certain point in time. Mandatory projects have the highest priority. Continuing projects started in a previous planning cycle must also be considered. Then the projects are ranked according to the level of meeting the criteria adopted in the planification phase. Decisions to initiate projects are made based on such a defined rank and the available budget (PMI, 2017c).

In Argentina, an investment plan is developed every year under the National Public Investment System (Sistema Nacional de Inversiones Publicas, SNIP; see Section 20.9.2) (Argentina MEFP, 2015). The components of the investment plan are projects. The initiation process covers all projects whose budget exceeds one per mille of the previous year's project portfolio budget. The development of the public investment plan begins in April with the establishment of project prioritization criteria jointly by the National Directorate of Public Investments and budgetary organizations (i.e., planification). Local authorities (Argentina consists of 23 provinces, the city of Buenos Aires, and Tierra del Fuego) and budgetary organizations are recognizing the need for projects. Similar analyzes, but concerning the needs of the central level, are performed by budgetary organizations. Project proposals are prioritized. The National Budget Office sets budget limits for individual projects. Consultations are held in the provinces, after which budgetary organizations make the final selection of projects. On this basis, the National Directorate of Public Investments publishes the National Plan of Public Investments. The National Budget Office sets project budgets. A provincial public investment plan is being developed. Budget organizations publish documentation of their projects, qualified for implementation. The National Directorate of Public Investments publishes its opinion on the Plan. Figure 20.3 schematically presents the Argentinian approach to project portfolio building.

Bid contests are a type of portfolio mode of project initiation (Morrison, 2015). Australian states compete with each other for central funds. In the first competition, carried out in 2008 as part of the national infrastructure review, 675 proposals were submitted, of which Infrastructure Australia selected 94 for further analysis. Finally, seven projects were awarded a budget from the Ministry of Infrastructure and Regional Development (Australia ANAO, 2010b).

A specific method of portfolio mode initiation of projects is the so-called **horizontal project initiation**. It consists in coming up with projects without precise references to strategic goals and then making a "horizontal" review of the existing strategic goals. This review tries to find those strategic goals most related to a specific project (Koenig, 2015). Those projects are selected for implementation that best pass the horizontal screening of strategic goals.

20.4.2.5 The Individual Modes

The individual mode of initiating public sector projects is the one in which, for each project initiative, a separate instance of the decision-making process concerning qualification for implementation is launched (Figure 20.4). Several major decision gates can characterize this process. We can distinguish:

- Direct initiation
- One-step process
- Two-stage process
- Processes with more decision gates
- Processes where the number of gates is set separately for each initiative

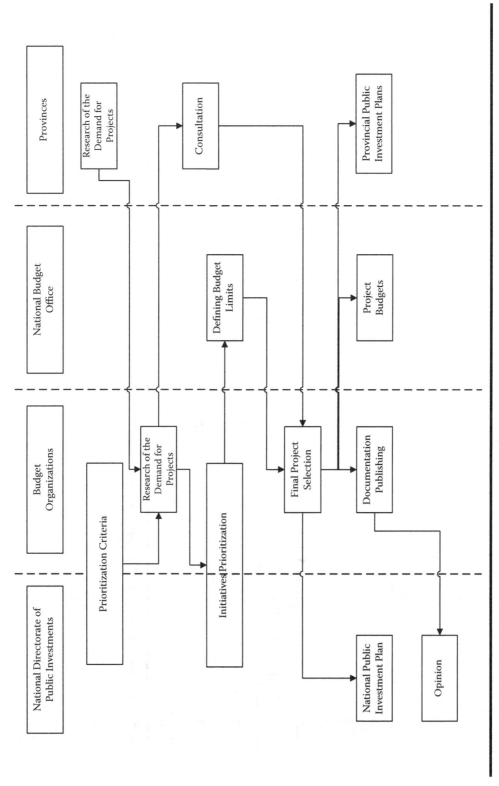

Figure 20.3   Initiating Public Sector Projects in Argentina. (Based upon Argentina MEFP, 2015.)

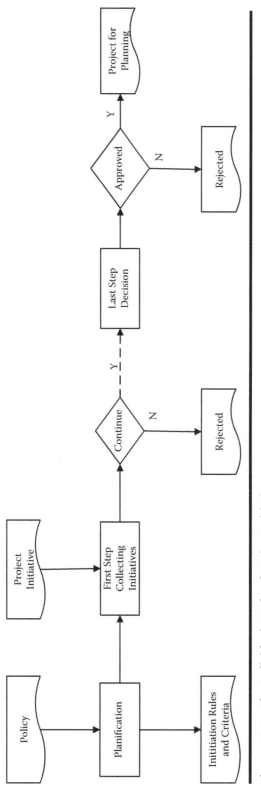

Figure 20.4 The Individual Mode of Project Initiation.

20.4.2.5.1 Direct Initiation

Direct initiation is the process by which submitting an initiative is equivalent to initiating it. An example of this mode is running projects under the Member of Parliament Local Area Development Scheme (MPLADS) in India. The purpose of MPLADS projects is the development of less developed areas of the country. Members of parliament have at their disposal a certain amount of money that they can spend on the development of the regions they represent (India MoSPI, 2015). Another situation in which direct initiation is used is mandatory projects, for example, ensuring the safety of mass events (Oliveira, 2014).

This mode of initiating public sector projects is also one of the inherent features of authoritarian-governed states, where decisions of governing individuals or political organizations are not subject to formal procedures. Decisions to implement projects are made randomly; large projects are carried out to show the leader's greatness so that undemocratic governments can maintain power or based on the interests of the entities initiating them (Shiferaw and Klakegg, 2012).

20.4.2.5.2 One-Step Process

In a one-step process of initiating public sector projects, the eligible body decides on the implementation once the initiative is submitted. The single-step approach does not mean that the process cannot be divided into components. Still, the execution of individual activities does not result in formal decisions being made.

In Iceland (Iceland Althingi, 2001), various project alternatives are considered during the pre-study phase regarding costs and benefits, and possible construction locations are indicated. The report of these activities also includes an initial design, if applicable. The information is presented to the Minister of Finance and the parliament for decision.

For projects of local importance in New South Wales counties, managers of functional areas (e.g., transport area) suggest project initiatives to directors; they make decisions based on budget availability and project priorities (Koenig, 2015). In the same state of Australia, managers responsible for road units identify the required level of expenditure, send information to the headquarters of the Road and Sea Services Board, these proposals are prioritized, and funds are allocated (Hodges, 2015).

20.4.2.5.3 Two-Step Process

A two-step process is one in which decisions about the initiation of a project are made as a result of two assessments, each of which can lead to the next phase of rejection. Technical and business assessments are frequently, but not always, applied types of evaluation. Different decision-making bodies may be involved in both evaluations. The two-step project approval process is in place in many countries, for example, in Hawaii (Hawaii Ige, 2015) or ICT projects of the Australian federal government (Australia DoF, 2016b).

The two-step processes can be further divided, taking the type of assessment performed in the first step as a criterion. The initial assessment may be on the business or technical aspects of the proposal submitted.

For the first variant, the goal and financing possibilities can be determined in the first phase, and preliminary declarations of commitment are obtained from entities planned for the project implementation. In the second phase, the project implementation and management methods are specified in detail (UK NAO, 2011).

In another variant, in the first phase, the project's objectives, need, and technical feasibility are assessed. The second phase is the decision on initiating, i.e., financing the project. The project management plan and other necessary project documentation are assessed. The assessment focuses on cost and may also include a detailed business case (Australia AGIMO, 2018).

The cabinet's objectives and overall business case are described and assessed in the first phase of the approval process for ICT projects in Australia. The government's decision is based on the Department of Finance and Deregulation opinion. A detailed business case is prepared in the second phase, considering inter alia, risk and mitigation measures, governance, and project budget (Australia DoF, 2016b).

The Norwegian system is also an example of this type of two-step process. In Norway, the Quality Assurance Scheme (QAS, Klakegg et al., 2008; Berg, 2012; Norway NTNU, 2013; Williams et al., 2010; Norway Christensen, 2008) applies to projects with an estimated value of over NOK 500 million. The purpose of introducing this scheme is to separate political and administrative decisions with a well-defined way of influencing these parties. There are two gates in this process: QA1 (Quality Assurance 1, Cabinet decision, where the need for the project is assessed) and QA2 (Quality Assurance 2, Parliament approval, where the project implementation strategy is assessed, among others).

Technology evaluation is the first to be done in, for example, Arizona, where state agencies' IT investment proposals must be approved by Arizona Strategic Enterprise Technology (ASET) before their initiation (Arizona ADoA, 2013). Projects with a budget above $ 1,000,000 must also be assessed by the IT Approval Committee (Arizona DoAT, 2013). A budget limit of $ 25,000 also applies to IT projects in Pennsylvania, where IT staff first approve projects at the Pennsylvania Office of Administration (Pennsylvania PoA, 2012).

As is the case in Maryland (Maryland DoIT, 2013b), the first decision point could be a request for permission to begin project planning to the Department of Information Technology, the Department of Budget and Management, and the Department of Legislation. If approved, a project plan is developed, then analyzed and approved by the same departments.

North Dakota has a governor's executive order to closely oversee contracts and major IT projects with a budget of more than $ 1 million (North Dakota Dalrymple, 2011). Before the decision on the implementation of the project by the applicant entity, the assessment of such projects is carried out by appropriate committees. The decision is sent to the Steering Committee, created for each large project.

20.4.2.5.4 More Complex Initiation Processes

An example of a three-step project selection process is the process used by the Ministry of Defense of Thailand (Thailand Puthamont and Charoenngam, 2007). The process consists of a concept phase, a technical design phase, and a final phase (Figure 20.5).

In India, initiating a project at the central government level consists of several stages. First, a Feasibility Study is drawn up by the proposing ministry. This document is sent for initial approval to the Planning Committee. After pre-approval, the ministry prepares a detailed project report, which is again sent to the Committee for evaluation. The ministry or the government (for the largest projects) decides to initiate a project (India MoF, 2003; India Planning Commission, 2010; India NITI, 2021; India MoSPI, 2013).

In Ethiopia, the initiation process begins with the ministry's submission of a proposal to the Ministry of Finance and Economic Development (MoFED). MoFED accepts (or rejects) this proposal. Then the proposing ministry prepares a detailed project plan. This plan is sent to MoFED. The decision to initiate is made by the government or parliament (Shiferaw et al., 2012).

In Peru, the project initiation process consists of three stages: preparation of the project profile, preparation of the preliminary Feasibility Study, and preparation of the final Feasibility Study. This phase ends with the approval of the declaration of feasibility by the authorized entity (Peru MEyF, 2014).

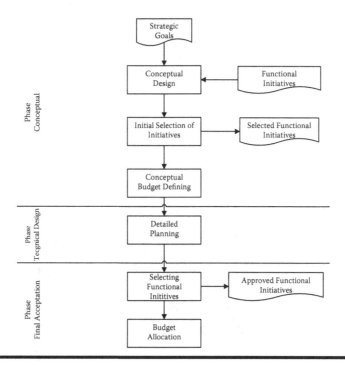

Figure 20.5 A Three-Step Process of Selecting the Strategic Project. (Based upon Puthamont and Charoenngam 2007.)

20.4.2.5.5 The Dynamic Project Initiation Process

The number of approvals does not have to be the same for every project, even within a specific Public Administration unit. For example, construction projects carried out in areas under conservation or ecological protection must obtain the conservator's approval of monuments or nature conservators. In contrast, in other areas, such consent is not required (Western Australia DoSD, 2013b). The project approval process can start with identifying the approvals needed and the issues to be resolved and setting a work schedule (at the milestone level) leading to the initiative's approval. Agencies involved in the approval process nominate those responsible for participation in the process. The agencies then evaluate the project in terms of formal and strategic requirements. The lead agency monitors the progress of the approval process, identifies, and supports problem-solving. If the proposal is rejected, the proponent may be assisted to modify the proposal for re-submission (Western Australia DoSD, 2013a). In Canada, the main task of the Major Projects Management Office is to manage the approval process specific to each infrastructure project (Canada MPMO, 2021).

20.4.2.6 Mixed Mode of Project Initiation

The mixed mode is one in which most projects are initiated in portfolio mode based on the implementation of the strategy. Still, when public intervention is necessary, it is also possible to initiate other projects (Figure 20.6). Such a project initiation process is possible, for example, in New South Wales (NSW Treasury, 2018b).

Projects can be initiated in a standard way, based on existing strategies (*basic projects*). They can also be started in exceptional situations to counter significant emerging problems that hinder

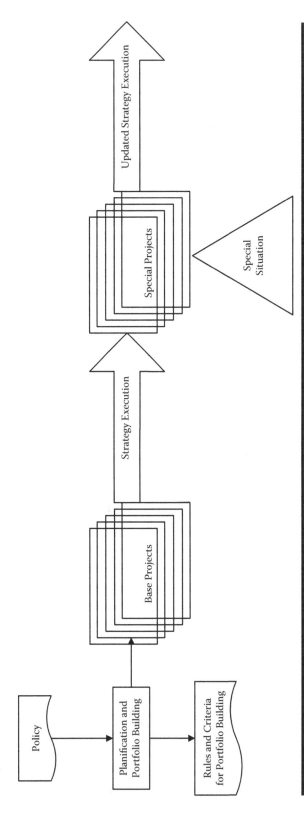

Figure 20.6 Mixed Mode of Project Initiation.

the implementation of the strategy or (less frequently) to exploit opportunities arising from the implementation of the strategy (*special projects*). Special projects can be financed directly from the resources allocated to the relevant government agencies for a given budget period or incorporated into the following budgeting periods.

In India, a Committee on Non-Plan Expenditure is created to analyze projects outside the plan. The results of this committee's analysis are sent to the Standing Committee on Finance, operating in the relevant public organization, for the decision to initiate projects (India MoF, 2014).

20.4.2.7 Initiation Monitoring

To ensure that projects will be appropriately selected, planned, and implemented, the project initiation process is verified by independent entities. Within the UK OGC Gateway Review Process (Point 20.5.1.1, UK OGC, 2007), the first four gates refer to the process of initiating them. These are the Strategic Assessment Gateway (for programs), Business justification gateway, Delivery strategy gateway, and Investment decision gateway.

In Australia, at the federal level, an implementation readiness assessment is performed for the initiatives that carry the highest risk (Australia DoF, 2013). The evaluation takes no more than one week and looks at the agency's ability to implement the programs and assess preparation for a specific program. As a result of this evaluation, the agency receives additional support on implementing the initiative.

Risk assessment is part of the project initiation. In Australia, each government agency must complete the Risk Potential Assessment Tool (Australia DoF, 2016a) form before submitting a proposal for an investment initiative to the government. Based on the questions in the questionnaire, the overall risk level (on a five-point scale) and the five most significant risks of the initiative are determined. For these risks, the proposing agency must identify ways to mitigate them. The risk of the initiative is then assessed after considering the proposed risk reduction measures.

In the initiation phase, the compliance of the proposals with state strategies and policies is also monitored (California CDoT, 2013a).

20.4.2.8 Stakeholder Participation

Even if there are objective measures of the project benefits, the communities for which they are implemented should accept them. The initiation process may start with identifying a group of stakeholders with whom to consult when designing project proposals. The most important include community members, local governments, residents, indigenous peoples, subcontractors, suppliers, consultants, state agencies, and landowners. The initiation process is the consultation process leading to the decision-making process: identifying stakeholders, informing, listening to opinions, exploring opportunities and innovations. Deciding is done jointly by consensus (Western Australia DPaC, 2002). This approach lengthens the process but reduces the risk of groups dissatisfied with implementing the project.

20.4.3 Planning

Planning is activities aimed at determining how a project will be carried out and documenting it in a project plan. Planning follows the initiation phase, but the transition from initiation to planning can be smooth. Planning activities may start in the initiation phase when specific information on how to execute the project is required to decide (Australia AGIMO, 2018; Thailand Puthamont and Charoenngam, 2007). The project plan describes the subsequent phases of a project.

While it is impossible to predict everything that may happen in the future with public sector projects, good planning and documentation of the plans is an important factor in the success of public sector projects. Each public project or project program should, by regulation (Scottish Government, 2013b; Colorado GASC, 2012) or, due to the procedures of the executing organization (Barkley, 2011), have an up-to-date, published implementation plan. In some organizations, recommendations emphasize project planning (Grabowski, 2014). Planning is better in countries with a long tradition of implementing projects essential to the functioning of the state (Flyvbjerg, 2018).

Public sector project plans can be created based upon documents such as PMBOK® Guide or Prince2®. Factors influencing the ability to achieve project goals in the planning phase are technical and organizational skills, project management skills, and knowledge of public policies (Gomes et al., 2012).

The structure of public project planning is shown in Figure 20.7.

20.4.3.1 Structure of Project Plan

The public sector project plan should be structured analogous or similar to project plans in other sectors. The complete plan should describe the activities in all management areas described in chapter 3. However, some solutions emphasize its specific elements.

In Colorado, mandatory elements of the project plan under the law include designation of a project manager, business case, business requirements, method of verification and validation, and financing strategy (Colorado GASC, 2012). Icelandic law indicates the detailed technical design, material list, cost estimate, schedule, and financial estimates for five years after the end of the project as the planning results to be included in the project plan (Iceland Althingi, 2001).

The most common requirement is to include their schedules and budgets in project plans (India MoPI, 1989; Poland Sejm, 2007). Northern Ireland's plans for public projects should describe its mandate, acronym, and definition, including vision, risk, and problem register, benefit profile, management strategy, stakeholder map, business case, quality and risk management strategies, and communication and benefit realization plans (NI DoF, 2014d). The British Columbia Ministry of Finance adds general management functions, products, and a detailed work plan to this list (BC MoF, 2012). As large projects and project programs may require the cooperation of many public organizations, their plans should describe the manner of this cooperation (USA Congress, 2010a). In Argentina, public sector project plans describe the rules of financing, rules for recruiting consultants, detailed requirements for individual functions, rules for using physical resources, accounting rules, and rules for conducting audits (Collia, 2014). Projects have baselines related to time and products. Some approaches talk about management strategy rather than management functions. The Brazilian method, in turn, determines the planning sequence: first, the scope of the project is specified, then the schedule, budget, quality criteria, team plan, communication plan, risks and responses to them, purchasing and contract plan (Brazil MPOeG, 2011). The process can result in a very detailed plan. The positive side of this approach is the very precise specification of the project product. The level of detail in project planning depends on its type, scope, number of contracts and stakeholders (Cota, 2014).

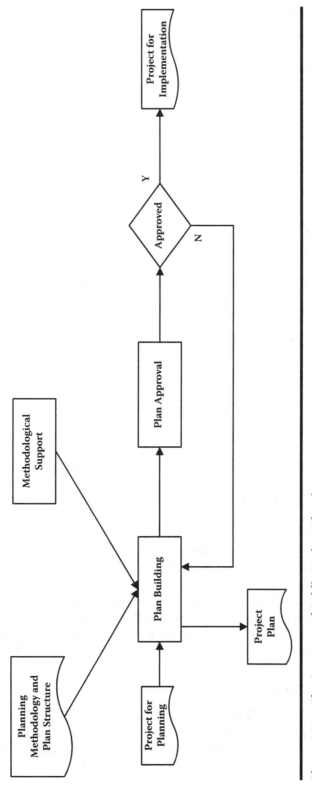

Figure 20.7 The Structure of Public Project Planning.

20.4.3.2 Developing a Plan

Project implementing organizations, L-PMOs, and external entities may be involved in project planning.

Consulting services in the planning and preparation phase of the project are, e.g., performed in California by the PMO of the Department of Transportation (California CDoT, 2013a), the Arkansas Department of Information Services (Arkansas DoIS, 2021), or by an analogous office in the California General Services Department (California DoGS, 2013). The role of the PMO can be defined as supporting planning. Planning can be done in agreement with a specialized government agency, as in Iceland, where the Government Construction Contracting Agency (GCCA, Iceland Althingi, 2001) plays this role. The same regulation requires the entity preparing the project to sign a contract for its planning with an external company. Specialized units (Cabinet Implementation Unit, CIU) in Australia encourage early and effective decision planning using project and project program management best practices and principles (Crawford and Helm, 2009). PMO may be responsible for some elements of the project plan, for example, placing there components that ensure the project's success, the durability of its products, and ways of their retirement (Colorado OoIT, 2013). The US Department of Energy's Office of Acquisition and Project Management does independent project costing for other organizational units (Lefman, 2014). In the United Kingdom, the Infrastructure and Project Authority creates an Integrated Assurance and Review Plan for major projects (HM Treasury and Cabinet Office, 2011).

Specialized entities can entirely carry out project planning. In Argentina, the Ministry of Planning is planning the most significant projects. Similar organizational solutions are applied in Arkansas (Arkansas DoIS, 2021), Bangladesh (Bangladesh DoPW, 2013), and Malaysia (Malaysia JKRM, 2013).

20.4.3.3 Project Plan Approval

Approval of the plan is equivalent to the final decision to start project execution. Before the project plan is approved, it may be reviewed. In India, the project planning status assessment is performed for the Public Investment Board (PIB) by the Infrastructure and Project Monitoring Division (India MoSPI, 2013). The organization that reviews the plan may be the Ministry of Finance (Iceland Althingi, 2001).

The project plan can be approved at different organizational levels. It may be approved by the body supervising the implementation of the project, i.e., its Steering Committee (Poland MS, 2014). It can also be approved by units within the jurisdiction where the project will be implemented, for example, by directors of a district office (Cowra Council NSW, 2015), directors in departments, and ministers in the Brazilian state of Bahia (Gaino, 2014). Major project plans may be approved at the highest organizational level: by the government or parliament (Norway NTNU, 2013).

The approved project plan may be modified as a result of internal (by project management) or external (by authorized bodies) control of the progress of work (Section 20.4.5).

20.4.3.4 Methodological Support

Governmental institutions develop recommendations, guidelines, and standards for project and project program planning. Such recommendations include standards such as PMBOK® Guide, Standard for Program Management (PMI, 2017b), or Prince2® (UK OGC, 2017). More detailed recommendations are also developed, for example, on project planning from the point of view of their sponsors (Australia ANAO, 2010), cost planning (Isakowitz 2002; USA GAO, 2020a; India Planning Commission, 2010), and project scheduling (USA GAO, 2012). A type of methodological support may be training conducted by international aid institutions, during which plans of real public sector projects are developed (Siles, 2014).

After project initiation, PMOs provide project plan preparation services (Malaysia JKRM, 2013; Pakistan Planning Commission, 2013), in particular, scope definition (California CDoT, 2013c), schedule and critical path (Arizona DoT, 2013, California CDoT, 2013a, Bangladesh DoPW, 2013), or budgeting (California CDoT, 2013a; Pakistan Planning Commission, 2013). In Georgia, support includes establishing governance structures and processes (Georgia GTA, 2014a). Standard schedules are provided (Missouri OoA, 2013a). For construction projects, technical projects are prepared, and construction sites are acquired (Malaysia JKRM, 2013; NSW PWA, 2021). The Office of Strategic Analysis and Estimation – a specialized unit of the Project Management Department of the Washington State Department of Transportation – provides services of estimation, risk analysis, and value engineering (Washington WSDOT, 2013b).

20.4.4 Execution

Execution is the process that brings a project to completion as defined in the project plan. Their primary effect is the production and delivery of project products.

Projects are carried out independently by PMO or are carried out independently by public agencies. There are several types of PMO participation in the execution of projects.

20.4.4.1 Project Execution by PMO

The complete way of L-PMO involvement in project execution is their full conduct (the G-PMOs usually do not execute projects by themselves). This is the approach, for example, of the Vermont Enterprise Project Management Office (Vermont DoII, 2010), Project Delivery Bureau (Vermont AoT, 2020), New York Enterprise Project Management Office (New York OoIT, 2013), California Project Management Division (California DoGS, 2013), two sections of the Arizona Department of Transportation Project Management Service Team (Arizona DoT, 2013), Program Development and Administration Division of Department of Transportation, Tennessee (Tennessee TDoT, 2021), Victoria Development (Victoria Development, 2021), Kuwait Mega Projects Agency (Kuwait MPA, 2021). The management of the largest infrastructure projects is performed by a specialized organization, especially if they are at high risk (Queensland QTT, 2013). Project Management Branch (PMB, Hong Kong ASD, 2013) manages all government building projects in Hong Kong. The Indian Central Public Works Department (India CPWD, 2016) carries out major projects in the areas of construction and infrastructure.

PMO may delegate project managers to public organizations where these people are responsible for project implementation. This approach is applied, for example, in New South Wales, where project management services can be performed by Public Works Advisory employees (NSW PWA, 2021), and in Western Australia, where projects are led by members of the Department of Finance Project Team (Western Australia DoF, 2020a).

The Virginia Technology Information Agency (VITA) manages IT projects in Virginia. Still, for some projects, that agency may delegate management to other public organizations – this is decided by its Program Management Division (Caudill, 2014).

20.4.4.2 PMO's Services for Projects

Project implementation services include advising or performing functions related to the implementation of public sector projects by entities outside the project team. The most important organizations dealing with this are also PMOs.

Services can cover any needed scope of activity without defining its detailed framework. PMOs perform such role, for example, in Hong Kong (Hong Kong ASD, 2013; Hong Kong ASD, 2021), Singapore (Singapore MoF, 2013), in the American states of Hawaii (Hawaii ETS, 2012), Arizona (Arizona DoT, 2013), Maryland (Maryland DoIT, 2013a), Missouri (Missouri OoA, 2013a), New York (New York OoIT, 2013), Vermont (Vermont DoII, 2010), California (California DoGS, 2013), and also in Brunei (Brunei DPW, 2021). PMOs operate similarly in Malaysia (Malaysia JKRM, 2013), New South Wales (NSW PWA, 2021), and India (India MoSPI, 2013). This is also how PMOs operate in the Australian state of Victoria (Victoria Development, 2021). In Northern Ireland, project management services are provided by the Centre of Expertise for Programme and Project Management (NI DoF, 2021). Collaboration can only involve smaller projects – where larger ones are directly managed by the PMO (Queensland QTT, 2013). In New Zealand, Major Projects advises projects to increase the likelihood of their success (New Zealand Infrastructure Commission, 2021).

PMO services may be limited to certain functions. For example, they support exchanging information between agencies, PMOs, and project managers (North Carolina OSCIO, 2013b). In Ontario, PMO supports creating project structures and identifying best project management practices (Tziortsis, 2014). In Peru, the General Directorate for Multiannual Public Sector Programming provides technological and functional support for organizations in the ministries (sectors), local governments, or provincial governments (Dirección General de Programación Multiannual del Sector Público, Peru MEyF, 2014). During the project implementation, PMOs provide document management services (Missouri OoA, 2013a), schedule, resources, and quality management (Malaysia JKRM, 2013). Many PMOs offer contract management services. Project-oriented PMO services include system interface management and data conversion (California CDoT, 2013c). PMOs may be responsible for applying project management methodologies (Fernandes, 2014). Very often, PMOs provide services related to project cost and communications management.

PMOs provide staff, particularly managers, for public sector projects. For example, public sector project managers are provided by Vermont's PMO (Vermont DoII, 2010) and Project Delivery Bureau Vermont (Vermont AoT, 2020), California Project Supervision and Consultation (California CDoT, 2013c), NSW Public Works Advisory (NSW PWA, 2021). Bahrain's Strategic Projects Directorate (Bahrain MoW, 2013) engages and manages project consultants. The PMO director may verify the assignment of a manager to each IT project (Maine MOIT, 2013a). Public sector projects sponsored by the Inter-American Development Bank (IDB) in South America are managed by IDB-contracted project managers.

One of the inherent features of projects, including public ones, is threatening them with risks. As a result of the risks, problems may arise, which in turn may necessitate changes to public sector projects. Risk mitigation may be the primary goal of a specialized organizational unit. This is the case of the Texas Quality Assurance Team (Texas SAO, 2013). MPAG New Zealand has developed a risk assessment profile form (New Zealand SSC, 2013). PMOs continuously and independently of the project team assess risks (New Zealand SSC, 2011). California CDoT (2013b) provides risk management services. Singapore CPPM (2011) identifies project risks and helps counteract them. Specialized services in estimation, risk analysis, value engineering, and project implementation support are provided by the Office for Strategic Analysis and Estimation (Washington WSDOT, 2013b).

PMOs provide problem management services (California CDoT, 2013a; UK IPA, 2021; India MoSPI, 2013). If a set of encountered problems threatens to terminate projects, actions known as project recovery are necessary. PMOs provide project recovery services (Vermont DoII, 2010). In justified cases, they may request project discontinuation (Arizona ADoA, 2013).

PMO work does not end at the end of the project. A PMO can support the benefits measurement and reporting process afterwards (Vermont DoII, 2010).

20.4.4.3 Reporting

Reporting is the systematic collection of information about the course of a project and its transfer to project stakeholders. The reporting results are then used for monitoring and control. Public sector projects have multiple layers of communications which complicate the reporting process. But accurate progress reporting is considered a success factor for public sector projects.

Project managers most often have two main reporting directions: PMO and the business owner, i.e., department director (Gavazza, 2014; Caudill, 2014). Public sector projects are also subject to scrutiny by various committees – the PMO provides them with the information they need (Naffah Ferreira, 2014). Many organizational solutions use PMO to collect project data (India NITI 2021; Vermont DoII, 2010; Arizona DoAT, 2013; Arizona DoT, 2013; Maryland DoIT, 2021a; New Zealand SSC, 2011; India MoSPI, 2013; Michigan MDoIT, 2003), which then deal with the further dissemination of project data. Arizona DoT (2013) manages project schedule, resource, and budget information.

Project progress reports are forwarded via PMO to their owners (New York OoIT, 2013). They can also be transferred to other superior organizational units, such as the CIO and bodies responsible for state finances (Montana SIT PMO, 2013b). UK IPA provides data to ministers responsible for project implementation (UK IPA, 2021), as does India MoSPI (India MoSPI, 2013). In Georgia, project implementation information is provided to the Georgia Governor's Advisory Body receives monthly reports of projects that are at risk or of community focus (Georgia GTA, 2014). In Victoria, Australia, the Governance and Business Services unit reported to several people, including the deputy secretary for Major Projects, the corporate deputy secretary, and the executive director of Major Projects Victoria (Victoria DSDBI, 2013). Periodic reports, in addition to data on the status of implementation of the scope, schedule, or budget, also contain comments, proposals for remedial measures, which in India are submitted to relevant ministers and government agencies by the Infrastructure and Project Monitoring Division (IPMD, India MoSPI, 2013). Data can be reported in aggregate form. The Ministry of Statistics and Programme Implementation in India reports central and sector data summarizing the status of implementation of programs (India MoSPI, 2009).

Collecting information on public sector projects can be centralized. The Max.gov portal (USA Government, 2010) is used in all policy programs, and it defines the standard of data exchange. In Peru, while implementing a public sector project, the implementing entity makes all information on project implementation available through the governmental PMO (Peru MEyF, 2014). One transport project reporting system operates in New South Wales (NSW Transport, 2021; Hodges, 2015). The system for collecting data on all public investment projects under SNIP (Section 20.9.2) also operates in Argentina (Argentina MEFP, 2021).

Reporting can relate to business issues, i.e., achieving goals or operational issues. In the United States, at the federal level, reporting of the achievement of most goals is carried out on an annual basis (USA Congress, 1993), and the priority ones every quarter (USA Congress, 2010a). In Peru, the results of assessing the project business effects are sent to the central PMO (Peru MEyF, ibid).

Operational reports describe the progress of work and compliance with the project plan. Project managers usually have a specific reporting schedule such as reporting project progress monthly (Naffah Ferreira, 2014), every three months (Cowra Council NSW, 2015), or annually (Fuertes and Herrera 2014). Project contractor reporting in a simple form can describe compliance with planned milestones (Clark, 2014; Astesiano, 2014). The traffic light metaphor can be used for project status reporting (Smith, D., 2015). Green means that the project is going according to plan, yellow means there are threats in the project, red means that the project has no chance of achieving its goals. The Action Committees (ACs) in Malaysia report to the heads of business units the progress of work on the projects (Malaysia Dahlan, 2015). Reporting may cover changes to the project that could affect its feasibility (Peru MEyF, 2014). Reports may describe the main obstacles to achieving goals and delivering programs (USA Congress, 2010a). PMOs can check that projects and project programs are implemented following the recommendations of appropriate, authorized bodies (New Zealand SSC, 2011). Operational reports can generally cover any area of project delivery. Project assurance (Subchapter 20.5) reports produced by the project assurance manager can describe the actions taken to increase the likelihood of project success (Georgia GTA, 2014b).

The state's statute may specify the scope of reporting on the implementation of public sector projects. For example, in the US state of Colorado, reporting on the implementation of public sector projects by entities outside the United States is required (Colorado OoSA, 2013).

20.4.4.4 Coordination of Cooperation

Public projects and project programs may be implemented by many public organizations. In such a situation, public organizations must cooperate, and this cooperation should be coordinated.

In Canada, an agreement is signed for projects supported by the Major Projects Management Office (MPMO), specifying the manner of involving government agencies involved in this work (signed by the deputy heads of these agencies), a work schedule, and a review schedule (Canada MPMO, 2013). Also, in the Brazilian state of Minas Gerais, a cooperation agreement (port. Acordo de Resultados) is prepared and signed when projects are implemented by multiple institutions (Prado, 2014). For the purposes of individual organizations, a description of the methods of cooperation with other organizations was prepared (USA Congress, 2010a).

Coordination activities can be assigned permanently to the PMO. This is the case, for example, in the state of Alaska (Alaska OPMP, 2013) and Hawaii (Hawaii ETS, 2012). Coordination of development programs and projects in Malaysia is the goal of Action Committees in ministries (Malaysia Dahlan, 2015). Coordination of projects or their aggregates may be delegated to an institution other than PMO. In the United States, the designation of a leading institution to achieve the federal government's priority goals is required by law (USA Congress, 2010a).

Detailed coordination procedures are, for example, the procedure for coordinating project reviews (Alaska OPMP, 2013) or coordinated procedures for cooperation between state bodies for the approval and review of projects and information on suppliers, implemented under the Statewide Project Delivery Framework (Texas DIR, 2013). The Missouri Project Management

Working Committee supports collaboration and information sharing between government agencies and other stakeholders to improve project management (Missouri OoA, 2013c).

20.4.5 Monitoring

Monitoring is an assessment of the project implementation status. In New Zealand public sector projects, monitoring is defined as the independent assessment of risks and work progress (New Zealand SSC, 2011). Monitoring can be performed based on previously collected and structured information (reports) or the basis of direct investigation of the project status (Pakistan Shah et al., 2011). A special material used by PMOs during monitoring may be the owner's opinions about the project (Iceland Althingi, 2001). The subject of monitoring may be compliance with the implementation of the project programs with their objectives (USA OMB, 2021; Colorado GASC 2012; Malaysia Dahlan, 2015) and management methods (USA OMB, 2021). The monitoring of project implementation is related to its baseline plan (BC MoF, 2012).

Each project should have a defined Key Performance Indicators (KPI) system to manage it properly. One of the primary project monitoring techniques devised initially for public sector projects is Earned Value Analysis (EVA; see Subchapter 3.5). Earned Value Analysis and its predecessor, Cost/Schedule Controls System Criteria (C/SCSC), have been used in the US Public Administration since the 1960s (Abba, 1997; Kwak and Anbari, 2011). Currently, the use of EVA is mandatory when acquiring complex systems by the US federal administration (USA GSA et al., 2019). EVA can be considered as a means of determining project performance indicators.

Monitoring is often combined, both in practice and in manuals, with control processes.

20.4.6 Control

Control is the issuing by an authorized parent entity, based on monitoring results, of project implementation orders and the enforcement of the implementation of these orders. Control uses the monitoring results. The result of the control review process may be request to change, for example, the project plan. Control is one of the main functions through which PMOs achieve their goals (Hawaii ETS, 2012). Supervision can be direct or indirect. Direct supervision is when the governing body itself determines the activities to be carried out by the public project or project program team or the organization implementing them. Indirect supervision is when the governing body requires the submission of an action plan, then approves the plan, which is ultimately executed by the supervised project, project program, or organization. Sometimes direct supervision is performed when the indirect supervision does not bring the intended results. There may also be mixed variants of control, where remedial measures are determined jointly by the controlled and controller.

The first instance that controls a project is its manager. One of the primary tasks of the project manager is to solve problems so that they do not have to be reported outside the project. If there are more severe problems, the project manager describes and justifies their reasons and determines corrective actions. The project program manager must determine remedies if program execution significantly deviates from the plan (USA GSA et al., 2019). The person responsible for defining project implementation recommendations, belonging to the organization implementing the project, may be the project assurance manager of a specific organization (Georgia GTA, 2014b).

The second level of control is the Steering Committee, which solves problems that cannot be solved at the operational level of the projects. Steering Committees of public sector projects make decisions related to the use and allocation of resources and the projects' external environment (Poland MS, 2014).

The next institution involved in controlling public sector projects is PMO. The Resource Management Offices operating within OMB (USA OMB, 2021) oversee the implementation of US government programs, in particular their budgets. Such tasks are also carried out by, for example, PMO in the California Department of Technology (California CDoT, 2013a), Centre of Expertise for Programme and Project Management in Northern Ireland (NI DoF, 2021), PMO IT in the US state of Maine (Maine MOIT, 2013b) or projects of US Department of Energy (Lefman, 2014). California CDoT (2013b) is authorized to control the implementation of state IT projects. The projects of the local agencies of central departments report to these headquarters. For example, Parkes County road projects in New South Wales are controlled by the headquarters of the NSW Maritime and Road Services Department (Hodges, 2015). PMOs practice to include advisers to projects whose main task is verification and control over project management (North Carolina OSCIO, 2013a; Maine MOIT, 2013b).

Major national projects may be subject to control by a governmental PMO. In the United Kingdom, major public sector projects are subject to the Infrastructure and Project Authority's control (UK IPA, 2016, 2021). In the United States, the regulatory authority for federal programs is the Office of Management and Budget (USA OMB, 2021), which performs high-level oversight. Central control is often conducted online. OMB in the United States specifies remedial measures for federal programs if they are not adequately defined at lower organizational levels (US OMB, 2021). A form of indirect central control in India is establishing responsibility for solving the cost and schedule issues by the Ministry of Statistics and Programme Implementation (India MoSPI, 2013).

Special positions in ministries may be created to control projects. In the Philippines, overseeing the work of the project management offices financed by FDI funds is the responsibility of the Undersecretary of State for PMO Activities (DPWH Philippines, 2021).

20.4.7 *Closing*

Closing of a project may occur as a result of the implementation of its plan, due to the inability to implement this plan, or the inability to achieve the business goals set for it. If performed in the ordinary, positive course, closing projects occur after transition of their products to operational processes (Subchapter 3.13).

The closing processes are the completion of purchases and contracts and the closure of the project, while their products are the formal closure of purchases and contracts and the closed project (Brazil MPOeG, 2011). Public sector projects may end prematurely. For example, six projects were abandoned in India at the Coal Ministry (India MoSPI, 2005). The body eliminating weak projects can be PMO, which stops attempts to implement unnecessary, redundant IT systems (Caudill, 2014). Another body that may suggest ending a project or program is the audit chamber. For example, the American GAO may recommend to Congress killing a program as a result of its audits (Wade, 2014). But sometimes, management does not close projects, even if they are poorly implemented. Projects favored by politicians and senior managers are very difficult to kill (Townsend, 2014).

Completing projects on time and budget is the primary responsibility of the project managers and their supervising IT Chief Information Officer (CIO) at the ministry in Minas Gerais (Naffah

Ferreira, 2014). Before the project closure, it may be necessary to obtain the opinion of the Project Portfolio Council, i.e., the body responsible for all projects in a given institution, after which the final report may be approved by the Steering Committee (Poland MS, 2014). PMO can support a project in obtaining approval of the final report (Cota, 2014).

At the end of the project, a review is performed to check the management methods (NI DoF, 2014c). The quality of execution is assessed in the project's final phase, where the project can be compared with other similar projects (Iceland Althingi, 2001).

After producing the project products, complex training and product implementation are performed (Townsend, 2014). After the execution, there is an operation phase: management of maintenance, use, personnel, and social issues (Chile MINVU 2009).

20.4.8 Evaluation

An element of completing public sector projects is their evaluation, i.e., checking their business results. The basis of the evaluation is the project baselines. The evaluation process consists of defining the scope, signing a contract, developing a detailed evaluation plan, evaluating activities, and producing a report (NSW Treasury, 2018a). In NSW, reviews are performed by the specialized Centre of Program Evaluation. A review may be required to verify that the project or program has been successful (NI DoF, 2014c). Evaluation can be performed in the post-investment phase (Peru MEyF, 2014). The evaluation components can be economic, social, and environmental factors (Chile MINVU, 2009). In India, the Development Monitoring and Evaluation Office (DMEO) evaluates programs, particularly major flagship ones, for their effectiveness, relevance, and impact (India DMEO, 2021). The evaluation may be performed by project financing institutions such as Inter-American Development Bank (Siles, 2014). A PMO can support the post-project benefit measurement and reporting process (Vermont DoII, 2010). It should be remembered that the assessment of the project results and impacts by public stakeholders may be subjective, not related to the evaluation results (Clark, 2014). Evaluation results are shared by the performing institution (India DMEO, 2021).

20.4.9 Variants of Project Life Cycles

The set of phases can be based on the above-described PMI's approach to public projects. This PMI-based set of phases is mainly valid in the US public sector and applied in some other countries. The set of phases of Brazil's IT projects, for example, is based too on PMI standards (Brazil MPOeG, 2011). Similar to the PMI approach is the process of implementing IT projects in the state of Vermont, where there are phases of exploration, initiation, planning, execution, and closing (Vermont DoII, 2010), while transport projects consist of the phases of submitting initiatives, selecting, defining, designing, and building (Vermont AoT, 1996). Other phases, incorporated into the appropriate places of public projects or project programs, include, for example, strategic planning and evaluation (Virginia VITA, 2017), aligning an initiative with business goals and gaining support for a plan (Washington WSDOT, 2013a), assessing the priority and funding opportunities of the project (Queensland QTT, 2013). In New Zealand, the first phase in which an investment decision is made is called **thinking** (New Zealand Infrastructure Commission, 2019).

The phase sets in some countries do not directly refer to PMI documents. In Pakistan, public sector IT projects go through the phases of initiation, planning and mobilization, direction and control, and closing (Pakistan Shah et al., 2011). In Switzerland, the phases of IT projects are initiation, concept, implementation, and deployment (Switzerland FIT SU, 2011). In the United Kingdom, a set of phases has been defined for agile projects: discovery, alpha, beta, live, retirement (UK HM Government, 2018). The very broadly defined phases of public sector projects in Peru are the pre-investment phase, the investment phase, and the post-investment phase (Peru MEyF, 2014). In Chile, on the other hand, the phases of the urban space project management process are planification, investment, acting, and evaluation (Chile MINVU, 2009). In Iceland, the phases of construction projects are pre-study, planning, construction, and completeness assessment (Iceland Althingi, 2001).

20.4.10 Agile Approach

The agile approach in the public sector, as in other sectors, is becoming more and more popular. Some administrations are increasingly turning to iterative and agile project life cycles.

The main agile-related issues in public organizations are:

■ Selection of projects to be implemented using the agile method
■ Ways of agile projects implementation
■ Implementing agile in the organization

20.4.10.1 Selection of Projects to Be Implemented Using the Agile Method

Agile has been a very fashionable approach recently. But even the greatest enthusiasts of this approach write that it should only be used "whenever appropriate" (Ambler, 2021). The most important public project in democratic countries – elections – cannot be credibly implemented iteratively (unless we consider the second round of elections the next iteration). Elections are not repeated until all voters are satisfied. This means that there is a question of identifying those projects that can be implemented in an agile way. Different organizations approach this issue in different ways.

The radical approach is to ban the sequential (waterfall) life cycle and mandate only the agile approach for IT projects (Maryland DoIT, 2021b).

A slightly more subtle way of predefining projects to be implemented this way is to identify the types of agile projects. Apart from software projects, they can be, for example, designing policies, project programs, or public services design (UK HM Government, 2016; Ontario Government, 2021a, 2021b). Agile can be recommended primarily for use in software projects, including using some of its elements in other types of projects. The Government Accountability Office suggests this approach for US federal administration projects (US GAO, 2020b). The use of agile is possible there because the Federal Acquisition Regulations (FAR, USA GSA et al., 2019) require government representatives to be innovative and do not prohibit the use of methods other than those defined in applicable regulations if they lead to better results.

Some organizations have developed guidelines to help you decide which approach is appropriate. The Government of Canada (Canada Government, 2021) proposes 20 questions divided into two groups (project characteristics, 14 questions and oversight, and governance, six questions) regarding the characteristics of the project and the environment based on which this decision should be made. Factors suggesting the use of the waterfall method are, for example, good definition and consistency of requirements, not being related to IT technologies, short (or very long) implementation time, legal or regulatory documentation is required, and finally – if the organization successfully implements projects sequentially. The use of the agile approach is supported by, for example, the use of IT technology, mean implementation time (about a year), the possibility of team co-location, low risk, financing for the entire duration of the project, with the possibility of making ongoing cash-flow decisions or bearing the organization failures in the implementation of projects using the waterfall method.

The other extreme approach is to use only the waterfall methods. It is difficult to find the requirement of using agile in construction or transport infrastructure projects. This is probably because these projects meet the conditions for applying the waterfall method (Canada Government, ibid). Construction projects, especially those affecting the natural environment, must have a precisely defined product before starting work, which is practically impossible to modify later.

20.4.10.2 Ways of Agile Projects Implementation

The agile approach is also a well-defined work methodology. The lack of methods and determining exactly everything during the project implementation cannot be called the agile approach (Maryland DoIT, 2021b). The agile approach does the same things as the waterfall (requirements gathering, planning, designing, building, testing), but simultaneously, not sequentially (Ontario Government, 2021a, 2021b).

Due to the specifics of the public sector, agile projects must have at least a specified scope and budget before deciding to implement. This information will then be used to assess whether the project has been successful. When defining the scope, first of all, business requirements, and rather not functional ones, should be taken into account. The detailed definition of project products should be avoided. Agility is about specifying these requirements during the course of the project. Projects must have their WBS (built in an agile-specific way) and a schedule baseline (USA GAO, 2020b), later appropriate for progress control. The agile contracts should be compatible with agile cadence, above all, frequent submission of new releases of products. An agile project must have its classic artifacts, including project charter, business, and non-functional requirements. In the beginning, you need to know what is to be delivered and what the schedule should be at the milestone level (Maryland DoIT, 2021b). A business case containing a project vision, final result, risk analysis, and cost-benefit analysis must be developed, as for traditional projects, to start the project (Texas DIR, 2021).

In IT-based projects, the SCRUM method is often used (Schwaber and Sutherland, 2017; Texas DIR, 2021; California CDoT, 2021). The scope and schedule of work can be determined by the hierarchy: user story, iteration, release, epic, project. Projects must use Earned Value Analysis to control work progress in an agile-adapted form: the basic unit of planning and control is the work effort of individual user stories (USA GAO, 2020b).

Another agile approach is used in the Australian state of Victoria (Victoria DPAC, 2016). The project is divided into two main, separate phases in which short sprints are used: conceptualization and implementation (separated by the procurement phase). There is no return to the conceptualization phase from the implementation phase, which means that the business scope of work is fixed at the entrance to the implementation phase, especially since this scope is included in the contract with an external contractor.

Another agile methodology can be used to design services (UK HM Government, 2016; Ontario Government, 2021a, 2021b). The project consists of four parts: Discovery, Alpha, Beta, and Live. The main problems, stakeholders, constraints, and risks are identified in the Discovery phase. In the Alpha phase, various prototypes of the solution are created and presented in short sprints – until the user accepts it. In the Beta phase, working solutions are created incrementally;

SCRUM can be used here. The phase ends when the minimum acceptable product is made. The creation of such a product begins the Live phase, in which the product is exploited and constantly improved.

The closure of an agile project is performed as in waterfall projects (Texas DIR, 2021). The product must be finally accepted, documentation and knowledge collected, systemized and stored, and the team should be disbanded.

20.4.10.3 Implementing Agile in the Organization

For agile projects to be successful in an organization, it must be prepared for this way of working. The agile approach means a certain mindset and should permeate the entire organization. The team must be built in accordance with agile principles and must have repeatable agile processes in line with those adopted for the organization. The decision to implement agile in an organization is a kind of generalization of the rules for selecting agile projects. If the organization is good at waterfall projects and its projects cannot be implemented in an agile manner, there is no way to implement agile. The environment of the organization should enable the use of agile. If the organization mainly implements software projects, then agile should be implemented, if infrastructure construction projects, rather not (USA GAO, 2020b).

The California Department of Technology (California CDoT, 2017) suggests a gradual, in line with real needs, transition of individual state agencies to the agile approach and highlights the cultural determinants of the implementation of such an approach, as well as the impact of experience, technology and the level of flexibility of the organization on implementation. The starting point could be a sequential model that creates software into which you can incorporate agile elements – for example, modular software development. SCRUM is the preferred agile method. The main phases of agile implementation in an organization include understanding this approach, planning its implementation, and working according to the new methodology. California DoT understands its role in agile implementation as providing knowledge that will enable individual state agencies to make their own decisions on this issue. Only actions that bring real benefits should be performed in the implementation. Implementing agile must not slow down service delivery (UK HM Government, 2016).

Implementing agile methods in many organizations, including public ones, has been prevalent recently. This is mainly due to the ability to quickly respond to changing or refined needs of users and the ongoing selection of a set of requirements to be implemented. But many criticisms of the sequential approach have no basis. For instance, the waterfall critics argue that all product is delivered in one shot with this approach. This is not true; the sequential approach allows you to divide each project into separate stages delivering specific products (cf. Figure 2.6 and Figure 2.7). The waterfall approach supposedly does not consider the user's reaction to the product. This is also not true – the pilot approach, which precisely aims to include product responses, was invented long before agile methodologies were developed. The waterfall is not geared to meeting customer needs, their critics say. Of course, it is difficult to imagine a sequential project for a client that would not consider his needs. Business analysis, which was also invented long before the agile approach, aims to identify his needs. Some critics state that the waterfall method only uses data to plan, measures after launch. On the contrary, the Earned Value Analysis, EVA, an integral part of every well-managed sequential project, requires the permanent collection of the project data.

But of course, rational, disciplined use of agile can bring benefits to many organizations that would not be achievable with traditional methods in many projects.

20.4.11 Classifications of Projects

Not all public projects have to be implemented in the same way in a given country. Mission-critical projects need to pay more attention to their management than local, small-scale projects such as school refurbishment or training for agency lawyers. Therefore, many governments have developed classifications based on which the processes, practices, and methods of project implementation adequate for their category are indicated. A set of management documents that must be created in the project may depend on qualifying for the appropriate category.

Projects can be considered national, perhaps the highest level of distinction encountered (NSW Parliament, 2021). Such projects give the green light to all activities. In such projects, the prime minister is empowered to make final decisions. The governor (head of the state) can regulate the national works. Breaking these regulations is punishable by law.

The estimated budget is most often used as the basis for project classification. The budget criterion is used in conjunction with other characteristics, for example, project reach (number of agencies involved), the technology used, state government and media interest, business criticality, number of product users, complexity, stakeholder support, project team size, and experience, complexity, risk, or duration.

In Michigan, projects can be divided into three categories: small (less than $ 1 million, performed for one agency), medium (meeting one of the following conditions: budget over $1 million, dedicated to multiple agencies, for non-standard technology, a representative of the Department of Information Technology is on the committee steering), and large (meeting one of the conditions: budget over $5 million, a project implemented with interest or on the order of state authorities, media involvement). There is a further breakdown for small projects: projects under $ 25,000, up to $ 250,000, and up to $ 1 million (Michigan MDoIT, 2004).

A project category can be defined by defining its characteristics based on answers to questions, for example, its criticality for business, number of product users, complexity, cost, duration, stakeholder support, team size and experience, and number of offices involved. Based on the answers to these questions, the project can be considered small, medium, or large (Montana SIT PMO, 2013a).

In Virginia (Virginia VITA, 2019), IT projects with a budget of less than $250,000 are considered internal agency projects and are not subject to the state management methodology. The other projects' risk level and complexity assessment are performed during their initiation. The first category includes projects at high risk and very complex. The second category is high risk and medium complexity or high risk and low complexity projects or medium risk and high complexity projects. The third category includes medium risk and medium complexity projects, medium risk, low complexity projects, and low risk and high complexity projects. The fourth category is low risk and medium or low complexity projects (Table 20.2). The higher the project category, the more formal methods must be applied.

Table 20.2 Classification of IT Projects in Virginia

Risk		Complexity		
		Low	**Medium**	**High**
	High	2	2	1
	Medium	3	3	2
	Low	4	4	3

Source: Based upon Virginia VITA, 2019.

Classification of projects may depend on the anticipated number of working hours needed. In Maine, work with less than 40 hours of labor is classified as a job request. A small project takes between 40 and 1,499 hours. Medium project: 1,500–9,999 working hours. Projects with a labor intensity of 10,000 or more are classified as large (Maine MOIT, 2013a).

In New South Wales, the essential characteristic of a project is the *project tier* (NSW DFSI, 2017). When determining it, the budget and priority of the government, the complexity of the interface (organizational, technical), procurement, and the capability and capacity of the agency implementing the project are taken into account. Project tier ranges from 1 (highest) to 4 (lowest). The highest tier covers with a budget of over $1 billion and all those exposed to the highest risk. Moreover, the project's designation as the highest risk level may be requested by many entities (prime minister, cabinet, minister …). Level 4 projects (below $ 50 million and low risk) do not require any mandatory action.

An extensive IT project classification system has been implemented in Maryland (Maryland Government, 2019). A major project is considered to be one that meets at least one of three conditions: has a budget of over $ 1,000,000, supports critical business functionality, or is designated as important by the Secretary because of its benefits or risks, its impact on society and local governments, or the public visibility. The Secretary may also designate any other project as the major one. A similar method of classification – defining the project by pointing them by authorized person functions in other governments (Texas DIR, 2013; NSW Procure Point, 2013).

The project category determines the approach to its management, for example, requirements for managers and other staff qualifications, the type of methodology used (simpler or more complex), the set of required documents, and the method of implementing project assurance.

In small organizations with few projects, one project management methodology may be sufficient. Developing their different versions would require more work than would not benefit from their use. On the other hand, in organizations implementing many different projects, the use of extensive methodologies for small projects would not pay off. Classifying projects solves these problems.

20.5 Project Assurance

Project assurance is activities, performed by teams independent from the project team, aimed at ensuring that projects are managed in a way guaranteeing achieving their goals. Assurance is one of the main elements of governance of projects (UK IPA, 2016; New Zealand Government, 2019). In the course of these activities, the project's status is analyzed, its weaknesses and strengths are identified, and ways of overcoming the existing problems are developed. The New Zealand government emphasizes independent project appraisal as an essential element of assurance (New Zealand Government, ibid). Assurance is essential in public projects because the project team and the project implementing organization are the holders but not the owners of the resources used – they are taxpayers. Project assurance can be considered an indirect form of control over their resources.

The assurance of projects can be performed in a planned manner as an organized process and on an ad hoc basis when there are signals of poor project performance or when otherwise authorized bodies wish to check the project's status. Special teams may be organized to implement assurance, for example, the Project Assurance team (e.g., New Zealand Government, ibid).

There are several main ways of project assurance. I describe them in the following sections.

20.5.1 Phases and Gates

Public agencies use a phase-gate structure for the governance of projects (Cooper, 1990; UK OGC, 2007). In general terms, each project or project program consists of phases separated by gates (Subchapter 2.4; Figure 20.8). Most governments apply the waterfall approach to their projects, but some have shifted to the agile approach, particularly IT. The phase-gate approach applies to both types of project life cycles. Each review analyzes and accepts the outputs of the phase (Victoria DPAC, 2016) and checks whether it is possible to achieve project or project program objectives (Barkley, 2011).

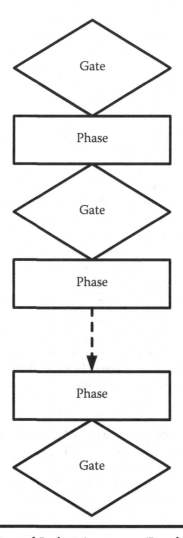

Figure 20.8   Phase-gate Structure of Project Assurance. (Based upon Cooper, 1991.)

20.5.1.1 UK OGC Gateway^TM Process

An example of a planned approach to project and program assurance is the Project Gateway Review Process defined by the UK Office of Government Commerce (UK OGC, 2007; UK IPA and Cabinet Office, 2021) shortly described below.

The zeroth gate of this process checks whether the program is needed and is likely to achieve its goals. The first gate verifies whether the business requirements can be met. After developing procurement and manufacturing strategy for products and an implementation plan, the procurement strategy is verified. In the next gate, after the entire business case and governance rules are in place, an investment decision is made and verified. It is verified whether the project is adequate to the situation. Before the completion of the delivery phase, the service readiness is verified, including organization readiness for product use, maintenance, and ownership. The Gateway Process extends to product use when its appropriate use and achieving expected business benefits are verified.

The UK OGC-defined process or its variations outside the UK Central Government is used, for example, in Australia at the federal level (Australia DoF, 2013), Victoria (Victoria DTF, 2019), Texas (Texas DIR, 2013), Queensland (Queensland Treasury, 2020), New South Wales (NSW Procure Point, 2013; NSW Treasury, 2017a), Northern Ireland (NI DoF, 2014b), and New Zealand (New Zealand Treasury, 2017). It is possible to develop different versions of assurance processes for different types of projects, e.g., IT, infrastructure, and recurrent projects (NSW Treasury, 2018a, NSW AO, 2015; NSW DFSI, 2017).

For some types of projects, versions of the gate review processes are different from those specified by the UK OGC. For example, after an investment decision, two additional gates are implemented for IT-related projects: checking the project charter and checking the detailed project plan (Canada CIOB, 2012). Other project life cycle structures that define the main control gateways are described in Section 20.4.9.

The most important public sector projects are usually subject to the gate review process. Various eligibility criteria for the process apply. For example, it covers infrastructure projects worth more than $ 30 million or projects with a budget of more than $ 30 million with an ICT component value of more than $ 10 million, and all project programs with a budget of more than $ 50 million and all high-risk projects (Australia DoF, 2013). The limit may depend on the level of project risk (NSW Treasury, 2018a). The Texas Project Delivery Framework is mandatory for projects with a budget of more than $ 1 million, lasting more than one year, involving more than one agency, or for projects designated by the relevant authorities (Texas DIR, 2013). In the Australian state of Victoria, this process must be applied to all complex, high-risk projects in areas such as policy, infrastructure, organizational change, procurement projects, and IT-based business change projects (Victoria DTF, 2019). In some solutions, the administration has the right to designate any project to perform the gate review process (NSW Treasury, ibid).

The process of gate reviews execution can also be improved. In Australia, ANAO performed an audit of how the gateway process was managed. As a result of this review, recommendations were made to improve this process (Australia ANAO, 2012).

By reviewing the gate, one can be sure that the project is being implemented using best practices and that the next phase will be successful. The lessons learned may be an additional product and benefit of the review.

20.5.1.2 Other Forms of Checking Project Status

Audits, reviews, and project evaluations are active forms of checking the status of public sector projects. If project status is analyzed with the active participation of project team representatives who can be involved in formulating the recommendations, this is called a review. If the project representatives and the documentation prepared by them are only a source of information, we are dealing with an audit. If monitoring concerns the business side of a project or project program, we are talking about its evaluation. Reviews are, for example, performed by Texas SAO (2013), UK IPA (2016), Vermont DoII (2010), Alaska OPMP (2013), Australia DoF (2021), Queensland Treasury (2020). Some PMOs conduct specialized audits, for example, in the area of quality management or the environment (NSW PW, 2013). The purpose of the review may be to ensure that quality and risk issues are adequately handled (Victoria DPAC, 2016). TBoCS Canada performs project risk assessment reviews necessary for their initiation (Canada TBoCS, 2019).

A special organizational unit may be created to supervise projects. In Maryland, a Project Management Oversight team (Maryland DoIT, 2009) is responsible for performing independent verification and validation, project reviews, and portfolio reviews.

20.5.2 Approaches to Project Assurance

An element of effective governance is the separation of project phases and the performance of reviews between them that determine the transition to the next phase. The above-mentioned example of different sets of gates shows that the most important issue is the very existence of well-separated phases and gates. The use of assurance increases the likelihood of early detection of project problems and applying remedial measures.

Assurance can be classified using the relation of the body performing this activity to the project as a criterion. In this regard, we can distinguish three types:

1. Self-assessment
2. Performed by the parent entity
3. Performed by external (independent) entities

20.5.2.1 Self-assessment

Self-assessment is the assessment of the project status by the entity implementing the project. Self-assessment is performed, for example, in Georgia IT projects. (Georgia GTA, 2014b). Such reviews may be for projects with a budget of less than $ 5 million. For an organization to be able to carry out the self-assessment, it must meet certain conditions: it must have its own PMO. The person who performs the review, known as the Project Assurance Manager, must be independent of the reviewed project. The institution must use the state GEMS Project Management Tool (Georgia GTA, 2014c). The institution must have successfully passed the governmental PMO review in advance. The Project Assurance Manager must have the appropriate qualifications: having a PMP® certificate issued by the Project Management Institute and completing the appropriate training set. It is also possible for someone who does not have a PMP® certificate to perform a review, but then he/she must undergo more governmental training. To know the project situation, the Project Assurance Manager must attend all meetings related to major projects. He/she also performs project documentation reviews (Georgia GTA, 2014b).

20.5.2.2 Assurance Performed by the Parent Entity

Public sector projects may be monitored by parent entities of the project implementing organization. The reviews are performed by the IMAP Investment Reviews team, which is a Treasury New Zealand organizational unit (New Zealand Treasury, 2021a).

Major Canadian projects are monitored by TBoCS (Canada TBoCS, 2019). In India, MP-initiated projects are monitored by a dedicated department of the Ministry of Statistics and Programme Implementation (India MoSPI, 2013), Twenty Point Program projects by the Twenty Point Program Department of the same ministry (India MoSPI, 2013), and major national projects are monitored every month by the Project Assessment and Management Department (India NITI, 2021; India MoSPI, 2013). In Pakistan, project monitoring is carried out by provincial planning and development departments or the Project Wing of the Planning Commission (Pakistan Shah et al., 2011). PW Pakistan (Pakistan Planning Commission, 2013) performs physical inspections, analysis of daily reports, and analysis of implementation reviews of all projects in business sectors.

Project monitoring is often the task of sector-specific PMOs; for example, IT projects are audited by the Vermont Project Office (Vermont DoII, 2010). Transportation projects are subject to the Vermont Department of Transportation's PMO audits (Vermont AoII, 1996). Projects may be monitored by specially appointed committees of the owning organization (Naffah Ferreira, 2014; North Dakota Dalrymple, 2011). For public aid projects, their implementation may be verified by a financing entity, for example, Inter-American Development Bank (Siles, 2014). The review result can be the basis for deciding whether to review a given project by a higher hierarchical body – then we are dealing with hierarchical reviews. In Georgia, the Georgia Technology Authority (GTA) runs the Critical Projects Review Panel (CPRP, Georgia GTA, 2014). The panel monitors the state's most important technology initiatives, decides which projects should be reviewed by the Governor's Office panel, and provides the Governor's Office with reports on selected projects.

In the United States, in large projects, the government party, together with the contractor, performs Integrated Baseline Reviews (USA GSA et al., 2019), the purpose of which is, among other things, to identify risks that threaten projects and define plans to mitigate them. The tool used for overseeing programs in the United States was the Program Assessment Rating Tool (PART[2], USA OMB, 2005, 2007). Based on the answers to the questions about its course, the program receives rating of Effective, Moderately effective, Adequate, Ineffective, or Results not demonstrated. Budgeting usually increases for the first two categories of programs and decreases for the next two.

20.5.2.3 Independent Assurance

We deal with independent reviews when they are performed by an entity or person not being a component of any institution directly or indirectly supervising the organization implementing a given project. Independent assurance is becoming the essential type in many governments. Independent reviews may be performed by institutions appointed to verify and validate projects or by external entities from the private sector (Norway NTNU, 2013; Shiferaw, 2013a). In the latter case, the role of a public organization is usually limited to selecting a monitoring entity and organizational and logistic support. The independent verification and validation system aims to assure a high level of trust as public sector projects are subject to public scrutiny. A system of independent verification and validation by an independent party (third party) also exists in the state of Colorado (Colorado GASC 2012; Colorado OoSA, 2013). Colorado state policy requires major projects to be assessed at specific points in their life cycle to determine if they are at high risk. If the risk is high, the project implementing organization must tender for independent project verification and validation. Independent verification and validation systems also exist, e.g., in the states of Georgia (Georgia GTA, 2014) and Maryland (Maryland DoIT, 2021a). In the United Kingdom, the project assurance has been the responsibility of the UK OGC (Crawford and Helm, 2009) and is currently performed by the Infrastructure and Project Authority (UK IPA, 2016).

In India, the Development Monitoring and Evaluation Office can independently evaluate programs with access to public funds or government guarantees (DMEO, India DMEO, 2021). Its results are publicly available.

Audits carried out by national audit chambers (Subchapter 19.3) are a specific method of independent monitoring.

20.5.2.4 Area Reviews

Area reviews are those in which one or more selected management areas are of interest.

During the implementation of projects, care should be taken to ensure that the planned benefits are achieved. Benefits management is one of the main processes in program delivery and should be integrated with other processes, including project management. In the course of implementation, benefits reviews should be performed for the possibility of obtaining benefits, and, if necessary, the set of them should be modified. To achieve the expected benefits, stakeholder buy-in, information management, and integration with organization processes are the most important (NSW Government, 2018).

A particular area that public organizations pay a lot of attention to in project reviews is risk management. Risks and risk management processes are analyzed, for example, in the reviews of the Australian state of Victoria (Victoria DTF, 2021) and Scotland (Scottish Government, 2013c).

Assurance may be focused on a set of red flags that may indicate risks to the project (Washington OCIO, 2011). These signals are defined for each success factor, e.g., management support, user engagement, project manager experience, clear business goals, limited scope, proper requirements management process, infrastructure, formal methodology, reasonable estimates, qualified team, contract negotiation and management, and implementation.

Quality management reviews are performed in the Australian state of Victoria (Victoria DPAC, 2018). Scottish public sector projects are reviewed on the progress of change from the perspective of cost, risk, priority, benefits, and strategic fit (Scottish Government, 2013c). The main documents and processes of the project relating to the strategic context, business case, governance and project management, implemented management processes, and project requirements may be analyzed (New Zealand SSC, 2011). In the Ministry of Justice in Poland, the circulation of information is monitored (Poland Sejm, 2009). PMO in reviews may be particularly interested in the schedule at the milestone level (Prado, 2014). In many countries, the financial aspects of projects are subject to special monitoring (Iceland Althingi, 2001). It can be checked whether projects and project programs are implemented in accordance with the recommendations of the appropriate, authorized bodies (New Zealand SSC, 2011).

20.6 Engagement of Stakeholders

A stakeholder is a person or group that affects the project, is affected by the project, or believes they are influenced by it. Public sector projects have more stakeholders than private ones (see Section 16.3.8 and Subchapter 3.10).

Stakeholder involvement is needed to ensure that projects are completed as expected by them. The Scottish Government mandates the identification and participation of stakeholders in projects and project programs (Scottish Government, 2013b). The way public sector projects are implemented, especially their criticism, often affects the government's image in the public's eyes. Errors in public sector projects affect more stakeholders than failures in private sector projects. Many people and bodies influence them and may have negative attitudes toward them, which may influence the popularity of the authorities (Lefman, 2014).

Ways of gaining stakeholder support should be considered from the very beginning of the project. Project proposers should prepare a stakeholder management strategy (Smith, D., 2015). Efforts should be made that the principles and processes of project implementation are appropriate for most stakeholders (Mayer, 2014). Negative stakeholders are of crucial importance for projects, and it is necessary to know the level of their impact and communication channels (Ross, 2014). It is essential to convince stakeholders of the validity and usefulness of the project and its outcomes. Stakeholder management plans often focus on negative stakeholders (Koenig, 2015) but should also include project-friendly subjects.

Good relationships can be the basis for conviction about the project. Building good relationships with stakeholders makes it easier to manage them and, in particular, to obtain the necessary information (Gavazza, 2014). On the other hand, awareness of projects by public entities contributes to the success of projects (Clark, 2014).

Stakeholder management tools include communication and stakeholder engagement in projects. If stakeholders are not informed about the project and its goals, the Not In My Backyard (NIMBY) effect may occur (Borell and Westermark, 2016; Smith, D., 2015). Communication with stakeholders is critical to the success of public sector projects. Ninety percent of public sector project managers' time is devoted to communicating with stakeholders (Gavazza, 2014). In Maryland, public project meetings, unless otherwise specified, are open to the public (Maryland OMAG, 2021). But there are barriers to external communication in public sector projects. Some information may not be disclosed to the public (New Zealand Infrastructure Commission, 2019).

Dialogue between project teams and stakeholders, involving stakeholders in the project, and taking their views into account are essential from the very beginning of the project definition (Vermont AoII, 1996; Gaino, 2014). In Western Australia, consultation occurs with stakeholders such as indigenous peoples, subcontractors, community members, suppliers, consultants, local governments, residents, state agencies, and landowners. The consultation process identifies stakeholders, informs them, listens to feedback, joint decision-making, and delegates authority (Western Australia DPaC, 2002). In India, stakeholder inclusion and involvement for implementation must be described when reporting the project for evaluation by the Planning Commission (India DoEA, 2011).

Various approaches can be found in the area of stakeholder engagement during project implementation. Some contractors of public sector projects believe that clients and other stakeholders should define the scope of the work and then allow it to act (Caudill, 2014). The opposite approach is that stakeholders learn the opportunities created by the project during their implementation. Therefore, stakeholder involvement is more important in the implementation phase than in the initiation phase (Siles, 2014). In this phase, controversial or conflict situations involving public stakeholders may arise. In such cases, project teams consult with representatives of the local community (Mayer, 2014). The final decision is made to guarantee the public interest (Koenig, 2015).

There are two categories of public project stakeholders who receive special attention: suppliers and subcontractors, and staff involved in project management. I devote the next subchapter to them.

20.7 Actors Management

In addition to project implementation organizations and their PMOs (Subchapter 20.1), the main actors involved in managing public sector projects are external companies (vendors) and project managers. External companies can perform specialized work (related to the production of products), or they can manage projects. Public organizations involve these actors in projects in various ways.

The inclusion of external entities in the implementation of projects takes place within the framework of the applicable regulations on public procurement (Subchapter 19.1). Such regulations usually define general rules of conduct in the scope of concluding and implementing contracts between the public and private parties, not only in the area of project implementation. The main elements of such regulations include defining the types of contracts depending on the type, importance, and complexity of works, defining the rules for admitting to tenders, determining the rules for awarding contracts, a description of dispute resolution methods, and legal remedies. Some regulations prefer domestic suppliers (India Lokh Sabha, 2012). These regulations create complex legal systems; their detailed analysis is beyond the scope of this book.

20.7.1 Suppliers

In the area of actor management, external companies are qualified to implement projects.

To facilitate the management of contracts by outsourcing their execution only to qualified companies, requirements are defined that must be met by companies implementing public sector projects. Such conditions may describe the experience and characteristics of the company – then we deal with the **direct certification** or indicate the certificates that companies implementing public sector projects must have – this approach is called the **indirect certification**. The first of these types is used, for example, in Hong Kong. In order to be entered into the register of qualified suppliers, you need to demonstrate qualified management personnel, experience in project implementation, and a good financial standing. The indirect certification is used, for example, in Australia at the federal level (Australia DoFD, 2012b; 2014): the condition to be included in the list of certified contractors of IT projects is holding CMMI® (SEI, 2010), OPM3® (PMI, 2018b), or P3M3® certification (UK OGC, 2010).

A specific type of certification is used in the USA. Tenderers participate in tenders pursuant to the Federal Acquisition Regulations (FAR, USA GSA et al., 2019). The FAR requires the contractor to have Earned Value Management System (EVMS) certification stating that this system complies with the EVMS standard (USA ANSI/EIA, 1998). However, the lack of such a certificate does not exclude the company from participating in the tender. It is possible to describe EVMS in tender documents and verify it by the government team implementing the tender.

Registers of qualified project contractors are maintained based on directly or indirectly formulated requirements. The Facilities Management, Design & Construction team (Missouri OoA, 2020) maintains the database of architects, engineers, and contractors. The Vermont Agency of Administration (Vermont AoA, 2021) maintains a list of qualified suppliers in the field of IT services and project management. In Australia, at the federal level, the Department of Finance and Deregulation maintains a list of providers accredited for ICT project management maturity (Australia DoFD, 2014). The Development Bureau (DB) of Hong Kong maintains a list of certified construction contractors eligible for public tenders (Hong Kong DB, 2021). The list is divided into three categories: A – possibility to participate in tenders with a value of up to HKD 30 million, B – up to HKD 75 million, and C – over 75 million. The list is not exhaustive – on the relevant page, you can find forms for submitting a company for the contractor status of public projects. Also, the agency Buy.NSW maintains a list of pre-qualified suppliers from which agencies buy goods and services (NSW Buy.NSW, 2021).

Public sector PMOs provide contract management services. NSW PW (2013) manages relations between the public and private sectors in the field of contract planning, tender management, and contract execution, including dispute resolution. Missouri OoA (2020) is also dedicated to supporting the selection of architects and engineers for projects.

Not only the abilities of suppliers but also of buyers can be assessed. NSW government procurement agencies fall into three categories (NSW Procurement, 2020). Non-accredited agencies can make the simplest purchases, assisted by accredited agencies in the cluster. Level 1 accredited agencies may complete purchases of no more than AUD 50 million with low risk and AUD 20 million with high risk. If the value is higher, it must be approved by a Level 2 or NSW Procurement agency. Level 2 accredited agencies can purchase without price limitation. In Canada, the level of project management maturity also determines the size of projects that can be implemented by a given public institution (Canada TBoCS, 2013a, 2013b).

Internet portals of government institutions inform about contracts. In Canada, such a portal for obtaining regulatory approvals is run by the federal MPMO (Canada MPMO, 2021). The government of New South Wales (NSW Procurement, 2020) informs about the projects to be implemented in the future.

20.7.2 Project Managers

In addition to companies, project managers significantly influence implementing public sector projects. Also, for them, as for companies, the requirements to be met to be able to participate in public sector projects are defined. Based on these requirements, registers of certified public project managers are maintained.

20.7.2.1 Requirements

In the United States, at the federal level, the Office of Personnel Management (OPM) has established guidelines for positions related to project management in the public sector (USA OPM, 2019). According to these guidelines, project managers must be able to define project products and services, develop and implement a project plan, coordinate and integrate project activities, manage resources, monitor work to minimize risks, implement quality assurance processes, manage problems, make project presentations, participate in end-of-phase reviews, define project documents and procedures, develop and implement a project product implementation plan. The guidelines also define the qualification levels of project managers.

There are five levels of qualification for project managers in Wyoming based upon possessed knowledge and their abilities. The lowest level allows one to perform simple management activities. The highest level allows planning and coordinating many projects simultaneously, managing projects with a budget of more than $ 1 million, and taking responsibility for their results (Wyoming WWP, 2013).

Virginia (Virginia VITA, 2018) has a standard for selecting and training project managers, including experience, personality, and formal traits. For internal agency projects (budget less than $ 250,000), the project manager must have a minimum of 1,500 hours of experience as a project team member and must have team building and leadership potential. Requirements for managers of more complex projects increase accordingly – up to 4,500 hours worked in projects, undergoing state training, having a PMI CAPM®[3] or PMP® certificate. The most important projects may be run only by people who have previously managed less-complex projects. People with higher education are preferred. The requirements for project managers also describe in detail their skills in individual management areas (e.g., project initiation, CBA, EVA, RoI, defining the scope, building WBS, creating a schedule).

The British government systematically approached the requirements for people involved in implementing projects. UK IPA has published the Project Delivery Capability Framework (UK IPA, 2018), which defines the competencies needed to perform project management roles and career paths in project management. According to this document, there are three groups of roles: leadership roles, project implementation specialists, and business analysis and change specialists. Leadership roles include the head of the profession (responsible for project management in government), Senior Responsible Owner, portfolio, project program, project, and PMO managers. Project Delivery Professionals operate in their areas such as business case, planning, configuration, and risk management. The required technical, behavioral, and leadership competencies are specified for each role. Possible levels of competency assessment range from lack of knowledge and experience to expert knowledge and experience.

Requirements for public sector project managers concern three areas: general project management skills and experience, management skills and experience specific to public sector projects (e.g., knowledge of applicable legal regulations), and knowledge of local realities. Human culture is also important in management (NSW Infrastructure, 2020).

20.7.2.2 Professional Development

Institutions involved in implementing public sector projects as statutory objectives have professional development in project management (Missouri OoA, 2013c; India MoSPI, 2013).

In Ireland, the Public Affairs Ireland agency organizes courses for the Public Sector Project Management certificate (Ireland PAI, 2013). This training is accredited by the state agency Quality and Qualifications Ireland (Ireland QQ, 2013), one of whose tasks is maintaining the national qualifications framework. Accreditation of training in project management proves the recognition of this specialty as an official profession in Ireland.

Washington State Human Resources Department (Washington WSHR, 2013) runs training sets that lead to several certifications, for example, project planner (81 hours of training), assistant/project coordinator (75 hours), project manager (150 hours), Preparation for obtaining the PMI CAPM® certificate.

In the United Kingdom, as part of the implementation of the project management policy, the Prime Minister's Office, together with Saïd Oxford Business School and Deloitte, established the Major Projects Leader Academy (UK Oxford University, 2012), and a project academy has recently been launched by IPA (UK IPA, 2020; UK HM Government, 2021). Starting in 2015, only graduates of this school with a Master's degree in project management have been eligible to conduct major projects in the United Kingdom. A project management academy (Victoria OPV, 2021a) has also been established in the Australian state of Victoria in collaboration with the University of Oxford.

Comprehensive training for project managers is provided in Michigan, covering the basics, soft skills, and advanced topics (Michigan MDTMB, 2004). Virginia runs a project manager training program (Virginia VITA, 2018). The program includes seven basic training courses, e.g., change management in an organization, team building, basics of project management, basics of portfolio management. There is a project manager Training Program (Kansas KITO, 2020) in Kansas. In Washington state, e-learning training covers change management, negotiation basics, risk management, and cost management using Primavera tools (Washington WSDOT, 2013a). Project management training is also organized by the Human Resources Department of Washington State (Washington WSHR, 2013). Traditional (Vermont DoII, 2010; Missouri OoA, 2013c; India MoSPI, 2013) or online training (Washington WSDOT, 2013a) is provided for project managers. It is also possible to increase knowledge through mentoring (Vermont DoII, 2010). The Washington State Department of Transportation hosts the Project Management Academy (Washington WSDOT, 2013c) focused on classroom teaching, case studies, experience sharing, and open discussions. Recently, the COVID-19 pandemic has caused an intense trend of transferring training to the Internet.

20.7.2.3 Certification

Admitting individual people to manage projects may be carried out separately for each project or may be performed based on certificates, i.e., a formal confirmation of their ability to manage projects issued by authorized bodies. Requiring a certificate may be a practice when selecting the person leading major public sector projects.

A frequently used solution is the requirement to have a PMP certificate issued by PMI (Maine MOIT, 2013a). In Arizona, only holders of the Project Management Institute's Project Management Professional (PMP®) certification who, in addition, must then be trained in state project management practices are authorized to manage major public projects (Arizona GITA, 2009). Similarly, in North Dakota (North Dakota Dalrymple, 2011), the major project manager must be PMP® certified and have experience running large projects. He/she must also know the specifics of North Dakota. In Michigan, the certificate is issued after passing the final examination completing a series of training (Michigan MDTB, 2004). In Kansas, IT project manager certification is obtained after completing three PMI-compliant project management courses and passing the appropriate exam (Kansas KITO, 2020).

The Federal Acquisition Institute (FAI), which is subordinate to the Office of Management and Budget, also issues project management certificates for procurement (USA FAI, 2021). FAI cooperates with PMI, among others, to properly recognize PMI's project management certifications for federal government procurement.

Another country with an extensive certification system related to project management is the United Kingdom. Organizations accredited by Axelos issue certificates related to Prince2® (UK Axelos, 2021), including Prince2® Foundation (Basic) and Prince2® Practitioner (Advanced). Also, APMG (UK APMG, 2021) issues certificates in project management, for example, Agile Program Management (AgilePgM®) or Project Planning and Control[TM4]. These certifications are required to perform many roles in UK public project programs (UK IPA, 2018).

In South Africa, there is a register of certified project managers maintained by the South African Council for Project and Construction Management Professions (RSA CPCMP, 2013).

Certifications may only cover part of the public sector project management process, for example conducting workshops on Investment Logic Mapping, ILM (New Zealand Treasury, 2019).

20.8 Knowledge Management

The fundamental resource necessary for implementing public sector projects (as well as other types of projects) is knowledge.

Project management uses knowledge previously gathered in a given project, organization, or global project implementation environment (Gasik, 2011). Projects also create knowledge that may be useful in the future. Knowledge management collects knowledge on project management and provides it to project management processes. This knowledge can be used to build and improve methodologies and systems for project implementation. During the project implementation, the earlier experience is delivered to the knowledge management processes for possible consolidation and later use in subsequent projects. This knowledge may be codified (e.g., in the form of project management standards) and may also be disseminated through contacts of communities involved in implementing public sector projects.

20.8.1 Knowledge Codification

Knowledge generated during the implementation of public sector projects should be stored. The storing of knowledge in written form is called codification. The unit of knowledge in the management area is best practice – a practice that gives good results and is recommended for future use. GPRA-MA (USA Congress, 2010a) requires identifying best practices related to project implementation. Some public sector project management policies require collecting knowledge (NI DoF, 2014c; Scottish Government, 2013a; Canada TBoCS, 2019). At the executive level, knowledge accumulation is required by the Office of Management and Budget (USA OMB, 2021). The Texas Department of Information Technology Resource (Texas DIR, 2013) and the Virginia Information Technology Agency (Virginia VITA, 2016) have identified a single location for collecting public sector project management knowledge. Special roles may be established responsible for documenting the acquired knowledge: knowledge managers. Storing and disseminating project knowledge can be the task of PMO. Such tasks are dealt with, for example, by India MoSPI (2013), the Center for Public Project Management in Singapore (Singapore MoF, 2013), and the Infrastructure and Project Authority (UK IPA, 2020). The most important element of disseminating knowledge is its application in subsequent similar projects – before starting the project planning, the knowledge obtained in previously implemented projects should be reviewed (Canada TBoCS, 2019).

Audit chambers are involved in knowledge management. Auditors gain knowledge about project implementation methods and the mistakes made. A panel of experts supporting the definition of public project management was appointed by the American Government Accountability Office.[5]

The British National Audit Office contributes to better project implementation by publishing the knowledge obtained during project audits. Such conclusions were, among others (UK NAO, 2020), ensuring alignment of projects with the strategic goals of the organization, better understanding of time and budget constraints, developing savings plans, ensuring management of dependencies with operational activities and other projects, considering operational planning from the beginning of the project.

Reports of audit chambers describing the methods of implementing public sector projects provide knowledge of public sector project management, which may be subject to further analysis (Kwak et al., 2014). Based on the knowledge gained from audits, audit chambers create guidelines for managing public sector projects (USA GAO, 2016, 2019, 2020a, 2020b).

20.8.2 Knowledge Exchange and Transfer

Knowledge about project management is promoted in various ways by public organizations. Internet forums are held to exchange knowledge between public sector project managers (New York NYS Forum, 2013). There are meetings (Durham RM, 2013), conferences (Western Australia ExpoTrade, 2020), and seminars (India MoSPI, 2013) for public sector project managers where they can make contacts and exchange knowledge. Such conferences are also forums for exchanging knowledge between the government and private parties. Government agencies have Communities of Practice dedicated to managing public sector projects (Tasmania DSS, 2021a). The American GSA maintains a website devoted to project management tools in historic building reconstruction projects (USA GSA, 2019). There are links to documents describing this type's effective project management tools – methodological (e.g., project manager checklist) and websites devoted to completed projects. As the use of these techniques is not obligatory for projects of this type, the website should be considered a tool for promoting rather than forcing the application of project management knowledge. Portals describe how to deal with the project separately for the project manager, owners, sponsors, or project team members (Michigan MDTMB, 2013). PMOs provide links to professional project management associations (North Carolina OSCIO, 2013a). The state of Virginia maintains an open-access portal of best practice and project knowledge (Virginia VITA, 2016). The Program Evaluation Organization performs scientific analyses of programs (India Planning Commission, 2010). Additionally, the foreign knowledge of programs evaluation is acquired by DMEO (India DMEO, 2021). A knowledge management method may also be sharing document templates used in project management (Tasmania DSS, 2021b).

Project management knowledge and skills should be acquired not only by project managers but also by other personnel employed in public organizations (NSW PSC, 2020a). This knowledge should apply to project planning, coordination, and control methods. It can be assessed on a scale from foundational to highly advanced.

20.8.3 Informing about Projects

Public sector projects typically have multiple stakeholders (see Section 16.3.8). Due to the multitude of stakeholders, it is essential to ensure an efficient, easily accessible channel of knowledge transfer between entities implementing projects and other stakeholders. To gather knowledge about project work progress, repositories of public sector projects are maintained. Web portals are mainly used to communicate with stakeholders.

PMOs maintain governmental repositories of IT projects (Vermont DoII, 2010). As a result, stakeholders are adequately informed about the progress of the projects. In Latin American countries cooperating with the Inter-American Development Bank, such repositories are maintained, which are the basis for informing the public about implementing public projects (e.g., BAPIN in Argentina, Argentina MEFP, 2021).

UK IPA requires ministries to publish data on major projects and annual reports on their websites (UK IPA, 2021). Portals of the most important government projects are maintained (Alaska ADoTaPF, 2013; Vermont DoII, 2010). Each project's owner, project manager, financing method, scope (state/regional), a short description, and the planned and executed financing level are provided. California hosts a portal for public IT projects implemented by the California CDoT (2013a). The portal contains information about the leading department, cost, significance level, complexity, and project status. In Montana, the portal is operated by the State Information Technology Services Division. Information about each project consists of start and end dates, status, percentage of completion (Montana SITSD, 2021a). In Florida, the Major Project Portal is hosted by the Real Estate and Management Team located at the Florida Department of Management Services (Florida DoMS, 2013). Basic data and information on work progress are presented for each project. In Nevada, information on major infrastructure projects, i.e., those with a budget of over US$ 100 million, is presented quarterly (Nevada NDoT, 2013). Some Canadian state governments, such as Saskatchewan (Saskatchewan MeO, 2013), British Columbia (BC Stats, 2020), and Alberta (Alberta EAAE, 2013), also run portals for public sector projects. These portals are used to disseminate information about the progress of work and possible contracts for subcontractors. Smaller administrative units can also run portals. The Maine Department of Transportation maintains Transport Infrastructure Projects Portal that presents projects in the tendering phase, projects in the implementation phase, and the tendering plan (Maine DoT, 2013).

Informing the public about implemented projects, their progress, and their status is one of the essential duties of public organizations implementing projects.

20.9 Methodologies and Systems

20.9.1 Project Management Methodologies

A methodology is a structured, comprehensive set of recommendations on how to perform a specific activity. The methodology covers principles, processes, activities, implementation methods, involved roles, inputs, outputs, and other elements. There are governance and management methodologies in the project area – although this term is used less frequently concerning governance, also because governance is considered a component of project management.

Project management methodologies are structured sets of recommendations on how projects are managed. If there are standards and methodologies in the country, the methodology should meet the expectations of the standard. For example, New York State developed the New York Project Management Methodology (New York SOT, 2013) based on the PMBOK® Guide standard. This methodology is valid for all projects carried out in the state of New York (IT, software, engineering, business development, etc.). Also, based on PMBOK®Guide, Michigan (Michigan MDoIT, 2004), Pennsylvania (Pennsylvania PoA, 2013), and Washington state (for transport projects, Washington WSDOT, 2013a) developed their methodologies. California Project Oversight and Consulting Division maintains PMI-based and CMMI®-based (SEI, 2010) policies and procedures in project management. The PMI standards are the basis for Vermont's IT project management methodologies, policies, standards, guidelines, procedures, tools, and patterns (Vermont DoII, 2013).

Management methodologies may supplement the governance of projects or be independent of it. If a specific management methodology has been adopted for the entire organization, it becomes an element of governance. If governance stops at the gate reviews level (Section 20.5.1), the management methodology is, in a sense, standalone or subordinated to governance. In this case, the governance process determines the sequence of gates and checks through which the projects pass, and the management methodology determines what the project should do to pass the gates successfully. For example, in Texas, each TPDF process gateway (Texas DIR, 2013) has the necessary processes, techniques, forms, and tools. With this approach, the management methodology complements the governance process.

Management methodologies may also include governance elements as their integral components. The governance process does not then exist independently. For example, elements of the Prince2® methodology are the definition of the project management hierarchy and the rules of transition between phases (UK OGC, 2017) – they are the elements of governance. In Montana, the Project Life cycle Framework (Montana SIT PMO, 2013c) methodology was developed as the main method, covering the business process cycle, project management cycle, purchasing cycle, and product development cycle.

In California, the Californian project management methodology was developed for IT projects (California CDoT, 2013a). Both government agencies and state vendors must use the methodology. The four main steps identified by this methodology are project initiation, planning, execution, and closure.

Missouri has developed its project management methodology (Missouri OoA, 2013b), one of the goals of which is to introduce a common language in this area. The methodology includes policies for individual management areas. The most important include configuration management (necessary to ensure that project baselines are managed and changes are controlled), planning, tracking implementation, requirements management, and risk management.

PDD Vermont maintains a transportation project implementation process (Vermont AoII, 1996), consisting of the selection, approval for execution, design, construction, and completion phases. In parallel, the Vermont DoII (2013) for IT projects defines the processes defining the business case and related metrics to measure project success. This shows that different methodologies can be used for different types of projects.

In Tasmania, guidelines for Tasmanian Government Project Management have been developed (Tasmania OeG, 2011). The guidelines contain two main sections: general project information and guidelines for managing the eleven project elements. The Tasmanian ICT Policy Board recommended applying these guidelines to all projects implemented by the Tasmanian government and its agencies.

In Pakistan, the Planning Commission developed a methodology for managing development projects (Pakistan Planning Commission, 2010). The methodology describes how to initiate projects in the context of the country's medium-term development plan and how to manage projects. The project life cycle according to this methodology consists of identification, Feasibility Study, preparation, evaluation, approval, implementation, and monitoring and evaluation.

Indian MoSPI (2013) developed a project implementation manual. The Victoria Department of Premier and Cabinet has developed a management framework used in their projects (Victoria DPAC, 2016). Queensland Treasury (2019) maintains major government policies such as the Project Assurance Framework. The PMO of the Virginia Department of Transportation maintains policies and procedures and ensures the application of best project management practices (Virginia VDoT, 2021).

Methodologies related to portfolio management are also defined, for example, regarding the project submission process, project evaluation, and project financing, including standards and criteria to ensure that projects are in line with their business objectives (Vermont DoII, 2010). New York OoIT (2013) prepares and applies systems for evaluating project initiatives.

A set of forms necessary for project implementation may be considered a kind of a fragmented methodology (RSA DoPW, 2013; Vermont AoDS, 2021). A set of forms (e.g., project brief, work progress report, checklists) and information necessary to be included therein define the set of activities required to be performed in the project.

Methodologies are an essential element of project management in organizations. The implementation of any form of project management methodology improves project management effectiveness. It demonstrates a higher level of project management capabilities compared to organizations where the management method is defined separately for each project.

20.9.2 Project Implementation Systems

The project implementation system is a set of organizational solutions introduced by the government for the implemented projects throughout their life cycle. The subject of such a system is each project. A component of such a system may be governance processes and project management methodologies, which become mandatory. An element of the system may be a database containing projects' descriptions and other information, for example, a set of gates through which projects have passed. Such systems apply to all projects or projects of a particular type performed in a given country, for example, technology or infrastructure projects.

In Argentina, there is a National Public Investment System (Sistema Nacional de Inversiones Publicas, SNIP, point 19.2.3.1). Its element is the IT system named the Bank of Public Investment Projects (Spanish BAnco de Proyectos de Inversión Pública, BAPIN), covering all public projects (Argentina MEFP, 2021). Projects are registered in BAPIN from the moment of submitting the initiative (see also point 20.4.2.4). The distinguished type of projects are projects implemented for the direct needs of Public Administration. The National Directorate of Public Investment (Spanish Direccion Nacional de Inversion Publica, DNIP; Argentina DNIP, n.d.) operates within the Ministry of Economy and Public Finance structure. DNIP is responsible for the functioning of SNIP, including project management methodologies in Public Administration. Each province in Argentina has the National Public Investment System equivalent, which covers all public sector projects in that province.

The National Project Management System (NPMS) has been implemented in Canada. It was developed by Public Services and Procurement Canada (Canada PSPC, 2019). NPMS is compliant with the PMBOK® Guide (PMI, 2017a). All IT and real estate projects must be implemented in accordance with the NPMS. NPMS requires each project to go through three phases: initiation, identification, and implementation (including completion). There are two versions of NPMS: full and lightweight.

In New Zealand, the investment management system comprises four phases of project execution: thinking, planning, executing, and reviewing. The system also includes other necessary components such as information, capabilities, and behaviors (New Zealand Treasury, 2021a). The system's functioning is supported by Investment Management and Asset Performance (IMAP) team (New Zealand Treasury, 2021b).

The earlier mentioned Norwegian Project State Model (Samset and Volden, 2013) may also be treated as a project implementation system (focusing mainly on project front-end).

Software applications support project management systems. PMOs maintain such tools. The Project Services team (NSW PW, 2013) runs a portal with web applications supporting project management, collaboration, communication, and information management. Victoria OPV (2016) supervises systems and technologies supporting project management. PMOs maintain a software tool of the Project Portfolio Management class and make it available to project teams (North Carolina OSCIO, 2013a; Virginia VITA, 2013). The Pakistani Planning Commission is responsible for maintaining the Project Management and Evaluation System (PMES, Pakistan Planning Commission, 2010). The Bangladesh Implementation Monitoring & Evaluation Division (IMED, Bangladesh MoP, n.d.) operates the Project Management Information System (PMIS). Michigan DoITS (2000) defined the applicable state standard for IT support in project management.

20.10 Model of Delivery Territory

The structures and processes described in the previous sections together form the area of Delivery of public projects.

Public projects work for the benefit of society, smaller communities, or public organizations. Public projects involve their stakeholders (Subchapter 20.6) who may belong to or come from outside these groups. The composition of public organizations' project portfolios is shaped by their portfolio management processes (Subchapter 20.3). Products and services are usually delivered to projects by external suppliers (Section 20.7.1), usually from the private sector. Projects are subject to the Management of Projects processes located in the center of the Delivery Territory.

Management of Projects (Subchapter 20.4) is based on the use and storage of knowledge. Management of Projects is implemented in project implementation systems (Section 20.9.2) supporting project management methodologies (Section 20.9.1). On the other hand, Management of Projects is influenced by two groups of entities: project managers (Section 20.7.2) and PMOs (Subchapter 20.1) which support and control this group of processes. A special process that is particularly important in public projects is Assurance (Subchapter 20.5), which tries to increase the chance of success of public projects.

Another special process of Management of Projects is Knowledge Management (Subchapter 20.8). Based on the project management knowledge, methodologies are built and management support systems are implemented. This knowledge is provided to project managers as well as directly to the Management of Projects processes. Project stakeholders are usually interested in the progress of project implementation which is also provided by the knowledge management processes.

PMOs (Subchapter 20.1) taking part in the implementation of many processes shape the methods of project management in public organizations. PMOs perform many functions. Firstly, it directly supports the implementation of the Management of Projects processes. PMOs also manage project management systems and methodologies. They support portfolio management in the organization as well as management of actors (Subchapter 20.7).

The structure of the Delivery territory is schematically shown in Figure 20.9.

Delivery Territory

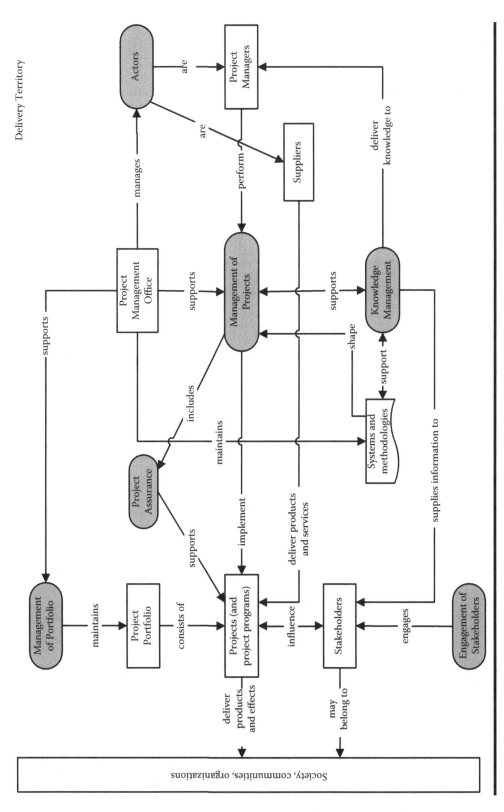

Figure 20.9 The Delivery Territory.

Notes

1. The GPM3® described in Chapter 12 is a tool for such assessment.
2. Retired in 2009.
3. CAPM® is a registered trademark of Project Management Institute.
4. AgilePgM and Project Planning and Control are trademarks of APMG International.
5. The author of this book was a member of this panel from 2014 to 2020.

Chapter 21

Support Area

Support are activities performed by entities external to projects, performed when the implementation of the project by its team gives worse results than if some external entity joined its implementation. Support is related to problems that could have occurred or had actually occurred. Support is provided by specialized organizational units. If one of PMO's tasks is building a schedule for each project, this is a standard service. However, if the construction of the schedule is the project team's responsibility, but there is an institution that can be turned to for help in particularly difficult cases, we are dealing with support.

There are two main types of support: problem-oriented (ex-post) and process-oriented (ex-ante).

21.1 Problem-Oriented Support

Problem-oriented support, sometimes called ex-post support, is solving problems that block the implementation of projects after they occur.

Project Monitoring Group (PMG) works in India. Projects (private or public) wishing to receive support from PMG must first register on the Cabinet Committee for Investments (CCI) electronic support platform. Private entrepreneurs submit them through the sponsoring ministry or by the ministry itself carrying out the project. The CCI reviews the application. The opinion is transferred to PMG, which decides on registration. The number of approved projects per month in 2014 reached 438 (India CSID, ibid). The projects proposed for rejection by the PMG are also transferred to the PMG. The registration enables the transfer of problems to PMG, the solution of which will be assisted by the Indian government. PMG supports problem-solving that should be solved by the Cabinet Secretariat and CCI (India CSID, 2014). It works with ministries, states, and investors through e-CCI. PMG is the single point of contact between the project teams and the state institutions that can help in the acceleration of the implementation of stalled projects. PMG itself does not solve problems but directs them to the appropriate ministries and monitors the problem-solving process, including meetings of CCI staff, ministries, and project representatives.

In the US state of Maryland, project support is one of the Oversight Manager (OM) functions. The OM is the single point of contact for project problem solving for the project manager. For example, when there is a contract problem, the project manager turns to the OM, who looks for the right person to solve the problem (Ross, 2014).

DOI: 10.1201/9781003321606-25

In Malaysia, the National Action Council (NAC, Malaysia Dahlan, 2015) is responsible for identifying and overcoming the constraints of development projects. PMOs also perform project recovery functions (Vermont DoII, 2010), which may be called performing direct interventions that address problems. British IPA ensures cooperation between ministries in solving problems in projects. The PMO can escalate conflicts to a higher authority (UK Cabinet Office, 2011).

21.2 Process-Oriented Support

Process-oriented support is sometimes called project facilitation or ex-ante support. Facilitation is guaranteeing support throughout the entire project process or its specific part – most often, it is project initiation – by a specialized entity. In this approach, the project is usually assigned a person to assist it.

The process-oriented approach is taking place, for example, in Alaska, at the federal level in Australia, in Western Australia, and Victoria state.

The minimum version to facilitate the implementation of projects is the support in preparing documents needed to obtain appropriate permits (California DoGS, 2013). When submitting an initiative in Alaska, a coordinator is assigned to assist the proposing entity in initiating the project, including obtaining permits (Alaska OPMP, 2013).

In Australia, at the Commonwealth level, the Government's Major Projects Facilitation Agency (MPFA, Australia DISER, 2021) exists at the Department of Industry, Science, Energy, and Resources. This agency provides services for projects carried out by private companies. Its operation shows how vital projects implemented by non-governmental entities are for the Australian government. In this way, private projects become partially governmental (Subchapter 16.2). The importance of the project for the development of Australia is decisive for granting the status of the major project (which enables the use of MPFA services). Each major project has a supervisor from Australia's central administration. MPFA provides services in the field of obtaining permits from state institutions required for the initiation and implementation of the project.

In Western Australia, the preparation of the most significant state initiatives may be facilitated by an employee of a specialized cell. His work aims to support the production of the main Project Definition Document (PDD). The level of involvement depends on the importance of the project. There are three levels of support. The first level is intended for projects that can be executed in an existing environment. The support consists of initial consultations and the presentation of the project to appropriate agencies. The second and most frequently used level is support for projects of significant or strategic importance. A project manager is appointed to assist the proposer with government-related aspects of defining PDD (e.g., site matters, regional issues, coordination, and interaction with government agencies). The proposer receives detailed information on the progress of works related to the project's approval. The third level of support is intended for specific, complex projects of strategic importance to the government. Here, a senior project manager is appointed who, in addition to the second level support, supports the initiative in coordinating government approvals and negotiating a Government Contract (Western Australia DoSD, 2013b; Western Australia DPaC, 2009).

A higher-level institution can support projects when they are threatened with problems. Advisors are assigned to projects, if necessary, whose main task is to verify and supervise project management. Advisors can, among other things, assess the ability of their managers to run a project, recommend decisions to the state IT director regarding the further course of the project, check whether the agency is prepared for the next steps of the project, identify risks, recommend corrective actions and possibly escalate them to higher management levels, be a mentor to the project manager, provide data necessary for inclusion in project plans, check tender documentation (North Carolina OSCIO, 2013a).

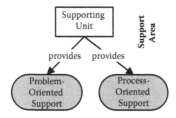

Figure 21.1   The Support Area.

It is also possible to appoint a "patron" of major projects – his task is to represent the project in the government forum and other communities (Victoria MPV, 2013). In the US state of Alaska, project initiatives are assigned a coordinator to help initiate the project, in particular obtaining the necessary approvals (Alaska OPMP, 2013).

The support area is schematically presented in Figure 21.1.

Chapter 22

Development Area

Projects succeed or fail. Therefore, it is legitimate to ask what determines the success of projects, including the public sector ones. Many researchers deal with the topic of project success in different ways. Also, public sector projects attract the attention of researchers. They try, among others, to answer the question of how to manage public sector projects in certain countries efficiently (Koenig, 2015; Bay and Skitmore, 2006; Simangunsong and Da Silva, 2013; Juchniewicz, 2012; Young et al., 2014; Prado et al., 2009; Prado et al., 2015; de Souza Silva and Gomes Feitosa, 2012).

If it turns out that public sector projects do not meet expectations all too often in a given country, governments may respond in different ways. In this chapter, I describe some practices aiming at improving the methods and effects of public project implementation.

22.1 Advisory Bodies

Public administration institutions establish bodies whose task is to advise on the development of project implementation processes. Advisory bodies play a fundamental role there. Such bodies can make improvements to project implementation processes on their own. On more important issues, usually governance issues, they advise governments that, within their powers, make decisions that make more significant improvements.

In the US federal government, based on the PgMIAA (USA Congress, 2015), the Program Management Policy Council (PMPC) was created for working, among others, to improve project programs and project management practices.

The Canadian Project Management Advisory Council (PMAC, Canada PSPC, 2009) was established to strengthen project management capabilities. PMAC's goal is to maintain and develop National Project Management System. To this end, it should disseminate best practices, identify similarities in procedures, tools, documentation, and review project practices. PMAC should also review project management methodologies to create the conditions for achieving project milestones on time. Innovative approaches to project management should also be explored. In 2015, PMAC was replaced by the NPMS Continual Improvement Team, having similar tasks and powers.

At the state level, a Project Management Methodology Group in North Carolina is dedicated to developing and improving processes, procedures, and benchmarks for projects (North Carolina OSCIO, 2013b). The advisory body may be responsible for defining state policies, procedures, and methodologies, including project management (Kansas Information Technology Executive Council, ITEC, Kansas OoITS, 2008a). In the Australian state of Victoria, the Advisory Board (Victoria OPV, 2021b) advises bodies involved in project management.

In addition to improving methodologies, advisory bodies also provide ongoing advisory services supporting project implementation, for example, advising on the use of project management methodologies and supporting the implementation of project management (e.g., Project Management Advisory Committee, Tasmania OeG, 2013). They also support exchanging information between agencies, PMOs, and project managers, try to remove barriers to project management, and support project managers in their home agencies (e.g., North Carolina Project Managers Advisory Group; North Carolina OSCIO, 2013b).

22.2 Development Directions

What is the desired target state for managing public sector projects? Countries wishing to optimize the benefits of public sector projects clearly define their strategic goals in this area and build action plans based on them.

The goals and methods of developing public sector project management systems are defined differently in different countries. Project management services development directions may be determined based on the results of customer satisfaction surveys of these services (North Carolina Mays and Bromead, 2012) or based on project management audits (Australia ANAO, 2011; Poland NIK, 2013). In most countries, such strategies are developed based on public sector project management status analyses, often in the broader context of a country's strategic development (Arizona Brewer et al., 2013).

Among the directions of development of public sector project management, there are general, business, managerial, operational, and knowledge-related goals.

General recommendations include developing a public sector project management strategy (Australia ANAO, 2011). A very general strategic goal may also be to strengthen the national way of managing projects and project programs (Arizona Brewer et al., 2013), increase the capability to implement ICT projects (Australia DoFD, 2012a), or develop and simplify processes and recommendations for project management (Canada TBoCS, 2010a; Poland NIK, 2013).

In Canada, the goal of the NPMS Continual Improvement Process (NPMS CIP, Canada PSPC, 2014) is to make NPMS the world leader in project management. The implementation of this process is mandatory for all institutions using the NPMS. By the assumptions of this process, all its documents are alive, which means they can be modified at any time. The same principle applies to processing web content. The NPMS Continuous Improvement Team leaders review each proposed change. If a change affects only one department, it can be approved based on this opinion by the responsible project implementation director. If the extent of the change is more significant, it must be approved by the Steering Committee for NPMS Continuous Improvement in the Public Services and Procurement Canada (PSPC). If a change could materially affect the entire NPMS, it must be reviewed and approved by the Project Management Advisory Board.

The group of business development goals includes planning project management to obtain the highest possible return on value (Washington WSDOT, 2013d). Projects must be linked to the country's strategy (Maine OIT, 2004). This group also includes achieving the appropriate effects of projects and project programs while reducing the risk for stakeholders and taxpayers (Canada TBoCS, 2019). Projects should be implemented in such a way as to achieve goals related to time and budget (Washington OCIO, 2011; Maine OIT, 2009; Canada TBoCS, 2010a, 2019).

The most often observed direction of project management development is the managerial direction. There should be an improvement in the management of individual phases, for example, planning (Australia ANAO, 2011; Kansas OoITS, 2008b) or completion of projects (Kansas OoITS, 2008b). Projects are to be managed transparently for their stakeholders (Arizona Brewer et al., 2013). Reports on the implementation of public sector projects are to be made available to the public (UK Cabinet Office, 2013). Documentation and reporting should be simplified (Arizona Brewer et al., 2013; North Carolina Mays and Bromead, 2012). Projects need to (better) manage risk and configuration (Kansas OoITS, 2008b) and contracts (Washington OCIO, 2011; Maine OIT, 2009).

The group of operational goals includes those that require the implementation of specific organizational solutions, for example, the creation of a university educating public sector project management staff (UK University of Oxford, 2012), the creation of the Major Projects Authority, or the definition of a portfolio of major projects, the implementation of which will be reported directly to the government (UK Cabinet Office, 2013). In the Canadian state of Ontario, a review of public sector project management by a dedicated team recommended introducing a gate review process (Grabowski, 2014).

The group of goals related to knowledge management includes planning to increase knowledge about project management, for example, through the implementation of training (North Carolina Mays and Bromead, 2012; Maine OIT, 2009), promoting knowledge about project management models (North Carolina Mays and Bromead, ibid). Recommendations for using knowledge gained in projects for future projects are to be developed (Australia ANAO, 2011). The development of knowledge about managing public sector projects may result from the activities of specially appointed teams (Norway NTNU, 2013).

22.3 Development Modes

The concept of the development mode refers to entities whose functioning is subject to optimization. There are two modes of development: the individual mode and the system mode.

22.3.1 Individual Mode

In the individual mode, improving public projects' implementation consists of improving their implementation in each organization separately.

In Australia, for institutions subject to the Financial Management and Accountability Act (Australian Government, 1997), P3M3® was adopted as a methodology for assessing and improving project management skills. These organizations must perform an annual maturity assessment and report the results to the Ministry of Finance (Australia DoFD, 2012a). Canada has implemented a policy that requires the continuous evaluation and improvement of project management in public organizations (Canada TBoCS 2010, 2019). The purpose of establishing an MPA in the UK was, among other things, to work with departments to build the capability of project and project program management.

In the United States, the Program Management Improvement and Accountability Act (USA Congress, 2015) mandated that each federal organization have a program management improvement policy and a management representative responsible for increasing the role and importance of program managers.

22.3.2 System Mode

In the system mode, the whole system of public project implementation in a given country under-stood as a whole is improved. For example, the government determines how project assurance in each organization is to be performed. The government can also improve processes unrelated to individual projects, such as streamlining the certification process for project managers.

The system-mode practice may be placing the order to improve project management in long-term strategies and plans. Such an improvement is required by the development strategy of Hawaii (Hawaii ETS, 2012). In India, the 12th five-year plan also mandated streamlining project management (India Planning Commission, 2011).

The system-mode practice described in PgMIAA is the establishment of the Program Management Policy Council at OMB. The Deputy Director of OMB is responsible, inter alia, for establishing a strategic program and project management plan in the federal administration. In the UK, the Programme and Project Management Council (PPMC) has been established to improve the way projects are carried out in the UK government. Currently, the functions related to improving the implementation system of British projects and programs have been taken over by the Infrastructure and Project Authority, analyzing and drawing conclusions from the implementation of the major British projects (UK IPA, 2021). The purpose of the advisory body (PMAC) may be to advise on improving project management methodologies (Tasmania, OeG 2013). In Scotland, a Project and Program Management Center of Excellence has been established with the task of, among other things, improving project and project program management capability (Scottish Government, 2013b). National audit chambers may also organize advisory bodies improving public project im-plementation – a group of experts developing recommendations for improving project management operates, for example, at the Government Accountability Office (USA Congress, 2015). In Canada, the Comptroller General, who is also the leader of the project management community, is working on developing project management practices (Canada TBoCS, 2019, 2021a).

Process practices of the system-mode streamlining of public project processes are visible, for ex-ample, in the project management methodology of the state of Michigan (Michigan MDoIT, 2004), which emphasizes gathering knowledge from the implemented projects to improve the implementation of future projects. In Norway, the Concept Research Program (Norway NTNU, 2013), located at the Norwegian University of Science and Technology in Trondheim (NTNU), supports and analyzes the development of the project initiation system. The Norwegian Ministry of Finance finances this pro-gram. It aims to improve resources' use and enhance the effects of large infrastructure projects.

The governmental PMOs are also responsible for improving project management systems and methodologies (India MoSPI, 2013; Missouri OoA, 2013c). Australia DoFD (2013) dealt with one aspect: ensuring the effectiveness of reviews, i.e., improving processes in this area. In Missouri, PMO owns a formal project management methodology improvement process (Missouri OoA, 2006). An element of process improvement is identifying best practices in the area of project management (Bangladesh DoPW, 2013). Bahrain SPD (2013) ensures the application of current engineering design standards, codes, management standards, value management practices, efficient project man-agement techniques, and quality management principles and ensures that project implementation complies with the highest international standards.

Maturity models can be used in both individual-mode and system-mode optimizations. The concept of the maturity model, its importance for the management of organizations, and the previous attempts to use maturity models in Public Administration are described in Chapter 24. A maturity model for the public project's implementation is described in Chapter 25.

The area of public project management development is presented in Figure 22.1.

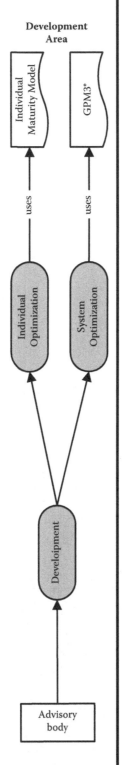

Figure 22.1   The Development Area.

Chapter 23

Governmental Project Implementation System and Managing It

23.1 Governmental Project Implementation System

Governments and their subordinate Public Administrations have developed effective project management methods. I have analyzed such practices in over 70 countries. Based on these practices, I defined four main public project management areas and territories: Delivery, Governance, Support, and Development of project management practices at the government level.

Chapters 19–22 contain a systematic, structured description of actually implemented (with the indication of the organizations implementing them) project management practices applied by governments in these areas and territories. They cooperate to ensure the effective and efficient implementation of public projects in the country. These areas were identified and characterized based on an analysis of practices actually carried out in governments.

The Governance Territory defines the main principles of implementing public projects in the state. These rules shape the manner of performing processes in other areas of public project implementation.

The Delivery Territory includes processes for implementing public projects (as well as project programs and project portfolios). This is the most extensive Territory. It consists of the areas of Management of Projects, Management of Portfolios, Engagement of Stakeholders, Project Assurance, Knowledge Management, and Actors Management. The Delivery Territory is influenced by all other areas and territories of government activities related to implementing public projects. This is what Governance, Support, and Development are for so that governments can implement their public projects in the best possible way.

The Support Area supports projects so that they can achieve success if they cannot guarantee its achievement by themselves.

The Development Area identifies the areas in which the implementation of public projects can be improved and indicates how these goals can be achieved. The development may concern Delivery, Governance, Support, and itself, too.

DOI: 10.1201/9781003321606-27

Presenting in a structured form solutions used in public project management at the government level can be a material that will significantly improve public project management. Governments can choose from among them for implementation those that will be particularly useful for them.

Since the entirety of the administration's activities, including project management, results directly or indirectly from actions or their omissions from the government's decisions, collectively, all these territories, areas, functions, and practices in the given country constitute the Governmental Project Implementation System (GPIS). This is the 2nd generation GPIS covering all practices related to implementing public projects in the state. It significantly extends the concept of 1st generation GPIS briefly mentioned in Section 20.9.2, which covered only project implementation processes, from initiation to evaluation. The 2nd generation GPIS includes many more elements than the 1st generation ones. The 2nd generation GPIS presented here also includes, for example, vendors and project managers, shaping governance, project management offices, developing project implementation or supporting project implementation, and other elements described above.

To improve public project management in the country, it is necessary to deal not only with the project process itself but also with those elements of the environment that support this process.

GPIS of the 2nd generation also covering its development processes is presented schematically in Figure 23.1.

Figure 23.2 shows the comparison of the 1st generation and the 2nd generation GPIS. The former is included in the latter or, speaking otherwise, the 2nd generation GPIS is a significant extension of 1st generation GPIS.

In the rest of the book, if I do not use the additional term, when I refer to GPIS, I will mean the full 2nd generation GPIS.

23.2 Governmental Project Management

Everything the government does with its GPIS, especially its building, maintaining, and improving, will be called Governmental Project Management (GPM).

The GPM was created for the government level as an analogy to organizational project management (OPM) (PMI, 2018a). OPM deals with the organization of all sectors. GPM is a higher management level than OPM – for public sector projects.

It is worth comparing GPM with lower levels of project management. The subject of project management is one project. OPM is responsible for the organization's projects, and GPM – for all public sector projects in the state. The project manager is responsible for project management, the organization's management board is responsible for OPM, and the government of a given state is responsible for GPM. As part of project management, the processes and structures of a single project are shaped. The processes, structures, and other factors enabling the implementation of projects (enablers) in a given organization remain within the scope of the impact of OPM. On the other hand, the object of GPM activities is the elements mentioned above shaping the methods of managing all public sector projects in a

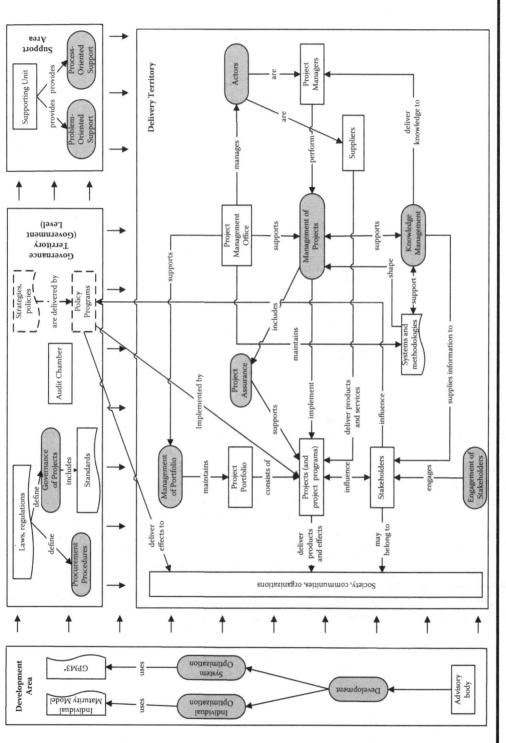

Figure 23.1   Governmental Project Implementation System.

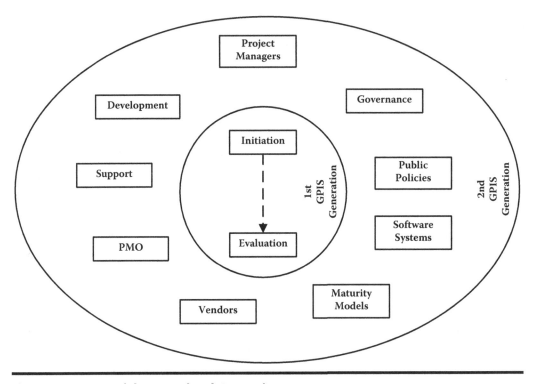

Figure 23.2   GPIS of the 1st and 2nd Generations.

given state. The higher the management level, the less direct the influence of the governing body is. The project manager's impact on project management is direct – he/she makes (together with the management team) all decisions. The organization's management board within OPM deals directly with the management of individual projects and shapes the organizational environment of management. As part of GPM, the government creates a project management environment; its impact on individual projects does not take the form of direct management (if the government deals directly with a specific project, it is an activity under OPM, not GPM). The subject of performance comparisons – and hence knowledge transfer – for projects are other projects, for organizations – other organizations, and for governments – other governments. Project management and OPM are intended for projects of any sector, while GPM is exclusively for the public sector.

The differences between project management, organizational project management, and Governmental Project Management are shown in Table 23.1.

Table 23.1 Levels of Project Management

	Project Management	*Organizational Project Management*	*Governmental Project Management*
Subject	Project	Project, project programs, organizational portfolios	Governmental Project Implementation System
Responsible	Project manager	Board	Government
Components	Project structures and processes	Organizational structures and processes	Laws, other regulations, institutions
Way of management	Direct	Direct/Indirect	Indirect
Benchmarking, knowledge transfer	Other projects	Other organizations	Other governments
Sector	Any	Any	Public

THE WAY AHEAD. HOW TO IMPROVE IT?

As a result of its development practice, the GPIS is becoming more and more advanced and efficient, i.e., more and more mature. It reaches higher and higher levels of maturity. The set of ever-higher maturity levels is called the "maturity model." This part describes how to create an increasingly mature GPIS. I use the Governmental Project Management Maturity Model (GPM3[1]®) for this purpose.

This part of the book begins with Chapter 24 explaining the maturity model concept. I will cover GPM3® in detail in Chapter 25 of this book.

Note

1. GPM3® is a trademark registered to Stanisław Gasik.

DOI: 10.1201/9781003321606-28

Chapter 24

Maturity Models

Economic growth is conditioned by the methods of organizing institutions responsible for development. Differences in levels of development are due to differences in the way institutions are organized and managed (Acemoglu and Robinson, 2008).

24.1 Ways to Improve Maturity in Organization (TBC)

Improving management methods, i.e., upgrading the level of development, can be done in several ways. The existing domain knowledge is always an indispensable element of improving management. It is stored both in documented form and in the minds of the people involved in the work (Polanyi, 1958, 1996; Nonaka and Takeuchi, 1995; Hansen et al., 1999). The level of systematic improvement of management is determined by the way this knowledge is incorporated into the processes carried out in the organization. Process improvement maturity has four main ways (levels) (Gasik, 2008). Each of them is more advanced than the previous one. Their description follows and continues in Subchapter 24.8.

24.1.1 Intuitive level

Management has always been improved by promoting people's knowledge in processes' implementation. Such improvement occurs daily, with direct development of better and better solutions. Usually, they relate to the implementation of specific functions or domain processes. Examples of practices that improve management at this level may be the work manager's instructions to those responsible for components of work: "pay more attention to determining how long your work will take" or "when receiving the work, pay more attention to whether the products have been well made." More advanced practices in this way of improving performance may be based on the use of knowledge about a specific area or function, for example, how to build a project schedule or how to involve stakeholders in the project. From the point of view of the impact on the entire managed system, this level of performance improvement is called **Intuitive** or local. This is level 1 of management improvement.

DOI: 10.1201/9781003321606-29

24.1.2 General Level

At some point in development, it was noticed that the improvement of the management level should be approached more systemically. An organization will perform better if a holistic approach is used to improve its functioning. General performance improvement systems have been developed, such as the ISO 9000 (ISO, 2015) and SixSigma (Tennant, 2001) approaches. The processes of the ISO 9000 standard in their basic form can be appropriately implemented in any organization. For example, the ISO 9000 standards describe general processes applicable to any organization. But it is not oriented toward specialized domain knowledge, for instance, regarding a specific set of processes or specific roles needed in an organization like project management processes or software development processes. This is level 2. **General** approach to improving management.

24.1.3 Specialized Level

Approaches based on knowledge in relevant fields try to fill the gap in the knowledge needed to perform a specific function, to operate in a particular area. The ISO 21500 series standards (ISO, 2012) were created directly based on ISO 9000 standards but for the area of project management. They cover the compendium of project management knowledge. Other holistic, specialized approaches to gathering project management knowledge include, for example, PMBOK® Guide and Prince2®, repeatedly cited in this book. The use of any of these methodologies is a more effective way of increasing management maturity in an organization than the use of general standards of the 2nd level. This is the 3rd level to improving maturity called **Specialized**.

But implementing specialized standards or methodologies creates problems. This is usually a tremendous job that takes a long time and consumes many resources. Attempting to implement all processes simultaneously may fail due to the need to control many new processes simultaneously. Andrews (2010) points out that reforms fail because all best practices are implemented simultaneously.

It is now time to introduce the concept of the maturity model. After its presentation, I will return in Subchapter 24.8 to classifying management improvement ways.

24.2 What Is a Maturity Model?

The maturity model is one of the tools widely applied for improving organizational performance. They have their roots in the works of Nolan (1973, 1979) and Crosby (1979). The first widely recognized maturity model was the Capability Maturity Model (CMM, Paulk et al., 1993) developed for software development. Since then, a plethora of maturity models has been published. Wendler (2012) lists 237 maturity models, but currently, this is undoubtedly not a complete list. Professionals recognize and accept project management maturity models and their usage (Albrecht and Spang, 2016).

There are several definitions of maturity models. Maturity models contain the best practices from a given area (Wendler, 2012). These practices are grouped into hierarchical sets of elements (maturity levels) describing the features of more and more effective processes (UK OGC, 2010). They describe typical patterns in developing organizational capabilities (Pöppelbuß et al., 2011). Each maturity level describes the progress in achieving excellence in a given process area (Marx et al., 2012). They characterize the effective processes at different levels of development of the entities implementing them (Pullen, 2007). Maturity levels build on each other (Wendler, 2012). Each higher maturity level describes more formal and sophisticated processes than the lower ones (Bititci et al., 2015). The maturity model also defines the breakpoints between the maturity levels and the methods of transition between these levels (Pullen, 2007). According to CMMI® (SEI, 2010), maturity models contain essential elements of effective processes and describe an evolutionary improvement path. All maturity models serve for assessing the management practices, processes, and competencies of organizations performing them. They are the basis for comparison in a particular business sector or other organizations (Tembo and Rwelamila, 2008). Maturity models may be applied for selected processes, process areas, or whole organizations (Becker et al., 2009).

Currently, the two best-known maturity models are probably the Software Engineering Institute Capability Maturity Model Integration (CMMI®) (SEI, 2010) and the Organizational Project Management Maturity Model (OPM3®[1]) (PMI, 2018b).

Maturity models, following the CMM/CMMI® formula, often consist of five levels: Initial, Managed, Defined, Quantitatively Managed, and Optimizing.

At the **Initial** level, organizations do not implement any well-defined practices. Success depends on the commitment and skills of the team members. At the **Managed** level, processes are implemented that deliver expected results based on the defined requirements. These processes are defined separately for each application in the analyzed area. The **Defined** level is the one at which the organization as a whole has a defined set of processes in a given area. Defined processes are tailored, when needed, to the specificity of instances of the work performed. The **Quantitatively Managed** level is the one at which the organization focuses on the effectiveness of processes and can manage them in such a way as to estimate and achieve the main parameters of work implementation, for example, time and budget. The **Optimizing** level is the one at which the organization implements processes ensuring continuous improvement of their work processes.

Each subsequent level of maturity contains the capabilities of the previous, lower levels. The general scheme of CMM-based maturity models is shown in Figure 24.1.

For instance, maturity models may be used to analyze the bidders' ability to perform specific work, such as project delivery. This idea prompted the US Department of Defense to commission Carnegie Melon University to work, which resulted in developing the first maturity model: the CMM. The use of the maturity model when selecting a supplier increases the probability of successful implementation of the purchased services. The Defined level is considered the maturity level, which significantly increases the likelihood of choosing the right tenderer.

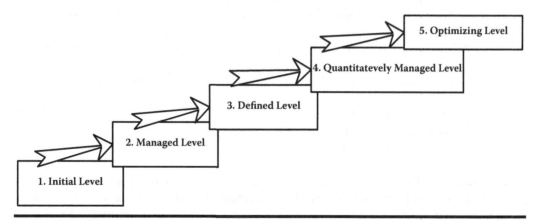

Figure 24.1 An Example of a Structure of Maturity Model.

The development and use of maturity models are robust and widely accepted tools for assessing an organization's ability. Interest in these methods has grown significantly recently. The benefits of using them justify the popularity of the maturity models. Higher project maturity levels result in higher project efficiency (PWC, 2012). The organizational maturity of project management improves the effects of project implementation (Shi, 2011; Crawford, 2006). In other words, projects require a well-prepared organization to succeed (Richardson, 2010). Several researchers, e.g., Besner and Hobbs (2008), Yazici (2009), Jiang et al. (2004), and Spalek (2014), showed that increasing an organization's project management maturity level positively influences the organization's performance. The purpose of maturity models is to describe organizational conditions to achieve excellence in performing processes (Wendler, 2012). The use of maturity models may enable organizations to achieve the benefits of improved capability at all levels of maturity (UK OGC, 2010). Applying maturity models turns immature processes or whole process areas into ones with high quality and effectiveness (Garzás et al., 2013). Maturity models can be used to indicate areas on which the management of an organization should focus its efforts (Pullen, 2007).

Along with the publication of documents on the comparison and analysis of maturity models (Pöppelbuß et al., 2011; Maximilian and Jens 2012; Wendler 2012), a field of research has emerged, now known as meta-maturity models (maturity analysis of maturity models). This interest is indirect proof of the importance of maturity models in the fields they are applied.

There are three main types of maturity models: vertical, horizontal, and network. A vertical maturity model applies to specific (public) service areas. The US Federal Aviation Administration (FAA) maturity model describing the service maturity of this agency (Ibrahim, 2000) or the Marshall eLearning systems maturity model (2005) fall into this category. In turn, horizontal maturity models are tools that can be applied to a specific aspect of each type of service. CMMI®, a process improvement tool, and OPM3®, a project management tool, are perhaps the best-known examples of horizontal maturity models: the project management assessment tool can be applied for any service field such as construction, agriculture, chemistry, military, information technology, etc. For the eGovernment area, examples of horizontal maturity models are found in Layne and Lee (2001), Almarabeh and AbuAli (2010), Baqir and Iyer (2010), Misra and Dhingra (2002), and United Nations (2002). The network maturity model integrates the service and tool dimensions mentioned above. Network maturity models define the process areas (vertical elements of the maturity model) and the maturity levels for these areas (horizontal elements). Examples of such models can be found in the works of Maranna (2011) and USA DoE (2012).

There are two perspectives of maturity models: the life-cycle perspective and the performance perspective (Wendler, 2012). The former shows how an organization should develop to achieve the final, excellent state. The latter describes the potential effects of achieving consecutive, higher and higher, maturity levels.

From an application sector perspective, there are three categories of maturity models:

1. Sector-independent maturity models,
2. Private sector maturity models (not discussed here),
3. Public sector maturity models.

A Public Administration managed by a government is a specific type of organization. Therefore, particular types of maturity models may also be applied to the Governmental level of project management. Maturity models have two primary uses: internal and external. Internally, they may be used as tools for evaluating and improving processes in the organization. At the Governmental level, they may be used for enhancing GPIS processes. The external usage at the organizational level is assessing the ability of potential contractors to perform their work. At the Governmental level, they may be used to determine by banks or by development-supporting institutions the government's ability to implement their investment projects successfully.

24.3 Sector-Independent Maturity Models

Maturity models are used in all sectors of activity: governmental, private, and non-governmental. Many models can be used in each sector, including the governmental one. Examples of project management sector-independent maturity models are CMMI® (SEI, 2010), Kerzner's model (Kerzner, 2005), and Crawford's maturity model (Crawford, 2006). More specialized project management sector-independent maturity models include the project risk management maturity model (Hillson, 1997), the stakeholder management maturity model (Bourne, 2008), and the contract management maturity models (Rendon, 2006; Jackson, 2007; BearingPoint, 2012). Examples of maturity models covering project aggregates are P3M3® (UK OGC, 2010) and OPM3® (PMI, 2018b) that describe, in addition to project management, project program management, and project portfolio management. Sector-independent maturity models have also been developed for specific types of projects, such as software development (Paulk et al., 1993), agile software development (Chetankumar and Ramachandran, 2009; Humble and Russell, 2009), new services development (Jin et al., 2014), construction (Sun et al., 2009), and many others.

Maturity models have also been developed for process activities[2]. This includes models that are applicable regardless of the business sector. Exemplary maturity models of this type are the business process maturity model (Rosemann and de Bruin, 2005), the resource management maturity model (Mahmood et al., 2015), the Big Data maturity model (Hansmann, 2017), the knowledge management maturity model (Teah et al., 2006; Lockamy and McCormack, 2004), the responsible science and innovation maturity model (Stahl et al., 2017), and the tertiary level institutions education maturity model (Marshall, 2005).

24.4 Public Sector Maturity Models

In addition to maturity models that can be used in any sector, there are maturity models that have been developed specifically for the public sector, including maturity models for projects and processes. An additional dimension of the classification of public sector maturity models is the organizational level. As stated earlier, for the public sector, one can consider an organizational level of project management higher than a single organization: the Governmental level.[3] This level is not helpful for private sector organizations, where the hierarchy of business power ends at the business owner level. Private companies most often compete with each other rather than co-operate. Like the sector-independent maturity models, the public sector maturity models can be divided into project-oriented maturity models and process-oriented maturity models. Based on bibliometric research, Moutinho and Rabechini (2020) distinguish the maturity of an organization as a well-defined area of practice and research in public project management. Table 24.1 shows the classification of maturity models for public sector projects, taking into account these two dimensions.

In the following subchapters, I describe some maturity models of 1, 3, and 4 types (Process management organizational level maturity models are less interesting for us).

Table 24.1 Public Sector Maturity Models Classification

	Projects	*Processes*
Organizational Level	1. Project management organizational level maturity models	2. Process management organizational level maturity models
Governmental Level	3. Project management governmental level maturity models	4. Process management governmental level maturity models

24.5 Project Management Organizational Level Maturity Models

The skills and capabilities of public organizations in the area of project management remain at different levels – some base their approach to project management solely on the abilities of project managers. At the opposite skill level, others have well-thought-out, efficient organizational systems. Organizations try to improve their project implementation skills. Organizational Project Management Maturity Models (sometimes called management capability standards) are used to improve skills. Upgrading the maturity level applies to both PMO and other Public Administration agencies.

Several governments have developed and applied project management organizational-level maturity models.

Canada has a project management policy in place (Canada TBoCS, 2019). Under this policy, the project management capabilities of public organizations are evaluated using the Standard for Organizational Project Management Capacity (Canada TBoCS, 2010a, 2013a, 2019). The main capability assessment form consists of 92 questions grouped into 12 areas. These are the areas of project management described in the PMBOK® Guide, the area of portfolio and investment program management, the area of supporting organizational structures, and the area of project management standards. A score is calculated based on the answers to these questions, which determines one of the five classes of the organization's ability (limited, supportive, tactical, evolutionary, transformative) in project management.

In the Greek public sector, a norm describing the maturity model for public sector institutions has been developed within the scope of the System for Assuring Managerial Capability (SAMC), the ELOT 1429 (Greece GOfS, 2009). This model specifies three levels of maturity. At the first level, the organization understands the need for project management and implements project management processes at the minimum Defined level. At the second level, the organization implements a consistent project management methodology for all projects. At the third level, the organization performs benchmarking, collects data that are the basis for decision making, and optimizes its project management processes.

The Australian Public Services Commission has defined Agency capability maturity levels (Australia APSC, 2012) that determine the capability of Australian federal-level public organizations to provide services. There are two areas in these guidelines for a projectized approach to management: Delivery and Implementation and Project Management. For each of these areas, there are five levels of maturity: Awareness, General Acceptance, Defined, Manager, and Leader/Excellence.

Researchers outside governments have also been actively developing public sector project management organizational-level maturity models.

Dettbarn et al. (2005) developed a CMM-based maturity model for capital project portfolio management for federal real properties. This model consists of five levels. At the first, ad hoc level, there are no stable processes. At the planned level, requirements and processes are defined for individual projects. At the Managed level, the investment portfolio is analyzed, and efficiency management processes are used when making a portfolio execution decision. At the integrated level, portfolio operations are used to allocate resources. Performance is rated, and feedback is obtained. The leveraged level requires continuous improvement of the portfolio management processes.

Jia et al. (2011) developed the PMOMIM-MCP maturity model for megaprojects managed by the state-owned Chinese Construction Headquarters. This model is based on OPM3®. The model contains two main components: the organizational sub-model (OMS) and the process sub-model (PMS). The OMS is defined in two dimensions: areas (structure, culture, technology, human resources) and SMCI (standardize, measure, control, improve). PMS has three dimensions: life-cycle groups (preparation, initiation, setup, delivery, and closure), PMBOK® Guide knowledge areas, and SMCI levels.

In Australia, the British P3M3® model (UK OGC, 2010) is used to assess project maturity. According to this model, each organization can be on one of five levels:

1. Awareness of the existence of project management processes
2. Performing repetitive processes
3. Existence of a process defined for the entire organization
4. Process management
5. Process optimization

The maturity assessment is performed in three main areas: project management, project program management, and project portfolio management.

The assessment of state agencies' ability to implement investments in New Zealand is very extensive (Investor Confidence Rating; New Zealand Treasury, 2016). Agencies are assessed in ten dimensions, three of which relate directly to project implementation: maturity according to P3M3®, project delivery (implementation of the scope within the planned time and budget), and obtaining benefits from projects. The remaining seven component assessments address issues less directly related to projects, such as resource management or the quality of an investment plan.

Other countries also work on the maturity of public organizations in the field of project management. UK IPA (2021) works with ministries to develop project and project program management capability within them. India MoSPI (2013) motivates Public Administration units to develop project supervision systems. An assessment of the organization's work processes is performed (Vermont DoII, 2010), for example, based on a simplified version of CMMI®. Customer surveys on satisfaction with the services provided are carried out. The review results are the basis for determining the directions of improving the work of the PMO (North Carolina OSCIO, 2013a).

24.6 Process Management Governmental Level Maturity Models

The creation of specialized models in this group began with the developing the e-government maturity model by Layne and Lee (2001). They defined four levels of maturity models for e-government. At level one (continuous), the government presents information online, and electronic forms are accessible to the public. At level two (transaction), services and forms are available online, and databases support online transactions. At level three (vertical integration), local systems are linked with higher-level systems. At level four (horizontal integration), all systems supporting different areas are integrated.

This application area is most commonly found in government-level process management maturity models. Fath-Allah et al. (2014) performed a comparative analysis of e-government maturity models. The most important result of their study is the identification, based on 25 models, of four main levels of maturity: presence (static, limited information accessible), interaction (forms accessible, communications with government by e-mail), transaction (online services, payments), and integration (one-stop-shop for everybody).

M-government maturity models are a kind of continuation of e-government maturity models. The m-government provides services through mobile devices. Maranny (2011) defines an m-government maturity model consisting of five stages and nine domains. The stages are Initial (information publishing), Enhanced (interaction), Reforming (transactions available), Enrichment (full integration of m-services), and Governance (transformation and participation). The nine domains are Technology, Infrastructure, Security, Application Services, Policy, Knowledge Management, Human and Organizational Factors, Privacy, and User Needs.

Also, a maturity model for interoperability in digital government has been developed for the broadly understood IT area, but not for its interface with the community (Gottschalk, 2009). Interoperability allows systems and organizations to work together. This model has five levels: computer interoperability, process interoperability, knowledge interoperability, value interoperability, and goal interoperability.

The maturity model for Health in All Policies (HiAP) (Storm et al., 2014) deals with another area of Public Administration. HiAP is a government approach that addresses health issues in all government policies. This model has six levels: unrecognized (no attention to the problem), recognized (recognizing the problem), considered (preparatory action on HiAP), implemented (HiAP investment), integrated (HiAP processes integrated with policy processes), and institutionalized (systematic improvement of HiAP quality).

Developing maturity models for processes implemented at the government level shows that such models can be created and used for this organizational level.

24.7 Project Management Governmental Level Maturity Studies

The first attempt to assess project management maturity at the government level was made by Baranskaya (2007). She did not develop a specific model but based her work on Kerzner's maturity model (Kerzner, 2005). But there is no information about its applications or further development. The Governmental level maturity models should not be mixed with national project management maturity models (Seelhofer and Graf, 2018), which cover all organizations in a given country, not only the public ones. – government-specific project management processes are not applicable there.

Hence, we may say that up to now, the 4th cell of Table 24.1 up to now is almost empty. But researchers asked questions about government maturity of project management using other tools and approaches.

In Australia, such studies were conducted by Young et al. (2014) using P3M3®. They assessed the maturity of public organizations and, on this basis, calculated the average maturity in the public sector. For projects, the average management maturity level was between 1 (benefit management) and 3 (risk management). The maturity of other areas (governance, financial, control, stakeholders, resources, generic) was at level 2. For project programs management, maturity was ranked at 1 (benefits, governance, stakeholders), 1 to 2 (risks, control, generic), and 2 (finance and resources). The highest level of maturity was found for portfolio management. In this area, governance, stakeholders, control, risks, benefits, and generic areas were at level 2. Levels 2 and 3 were found for finance and resources.

Prado and Andrade (2015) distinguished the public sector as the subject of part of their project management maturity studies in Brazil. They also averaged maturity in public organizations to obtain the score for the whole public sector. The Prado Project Management Maturity Model

(Prado-PMMM) has five levels, from 1 (Initial) to 5 (Optimized). The model covers seven dimensions: Competence in Project and Program Management, Competence in Technical and Contextual Aspects, Behavioral Competence, Methodology Usage, Computerization, Usage of Structural Organizational Structure, and Strategic Alignment. The average score for all examined public organizations was 2.5.

In New Zealand, KPMG conducted public sector organizations' project management maturity studies (KPMG, 2011) using P3M3®. The study found that 80% of organizations are at level 2 or lower on a scale of 0–5, with 50% at levels no higher than 1.5.

In Ghana, maturity studies were conducted (Ofori and Deffor, 2013) in which public sector organizations were also identified. A questionnaire verifying the achievement of level 2 in the Kerzner (2005) model was used in this study. The study found that public organizations in Ghana remain on the embryonic level of maturity on average.

Kerzner's maturity model was also used in the Brazilian state of Pernambuco to study project management maturity in public sector organizations (de Souza Silva and Gomes Feitosa, 2012). The study found that public organizations in the state of Pernambuco are below level 2 (common procedure) in Kerzner's model.

Other studies of this type were undertaken, among others, by Koenig (2015) in New South Wales, the state of Australia, Bay and Skitmore (2006) and Simangunsong and Da Silva (2013) in Indonesia, and Juchniewicz (2012) in Poland.

All such research aims to determine the level of advancement of project management in a particular jurisdiction. The general scheme of these studies is in all cases similar: a specific model of organizational maturity of project management is adopted, e.g., P3M3® (UK OGC, 2010), Prado MMGP (Prado, 2016), or PMMM (Crawford, 2014). Then the maturity level is assessed for the set of organizations under study. Here, the basis for evaluating maturity in a given jurisdiction (of a given government) are statistics obtained from collecting data in the surveyed organizations, for example, mean, statistical deviation, or tests comparing maturity levels between specific groups. I will call this approach the **bottom-up maturity assessment** of GPIS/GPM. For such research to have not only academic significance, it should be the basis for making decisions that improve public project management in a given area or country. But the maturity models used for this purpose were created at the level of individual organizations; they are part of organizational project management. Models such as CMMI®, OPM3®, or P3M3® do not assess the maturity of project management in the organization by averaging the parameters observed in the implementation of individual projects but by confirming the implementation of specific practices by the organization. Averaging the results of the maturity survey of individual organizations at a higher level does not give an accurate picture of project management at this higher level. For example, Young et al. (2011) identify risk management, portfolio management, and project program resource management maturity targets for each surveyed agency, ignoring central government practices in achieving this maturity. The quality of project management in public organizations depends not only on the decisions of the boards of these organizations but also on the management framework established at the government level, concerning, for example, how to manage public contracts or the rules for hiring and firing public service personnel.

Not considering the governmental practices in shaping public project management is a significant gap that this book is trying to fill.

24.8 Ways to Improve Maturity in Organization (continued)

Now I can go back to the beginning of this chapter (Subchapter 24.1) and complete our discussion of the maturity levels of organizational management improvement. The highest, fourth

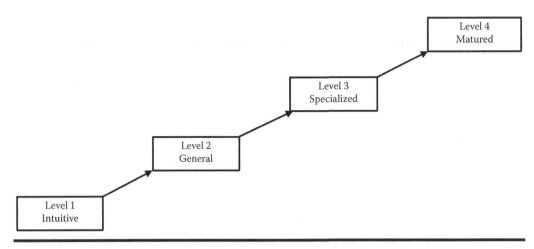

Figure 24.2 Project Management Maturity Meta-Model. (Based upon Gasik, 2008.)

level of management improvement using domain knowledge and indicating the order of implementation of practices is based on maturity models.

This approach I call **Matured** level.

The three previously presented ways to improve management, and the maturity model approach form the Project-Management Maturity Meta-Model (PM4). PM4 is shown schematically in Figure 24.2.

The government level of project management deserves its maturity model.

In the next chapter, I describe a maturity model that can be used for improving GPIS processes.

Notes

1. The author of this book was a member of the team developing OPM3® 2nd Version (PMI, 2008a).
2. Projects are a special kind of process. The full wording should be: maturity models for non-project processes, but for simplicity, I use the term process management maturity model instead.
3. Government is a special type of organization. Here, too, we use the term "government" for simplicity of writing.

Chapter 25

Improving Governmental Project Implementation Systems with GPM3

There are countries with no regulations on public project management; also, many countries do not have institutions supporting the management of public sector projects. In other countries, however, such regulations and institutions operate. Thus, there is a specific continuum of the government's influence on the public sector projects it implements. Many countries have improved their ways of implementing public sector projects. The history of project management, particularly in its early years, has been the history of (streamlining) public sector project management (Morris, 1994; Archibald, 2008; Lenfle and Loch, 2010; Kwak 2003). Contemporary public sector project management systems in the United States at the federal level or in some of their states, India, Canada, the United Kingdom, Norway, or Australia, are much more complex and mature than fifty or hundred years ago. This creates a temptation to systematically compare project management solutions at the government level using the maturity model. But since such a model does not exist so far, it is necessary to build it. I have created a maturity model that includes GPM and the results of its operation – GPIS. This will be called Governmental Project Management Maturity Model (GPM3®), to which this book continues.

Capability Maturity Model Integration (CMMI®™, SEI, 2010) was selected as inspiration for building GPM3®. CMMI® is recognized as a type of template (Wild and Lindner, 2017) and has been the basis for developing many other maturity models.

When building GPM3®, I adopted the principle that the primary entity whose maturity is analyzed is the government administration of a given country as a whole. At the same time, its components are individual subordinated organizations: agencies, institutions, and entire departments. The concept of Organization in CMMI® thus corresponds to that of Government in GPM3®. The concept of a project from CMMI® corresponds to a public organization (ministry, department, another public agency). These dependencies are shown in Figure 25.1.

DOI: 10.1201/9781003321606-30

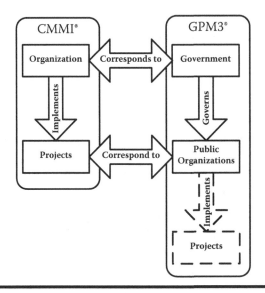

Figure 25.1   CMMI® and GPM3®.

GPM3® consists of five maturity levels:

1. Initial
2. Local
3. Governmental
4. Support
5. Optimizing

The following subchapters describe levels of GPM3® maturity.

Table 25.1   The Correspondence of CMMI® and GPM3® Maturity Levels

CMMI	GPM3
Initial	Initial
Managed	Local
Defined	Governmental
Quantitatively Managed	Support
Optimizing	Optimizing

25.1 The Initial Level

25.1.1 Definition

At the Initial level of GPM3 level, there is no awareness of the importance of project management for the successful functioning of the administration and the country. Governments sometimes use the term "project" but fail to recognize the importance of their projects for the proper functioning of the state. Projects are considered to be some form of technical activity. Governments believe that the efficiency of projects depends primarily on technical skills.

This level corresponds to the Initial level of the classic CMMI® model if we replace the CMMI's concept of the project with the concept of "public organization" and the CMMI's concept of organization with the concept of government. Then the characteristics of this level will look like this: organizations implement their projects in an ad hoc and chaotic manner. The government usually does not provide a stable organizational and legal environment for project implementation. Project management success in these countries depends on people's competencies and heroism, not on proven processes. Public projects are often abandoned in such countries and exceed the schedule and budget. An exemplary description of such a reality is presented in the report *Public Sector Projects in Poland* (Gasik et al., 2014; Gasik, 2017).

25.1.2 Exemplary Practices

The most common form of government involvement in projects of this type is allocating resources and waiting for products (Judah, 1857; Kozak-Holland and Procter, 2014; Kwak et al., 2014). There are no institutions specialized in project implementation. National audit chambers are not qualified in project management and avoid project audits. And if they do, they focus on the technical side and individual activities and not on complete project management processes (Poland NIK, 2014). Public sector projects may be implemented based on general public procurement regulations (if any), which do not consider the specific nature of projects.

At a late stage of this level of advancement, regulations concerning individual projects may be adopted, but not referring to their management or relating to a minimal extent, usually by requesting the delivery of management products (schedule, reports), but without requiring the use of management processes (Poland Sejm, 2009, 2016). All countries remained at this level of maturity more or less until the Cold War.

25.2 The Local Level

25.2.1 Definition

The Local level is the one at which individual projects and public sector institutions begin to apply project management practices. At this level, regulations, guidelines, and standards are created for use by organizational units at a level lower than the government as a whole. The Local level may correspond to the Managed level in the CMMI® or P3M3® models if all government institutions adopt a structured approach to project management. However, this situation usually does not occur. After introducing effective project management in many government institutions, the government decides to cover the entire group of institutions under its responsibility by GPIS.

25.2.2 Exemplary Practices

Entering this level usually starts with an interest in managing individual projects. In the USA, in 1956, the Navy began implementing the Polaris project, for which the PERT technique was developed (Lenfle and Loch, 2010) or large civil construction projects, e.g., the Australian Sydney Opera House (Kouzmin, 1979). The United States entered the local GPM3® level with the Polaris project. This level was further developed when Robert McNamara introduced the US DoD Program Planning and Budgeting System (PPBS; Hitch, 1965) after 1960, emphasizing up-front analysis, planning, and control of projects. This was a fundamental change in the definition of projects: the description of the system had to be known before making a decision (Lenfle and Loch, 2010). One could say that McNamara and Hitch invented the waterfall approach to project implementation. In the United States, the first order to use a specific methodology across an organization was the DoD recommendation for the Cost/Schedule Controls System Criteria (C/SCSC) published in 1967 (Abba, 1997).

Today, many countries are located at the Local level, with different project and project program management practices for individual projects, agencies, entire departments, or types of projects, but that does not cover a given government's whole area of operation. In Brazil, the Central Bank has developed an integrated project management methodology (Brazil MPOeG, 2013) that applies only to this bank. The Office of Project Management & Permitting operates in Alaska, supporting project management at the Department of Natural Resources (Alaska OPMP, 2013). The National Road Administration has published Project Management Guidelines for constructing Irish roads (Ireland NRA, 2010). In Chile, the Ministry of Housing and Urban Development has published Recommendations for Project Management (Chile MINVU, 2009) under its jurisdiction. Canadian MPMO supports implementing the most important Canadian projects related to natural resources (Canada MPMO, 2021).

Some governments implement practices related to a specific project type. The California Project Management Methodology (California CDoT, 2013a) was developed for technology projects in California. The IT project management methodology has been published in Kansas (Kansas OoITS, 2008a). Government ministries create public sector project management offices, e.g., Vermont AoT (2020) dealing with transportation projects, Project Management and Development Branch in Real Estate Services Division in California (California DoGS, 2013), or Vermont DoII (2013) dealing with IT projects. There is a strategic plan for IT in Arizona (Arizona Brewer et al., 2014). One of his goals is to strengthen the management of IT projects and project programs in this American state. Some governments issue requirements or guidelines for using the agile approach, which is primarily applicable for software projects (Maryland MDoIT, 2021b; Canada Government, 2021; US GAO, 2020b), but its practices may be implemented in other project types.

We can come across a situation where practices from higher maturity levels are performed in a specific country for specific areas like IT or construction. However, as they target only a particular project type, these practices are part of the Local rather than the next, Governmental level of project management maturity.

Generally, Local-level practices relate to the Delivery Territory of the GPIS model (Chapter 20).

25.2.3 Benefits

At the Local level, the benefits of project management are only found in organizations and agencies that implement project management practices. The knowledge accumulated while the government is at the Local level is necessary to take project management maturity to a higher level. The government is convinced by the example of successes in lower-level organizations that introducing organized project management brings benefits and then introduces processes in this area for its entire area of operation, moving to a higher level of maturity.

25.3 The Governmental Level

25.3.1 Definition

The Governmental level is the one at which governments are actively involved in shaping the approach to project management for the entirety of their agencies. The government enacts laws and regulations. In particular, it may establish government-wide project management agencies (Governmental PMO). Generally, the government performs governance processes (Chapter 19) concerning the entire portfolio of public projects. At this GPM3® maturity level, these governance activities mainly include the Delivery Territory (Chapter 20) practices. Governmental-level practices may overlap with Local-level practices. For example, an analogous project initiation process may be defined and executed in a particular organization by the decision of that organization's board of directors or by government regulation. The difference between these levels is in the extent to which they are applied: local or government-wide. The factor that distinguishes the Local from the Governmental level is that all public projects (usually above a certain budget threshold (Section 20.4.11) are covered by the governance and project management processes.

25.3.2 Exemplary Practices

Institutions supporting the management of all projects and project programs operate at this maturity level. In the United States, the Office of Management and Budget overseeing the implementation of project programs and major projects (USA OMB, 2021) reports directly to the President. In the Australian state of Victoria, there was the position of Minister for Major Projects managing Major Projects. In the UK, the Infrastructure and Project Authority operates within the structures of the Prime Minister's Office (UK IPA, 2016). There is a Strategic Projects Directorate in Bahrain (Bahrain MoW, 2013). In Singapore, there is a Center for Public Project Management (Singapore CPPM, 2011). In India, several federal institutions are involved in managing public sector projects at the federal level. NITI (India NITI, 2015) is responsible for selecting projects for implementation, and India MoSPI (India MoSPI, 2013) monitors project implementation.

Project management policies, guidelines, standards, and regulations are published. Probably the first regulation that applied to all programs of a given country was Circular A-109, issued in 1976 by the American OMB (1976). This document, among others, extended the application of the DoD-proven C/SCSC method to federal projects of other departments.

Canada has a Project Management Policy (Canada TBoCS, 2010a, 2019). Standards are adopted for public sector projects. PMBOK® Guide has been recognized as a standard by American National Standards Institute in the United States. The British government commissioned Prince 2® (UK OGC, 2017) for use in government projects in the United Kingdom. A group of experienced project managers developed a project management methodology in New York state based on the PMBOK® Guide. This methodology is valid for all projects carried out in the New York state (IT, software, engineering, business development, etc.). In Tasmania, Tasmanian Government Project Management guidelines have been developed (Tasmania OeG, 2011). The Tasmanian ICT Policy Board recommended applying these guidelines to all projects carried out by the Tasmanian government and its agencies. The Michigan project management methodology was developed in Michigan (Michigan MDoIT, 2004). The methodology is based on an early version of the standard of the Project Management Institute (PMI, 2004). This methodology is very extensive, equipped with a set of forms. In Scotland, governance rules for all government projects and project programs have been published (Scottish Government, 2013b).

Regulations and standards may apply to specific phases or elements of project management.

At the Governmental level, there are defined rules for including projects in the portfolio of individual institutions or the government project portfolio, managed in accordance with specific rules. The method of building the portfolio of programs is dealt with in GPRA (USA Congress, 1993). The method of selecting projects for implementation was defined in India (India PMD, 2012). Norway has the Quality on Entry procedure (Magnussen and Olsson, 2006; Klakegg et al., n.d.) that describes how to initiate the country's major public sector projects. New Zealand has guidelines for the assurance of major projects, project programs, and portfolios (New Zealand Government, 2019). New South Wales has developed evaluation guidelines for all public sector projects of appropriate value (NSW Treasury, 2018a). Queensland has Project Assessment Framework (Queensland Treasury, 2015) whereby all projects are assessed in the appropriate phases (Gates), from needs assessment, through implementation to service delivery and benefit realization. In the Australian state of Victoria, the Department of Jobs, Precincts, and Regions oversees the implementation of major projects (Victoria DJPR, 2020).

Regulations concerning the education and qualifications of public sector project managers are introduced. The Government Projects Academy was established in the UK to educate government project managers (UK HM Government, 2021; previously, they were trained by the Said Business School at Oxford University, UK Oxford University, 2012). In New South Wales, recommendations for the qualification of public project managers have been developed and published (NSW PSC, 2020a)

At this level, national audit chambers play a significant role. They are competent in performing audits of public sector projects. They can also develop recommendations for public organizations on implementing projects (USA GAO, 2012, 2016, 2020a; Australia ANAO, 2010a).

25.3.3 Benefits

The transition from the Local to the Governmental level is the essential change in the approach to project management in governments.

At this level of maturity, the government recognizes the importance of project and project program management for the country's development. All projects, project programs, and the entire project portfolio of a country are more likely to succeed. Best practices are distributed to all organizational units. Due to the standard language resulting from the existence of general government institutions, processes, and methodologies, it is possible to exchange knowledge between individual organizations. The government has the tools to shape the way it manages its projects.

25.4 The Support Level

25.4.1 Definition

The Support level is the one in which the government side not only defines project implementation processes and leads organizational units dealing with projects but engages in activities that increase the chances of success of these projects, for example, by removing problems facing projects. The Support level practices differ from Local and Government level practices regarding timing and reasons for their implementation. At the Support level, services are provided when and if projects are at risk or are expected to encounter implementation difficulties. Support level practices may be implemented at the request of the project team. They are not a part of the standard implementation processes.

25.4.2 Exemplary Practices

Support level practices are divided into two main groups: problem-oriented and process-oriented. The first group of these practices supports projects when they encounter problems. The second group consists in assigning a supporting person for the duration of the entire project process or its part.

Support level practices are described in Chapter 21 dedicated to the Support area.

25.4.3 Benefits

The governmental side not only passively waits for the implementation of public sector projects but also transfers the best knowledge, practices, and abilities to projects. Eliminating problems facing projects and involving professionals from specialized organizational units increases the chances of success. Their knowledge and experience may support the implementation of public projects at various stages of their implementation. Involvement in the project implementation process and problem-solving increases government experts' knowledge that can later be used to improve project implementation processes.

25.5 The Optimizing Level

25.5.1 Definition

At the Optimizing level, there is a mandate from the Government level to improve public sector project management processes and systems continuously. Streamlining becomes an element of governance. At the Optimizing level, two sub-levels can be distinguished: individual and system. The individual level is achieved by governments that apply the practices of the individual mode of the Development area (Section 22.3.1). On the other hand, the system level is achieved by governments that use the practices of the system mode of development for this purpose (Section 22.3.2).

25.5.2 Exemplary Practices

Practices of improving project implementation, both for the individual level (Section 22.3.1) and the system level (Section 22.3.2), are described in Chapter 22 on project management development. What has not been described there is the use of maturity models. The following two sections are devoted to this subject.

25.5.2.1 Individual Optimizing Level

For the individual improvement of project management, maturity models may also be used. In Australia, P3M3® was adopted for organizations subject to the Financial Management and Accountability Act (Australian Government, 1997) as a methodology for assessing and improving project management skills. These organizations must perform an annual maturity assessment and report its results to the Ministry of Finance (Australia DoFD, 2012a). Canada has implemented a policy that requires the continuous evaluation and improvement of project management in public organizations (Canada TBoCS, 2010a, 2019). The maturity model used in assessing the advancement level of public organizations in Canada is the Organizational Project Management Capacity Assessment Tool (Canada TBoCS, 2013a, 2013b).

25.5.2.2 System Optimizing Level

To date, no maturity model used in practice to improve GPIS / GPM can be identified. GPM3® is the first attempt to create such a model. GPM3® is an example of a top-down approach to developing and assessing the level of GPIS maturity, that is, the one that recognizes the essential, dominant role of governments in shaping GPIS. The main focus of GPM3® is GPIS, which is the way a government approaches project management.

25.5.3 Benefits

At the Optimizing level, the management of public projects is constantly improved. There are operations processes that use the knowledge gathered from previously completed projects to improve project management in individual organizations or GPIS as a whole. Therefore, the likelihood of success is even greater than at the Support level. The processes of supporting the implementation of projects and practices from lower maturity levels can also be improved at this level – which justifies considering the Optimizing level as the highest level of GPIS advancement.

The Governmental Project Management Maturity Model is shown schematically in Figure 25.2.

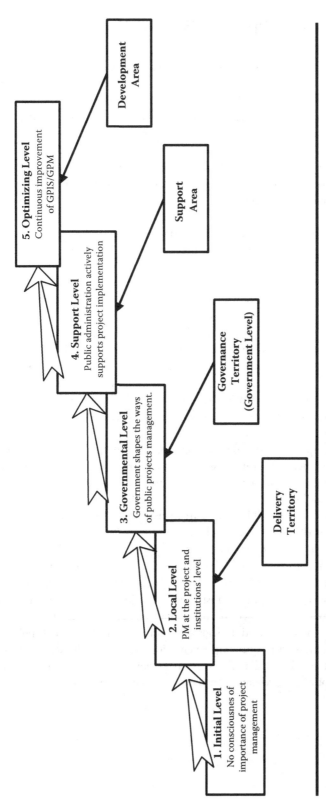

Figure 25.2 Governmental Project Management Maturity Model.

25.6 The Role and Applications of GPM3®

GPM3® is a maturity model that shows how to improve public project implementation. Project management tries to answer the project manager's question about managing a single project. Organizational project management tries to answer the question of the organization's management board, how to organize a company to efficiently manage its projects, project programs, and project portfolios. Governmental Project Management tries to solve the problem governments face: how to organize the management of public sector projects and project programs in a given country to ensure the stable functioning and development of the country in line with its capabilities. OPM3® has contributed to improving project management in individual organizations, improving their methods of operation. Similarly, GPM3® may contribute to improving the functioning of state administrations, affecting the well-being of entire societies.

The Governmental Project Management Maturity Model, like other maturity models, may have three types of applications: descriptive, prescriptive, and comparative (de Bruin et al., 2005; Pöppelbuß et al., 2011).

The descriptive application of GPM3® generates an assessment of the current state of the GPIS in a given administrative unit. Therefore, it can be used to assess the level of effectiveness of public investment as made, for example, by the International Monetary Fund (Dabla-Norris et al., 2010).

The prescriptive application shows the roadmap for the improvement of GPIS. Thus, it can be a tool that helps improve the manner in which Public Administration functions, affecting the well-being of whole societies. For instance, it may be used by organizations specialized in supporting supranational development and aid organizations that are currently more focused on managing individual projects than organizing the GPIS. With the GPM3®, those organizations will be able to define characteristics and roadmaps to develop project management capabilities tailored to a given administrative unit.

The comparative application enables the comparison of the GPISs between sibling administrative units. For instance, in a country with a federal structure consisting of autonomous states, it will allow a comparison of the maturity of GPIS in these states. GPM3® could become the basis for defining the area of knowledge that, tentatively, could be called Comparative Governmental Project Management – which will become part of the long-established discipline of Comparative Public Administration (Riggs, 1954; Heady, 2001; Jreisat 2011).

GPM3® may be the basis for many directions of further work. A precise methodology for assessing GPIS/GPM maturity should be developed. In the future, to better understand its maturity levels, dimensional maturity models may be developed to evaluate the level of maturity in specific aspects, such as legislation, institutions, project manager qualifications, stakeholder engagement, and public sector project portfolio management. Examples of maturity levels in the area of legislation may be no legislation, legislation on public procurement, public sector projects, and legislation on improving project implementation. However, the definition of these levels is beyond the scope of this book. Based on practices defined in this book, it will be possible to introduce the concept of p-government in the future, i.e., a government that efficiently manages its projects.

Summary

The modern world is more and more project-oriented and projectified. The same is happening with government-led public administrations. Projects are an integral component of Public Administration. The implementation of public policies is based on the implementation of projects. Project management is an essential component of the knowledge and skills of every public manager. Governments can shape the modes of operation in their subordinate administration. They can also shape how public projects are implemented – although not all of them do it or do it ineffectively. There are two reasons for this situation. First, governments may not be aware of the importance of project management for state and national development. Second, they may not know how to do it. The book tries to overcome both of these problems. First, it shows the place of projects in implementing public policies. Second, it describes the project management practices from many countries structured by constructing the Governmental Project Implementation System (GPIS). Two generations of GPIS may be defined: the first covers only processes from the idea arising and project initiation to product implementation and outcome evaluation (some governments have already implemented such GPISes). The second, more comprehensive, also covers many other activities necessary to ensure success in implementing public projects, for example, managing project actors, knowledge management, structures supporting project management, or government-level project governance. Barkley (2011) expected the American knowledge about managing government programs to be transferred to other countries. But the knowledge of GPIS and GPM was developed in many other countries as well, so Barkley's expectation may be extended to the transfer of knowledge about government project management practices from other matured in project management countries to less matured ones.

For practitioners, the essential value of the book is a structured presentation of effective Governmental Project Implementation System and Governmental Project Management practices.

Projects are the primary means of introducing changes and performing statutory activities in Public Administration. Therefore, knowledge about project management should be an integral part of Public Administration. However, it seems that this field of social sciences and Project

Management are developing separately. The book tries to fill this gap by presenting to the Public Administration community the most important concepts in the area of Project Management (like a project, project program, project portfolio, project families, project life-cycle) and, *vice versa*, Public Administration concepts (like government, state, administration, public policy, quality of government, state capabilities) to the project management community. This also includes describing the place of projects management in public policy implementation.

> I believe that this will contribute to bringing the research areas of Public Administration and Project Management closer together.

So far, the discipline of Public Administration has focused on operation-based policies whose main component was continuous operations (e.g., tax policies, health policies, education policies) and neglected project-based policies whose essential element is project implementation (e.g., infrastructure policies, natural resources policies, some military policies). Much of the results of public policy research were not adequate for the latter type. The book introduces the division of public policies (and their instruments) into two main types: operation-based and project-based.

> The defining of project-based policies will enable the study of their specific features and thus contribute to the development of Public Administration.

The book also contributes to the development of the theory in other ways. The concept of a program is one of the main concepts in both the areas: of Public Administration and Project Management. From the point of view of knowledge management, it is the so-called boundary object (Fong, 2003, 2005). Boundary objects tend to be understood differently by the parties involved. The problem of different meanings of such objects can be solved in several ways. One is to identify common features known on both sides and features specific to each knowledge community. Identifying different characteristics can lead to the definition of sub-types of boundary objects which removes misunderstandings between the communities involved. For the boundary objects of a program, features specific to Public Administration and Project Management were indicated, which led to the separation of the concepts of "policy program" and "project program."

> Defining policy programs and project programs as sub-types of the general concept of a program will facilitate its better understanding and cooperation between the Public Administration and Project Management communities.

Researchers have addressed various aspects of government actions regarding project management until now. An essential step in developing knowledge about specific issues is understanding that previously separately conducted research concerns one well-defined field. For example, in the area of project management, an important milestone was the definition of the concept of organizational project management which covers issues related to project management at the organization level. The introduction of OPM resulted in the intensive development of knowledge about project management at the organizational level. The Governmental Project Management (GPM) introduction should result in an analogous development of project management knowledge at this level. Governments and organizations supporting the development of states may be interested in this knowledge.

Maturity models are becoming more and more popular; however, they are not used in the Public Administration field. Fitzpatrick et al. (2011) analyzed the content of scientific articles on Comparative Public Administration. None of these papers used maturity models. The book proposes the Governmental Project Management Model (GPM3®) dealing with GPIS and GPM, a part of the administration's activities. According to GPM3®, the Public Administration is not interested in project management at the Initial level. At the second level of maturity, the Local one, individual organizations develop their project management methods. The third level of maturity is the one in which governments define the approach to project management in their organizations. Next, governments not only define management processes but also directly support the implementation of public projects. The highest level of GPM3 maturity is where governments systematically improve project management in their jurisdictions. GPM3®, like other maturity models, will be able to be used as a tool to evaluate the implementation system of public sector projects and define directions for the improvement of these systems. Both of these tasks can be performed by international (or supranational) organizations (e.g., UN Development Program, UN Office for Project Services, World Bank, or Inter-American Development Bank) supporting the development of governments as a whole and their agencies. The tools described in this book may also be used directly by national governments to improve their project management capabilities.

> Governmental Project Management is a vital area from the point of view of countries' development because it deals with the essential tool for implementing changes – projects.

Models that abstractly describe phenomena become the basis for creating theories, i.e., explaining the observed phenomena. GPIS and GPM3® can become the basis for asking questions and making theories explaining the phenomena occurring in the GPM area. When analyzing GPIS, you can ask questions like: How does the implementation of actor management affect the effectiveness of projects? How does the performance of GPIS and GPM affect the implementation of public policies? The existence of GPM3® will allow, for example, research to what extent achieving successive maturity levels improves the effectiveness of projects implemented in a given administration.

> From the theoretical point of view, the book proposes and creates the basis for researching the area of Governmental Project Management.

The book also contributes in other ways to the development of theories in the area of project management. The three existing models of differences between public organizations and organizations in other sectors (the generic model, the core model, and the dimensional model) did not adequately describe these differences at the level of projects – that are a type of organization. The book, therefore, introduces the layered model of differences at the level of projects. Also, the maturity meta-model has not been made public by the author so far. The meta-model helps you understand the role of maturity models in improving project management and points out that formal maturity models are not the only way to improve project management maturity.

Learning about Public Administration is empty if there is no comparative component (Dahl, 1947). Comparative Public Analysis (CPA) is a branch of science that compares different entities, such as countries, in terms of their concepts related to Public Administration (Fitzpatrick et al., 2011; Jreisat, 2011, 2012; Dowding et al., 2016; Heady, 2001). The theoretical result of CPA is the construction of models explaining the identified differences (Wilson, 2011). In practical terms, CPA makes it possible to identify suitable solutions from one compared entity and transfer them to other entities. Comparative studies can be performed at different levels – for example, in organizations, governments, or countries. Researchers compare project management solutions in different countries (Crawford et al., 2007; Koops et al., 2016; Pilkaite and Chmieliauskas, 2015; Williams et al., 2010; Samset et al., 2016; Volden and Samset, 2017). GPIS, GPM, and GPM3® make the basis for a systematical comparison of project management practices at the government level. Comparative Governmental Project Management (CGPM) emerges.

The main part of this book is devoted to Comparative Governmental Project Management, and I hope it will facilitate its further development.

References: Source Materials

Alabama OoIT (2017). *Guideline 400G1: IT Project Governance.* Montgomery: Office of Information Technology. https://oit.alabama.gov/wp-content/uploads/2017/09/Guideline-400G1-IT-Project-Governance.pdf. Accessed June 2021.

Alaska ADoTaPF (2013). *Alaska DOT&PF Statewide Project Information.* Juneau: Alaska Department of Transportation and Public Facilities. http://dot.alaska.gov/project_info/index.shtml. Accessed May 2021.

Alaska OPMP (2013). *Office of Project Management and Permitting.* Juneau: Department of Natural Resources. http://dnr.alaska.gov/commis/opmp/. Accessed May 2021.

Alberta EAAE (2013). *Alberta Major Projects.* Edmonton: Government of Alberta. https://majorprojects.alberta.ca/. Accessed May 2021.

Argentina Congreso de la Nacion (1994). *Ley 24.354. Sistema Nacional de Inversiones Públicas.* Buenos Aires: Congreso de la Nacion Argentina.

Argentina DNIP (n.d.) *Dirección Nacional de Inversión Pública.* Buenos Aires: Gobierno Argentino. https://www.argentina.gob.ar/jefatura/evaluacion-presupuestaria/dnip. Accessed April 2021.

Argentina MEFP (2015). *Plan Nacional de Inversiones Publicas 2015-2017, Subsecretaria de Coordinacion Economica y Meyora de la Competitividad.* Buenos Aires: Direccion Nacional de Inversion Publica, Ministerio de Economia y Finanzas Publicas. https://www.argentina.gob.ar/sites/default/files/texto_plan_2015-17.pdf. Accessed May 2021.

Argentina MEFP (2021). *Banco de Proyectos de Inversión Pública BAPIN.* Buenos Aires: Direccion Nacional de Inversion Publica, Ministerio de Economia y Finanzas Publicas. https://www.argentina.gob.ar/dnip/bapin. Accessed May 2021.

Arizona ADoA (2013). *Project Investment Justification (PIJ) A Statewide Standard Document for Information Technology Projects.* Phoenix: Arizona Strategic Enterprise Technology, Arizona Department of Administration. http://aset.azdoa.gov/content/project-investment-justification. Accessed May 2021.

Arizona Brewer, J. K., Smith, S. A., & Sandeen, A. V. (2013). *2013 Statewide Strategic IT Plan, A Plan for The Future, Arizona Strategic Enterprise Technology.* Phoenix, Arizona: Department of Administration. https://aset.az.gov/sites/default/files/2013%20Arizona%20Statewide%20Strategic%20IT%20Plan%5B2%5D.pdf. Accessed May 2021.

Arizona DoAT (2013). *Information Technology Authorization Committee (ITAC).* Phoenix: Arizona Department of Administration Technology. https://aset.az.gov/governance/it-authorization-committee-itac. Accessed May 2021.

Arizona DoT (2013). *Project Management Services.* Phoenix: Department of Transport. http://www.azdot.gov/business/ManagementServices. Accessed May 2021.

Arizona GITA (2009). *Statewide Policy. Project Management Certification.* Phoenix: Government Information Technology Agency. https://aset.az.gov/sites/default/files/p335projectmanagementcertpolicy_1.pdf. Accessed May 2021.

Arkansas DoIS (2021). *Project Management.* Little Rock: Department of Information Services. https://www.transform.ar.gov/information-systems/products-services/professional-services/project-management/. Accessed May 2021.

Astesiano, H. (2014). *Argentina Grupo Provincia Projects*. Buenos Aires: Interview by S. Gasik on 2014.09.25.

Australia AGIMO (2018). *ICT Investment Approval Process*. Canberra: ICT Australian Government Information Management Office, Department of Finance and Deregulation. https://www.finance.gov.au/government/assurance-reviews-and-risk-assessment/ict-investment-approval-process. Accessed May 2021.

Australia ANAO (2010a). *Planning and Approving Projects – An Executive Perspective Setting the foundation for results. Better practice Guide*. Canberra: Australian National Audit Office.

Australia ANAO (2010b). *Conduct by Infrastructure Australia of the First National Infrastructure Audit and Development of the Infrastructure Priority List Infrastructure Australia*. Audit Report No. 2 2010-11. Canberra: Australian National Audit Office.

Australia ANAO (2011). *Management of the Implementation of New Policy Initiatives*. Audit Report No. 29 2010-11. Canberra: Australian National Audit Office.

Australia ANAO (2012). *Performance Audit Administration of the Gateway Review Process*. Audit Report No. 22 2011-12. Canberra: Australian National Audit Office.

Australia APSC (2012). *State of the Service 2010-11. Appendix 4 – Agency Capability Maturity Levels*. Canberra: Australian Public Service Commission. http://www.apsc.gov.au/about-the-apsc/parliamentary/state-of-the-service/state-of-the-service-2010/appendices/appendix-4-agency-capability-maturity-levels. Accessed May 2016.

Australia DISER (2021). *Major Projects Facilitation Agency*. Canberra: Department of Industry, Science, Energy and Resources. www.business.gov.au/expertise-and-advice/major-projects-facilitation-agency. Accessed May 2021.

Australia DoF (2013). *Gateway Review Process*. Canberra: Department of Finance. https://www.finance.gov.au/government/assurance-reviews-and-risk-assessment/gateway-reviews-process. Accessed April 2021.

Australia DoF (2016a). *Risk Potential Assessment Tool General Guidance*. Canberra: Department of Finance. https://www.finance.gov.au/publications/resource-management-guides/risk-potential-assessment-tool-general-guidance-rmg-107. Accessed May 2021.

Australia DoF (2016b). *ICT Investment Approval Process*. Canberra: Department of Finance. https://www.finance.gov.au/government/assurance-reviews-and-risk-assessment/ict-investment-approval-process. Accessed May 2021.

Australia DoF (2021). *Assurance Reviews and Risk Assessment*. Canberra: Department of Finance. https://www.finance.gov.au/government/assurance-reviews-and-risk-assessment. Accessed September 2021.

Australia DoFD (2012a). *Capability Improvement Plan 2012–2013*. Canberra: Portfolio Management Office, Australian Government, Department of Finance and Deregulation.

Australia DoFD (2012b). *Organisational Capability in ICT Investment*. Canberra: Department of Finance and Deregulation. https://silo.tips/download/organisational-project-management-maturity-assessment. Accessed May 2021.

Australia DoFD (2013). *Implementation and Performance Improvement Division, Governance & Resource Management Group*. Canberra: Department of Finance and Deregulation. http://finance.gov.au/about-the-department/governance-and-resource-management-group.html. Accessed September 2013.

Australia DoFD (2014). *Prequalified Multi-use List Tenders*. Canberra: Department of Finance and Deregulation. https://www.aph.gov.au/-/media/Committees/rrat_ctte/estimates/bud_1415/infra/Answers/attachments/QoN37A_B.pdf. Accessed May 2021.

Australia DPMC and ANAO (2006). *Implementation of Programme and Policy Initiatives Making Implementation Matter. Better Practice Guide*. Canberra: Australian National Audit Office.

Australia Snowy Hydro (n.d.) *Snowy Hydro. Our Company*. Coma NSW: Snowy Hydro. https://www.snowyhydro.com.au/about/our-company/. Accessed June 2021.

Australian Government (1957). *National Capital Development Commission Act 1957*. Canberra: Australian Government.

Australian Government (1997). *Financial Management and Accountability Act 1997*. Canberra: Australian Government.

Australian Government (2021). *Major Projects Facilitation Agency*. Canberra: Government of Australia. https://www.business.gov.au/Expertise-and-Advice/Major-Projects-Facilitation-Agency. Accessed June 2021.

Bahrain MoW (2013). *Strategic Projects Directorate.* Manama: Ministry of Work. http://www.works.gov.bh/English/WhoWeAre/Pages/profileinfo.aspx?profId=17. Accessed May 2021.

Bangladesh DoPW (2013). *Project Circle.* Dakka: Department of Public Work. http://www.pwd.gov.bd/index.php?option=com_content&task=view&id=363&Itemid=393. Accessed September 2013.

Bangladesh MoP (n.d.) *USER Manual (End User) of Project Management Information System (PMIS) Application.* Dakka: Ministry of Planning. Implementation Monitoring & Evaluation Division (IMED). https://imed.portal.gov.bd/sites/default/files/files/imed.portal.gov.bd/page/2ed1f2fc_fd2b_4d5f_b0ef_40ed6cc52835/PMIS_User_manual.pdf. Accessed May 2021.

BC MoF (2012). *Information Management and Information Technology Management.* Victoria, BC: Ministry of Finance British Columbia. https://www2.gov.bc.ca/gov/content/governments/policies-for-government/core-policy/policies/im-it-management. Accessed May 2021.

BC Stats (2020). *Major Projects Inventory.* Victoria, BC: British Columbia Stats. https://www2.gov.bc.ca/gov/content/employment-business/economic-development/industry/bc-major-projects-inventory. Accessed May 2021.

Botswana CEDA (2020). *Citizen Entrepreneurial Development Agency.* Gaborone: Government of Botswana. https://www.ceda.co.bw. Accessed May 2020.

Brazil MPOeG (2011). *Metodologia De Gerenciamento De Projetos Do SISP.* Brasilia: Ministério do Planejamento, Orçamento e Gestăo.

Brazil MPOeG (2013). *Metodologia de Gerenciamento de Portfólio de Projetos do SISP.* Brasilia: Ministério do Planejamento, Orçamento e Gestăo.

Brazil Presidência da República (1993). *Lei № 8.666, de 21 de Junho de 1993.* Brasilia: Presidência da República Brasil, Brasilia.

Brunei DPW (2013). *Department of Public Works, Bandar Seri Begawan.* Brunei: Department of Public Works. http://www.mod.gov.bn/pwd/SitePages/Project%20Management.aspx. Accessed May 2021.

California CDoT (2013a). *California Project Management Methodology.* Sacramento: California Department of Technology. http://www.cio.ca.gov/Government/IT_Policy/SIMM_17/index.html. Accessed May 2021.

California CDoT (2013b). *IT Project Tracking.* Sacramento: California Department of Technology. https://cdt.ca.gov/policy/it-project-tracking/. Accessed May 2021.

California CDoT (2013c). *Project Delivery.* Sacramento: California Department of Technology. https://cdt.ca.gov/project-delivery/. Accessed May 2021.

California CDoT (2017). *Understanding Agile.* Sacramento: California Department of Technology. https://projectresources.cdt.ca.gov/wp-content/uploads/sites/50/2019/10/UnderstandingAgile.pdf. Accessed September 2021.

California CDoT (2021). *California Agile Framework (CA-Agile).* Sacramento: California Department of Technology. https://projectresources.cdt.ca.gov/agile/. Accessed September 2021.

California DoGS (2013). *Project Management Branch, Real Estate Services Division.* Sacramento: Department of General Services. https://www.dgs.ca.gov/RESD/About/Project-Management-and-Development-Branch/Project-Management. Accessed May 2021.

Canada CIOB (2012). *A Guide to Project Gating for IT-Enabled Projects, Chief Information Officer Branch.* Ottawa: Treasury Board of Canada Secretariat. http://www.tbs-sct.gc.ca/itp-pti/pog-spg/irp-gpgitep/irp-gpgiteptb-eng.asp. Accessed April 2021.

Canada Government (2007). *Cabinet Directive on Improving the Performance of the Regulatory System for Major Resource Projects.* Ottawa: Government of Canada. https://www.ceaa.gc.ca/050/documents_staticpost/cearref_21799/83452/Vol1_-_Part03.pdf. Accessed May 2021.

Canada Government (2020). *Treasury Board Submissions.* Ottawa: Government of Canada. https://www.canada.ca/en/treasury-board-secretariat/services/treasury-board-submissions.html. Accessed May 2021.

Canada Government (2021). *Is Agile Right for My Project? (draft).* Ottawa: Government of Canada. https://canada-ca.github.io/digital-playbook-guide-numerique/views-vues/agile/en/agile-use-when.html. Accessed September 2021.

Canada MPMO (2013). *Project Agreements.* Ottawa: Major Project Management Office. http://mpmo.gc.ca/projects/9. Accessed May 2021.

Canada MPMO (2021). *MPMO Mandate*. Ottawa, Canada: Government of Canada. https://mpmo.gc.ca/8. Accessed April 2021.

Canada Parliament (1990). *Hibernia Development Project Act*. Ottawa, Canada: Senate and House of Commons of Canada.

Canada PSPC (2009). *Project Management Advisory Council*. Ottawa: Public Services and Procurement Canada. http://www.tpsgc-pwgsc.gc.ca/biens-property/sngp-npms/ccgpm-pmactor-eng.html. Accessed May 2021.

Canada PSPC (2014). *NPMS Continual Improvement Process*. Ottawa: Public Services and Procurement Canada. https://www.tpsgc-pwgsc.gc.ca/biens-property/sngp-npms/amelioration-improvement-eng. html. Accessed June 2021.

Canada PSPC (2019). *National Project Management System*. Ottawa: Public Services and Procurement Canada. https://www.tpsgc-pwgsc.gc.ca/biens-property/sngp-npms/index-eng.html. Accessed May 2021.

Canada TBoCS (2007). *Assessing, Selecting, and Implementing Instruments for Government Action*. Ottawa: Treasury Board of Canada Secretariat. https://www.canada.ca/en/government/system/laws/developing-improving-federal-regulations/requirements-developing-managing-reviewing-regulations/guidelines-tools/assessing-selecting-implementing-instruments-government-action.html. Accessed May 2021.

Canada TBoCS (2008). *Project Charter Template. An Enhanced Framework for the Management of Information Technology Projects*. Ottawa: Treasury Board of Canada Secretariat. https://www.canada.ca/en/treasury-board-secretariat/services/information-technology-project-management/project-management/project-charter-template.html. Accessed June 2021.

Canada TBoCS (2010a). *Policy on the Management of Projects*. Ottawa: Ottawa: Treasury Board of Canada Secretariat. http://www.tbs-sct.gc.ca/pol/doc-eng.aspx?id=18229&section=text. Accessed May 2021.

Canada TBoCS (2010b). *Standard for Project Complexity and Risk*. Ottawa: Treasury Board of Canada Secretariat. http://www.tbs-sct.gc.ca/pol/doc-eng.aspx?id=21261. Accessed May 2021.

Canada TBoCS (2013a). *Organizational Project Management Capacity Assessment Tool*. Ottawa: Treasury Board of Canada Secretariat. https://www.canada.ca/en/treasury-board-secretariat/services/information-technology-project-management/project-management/organizational-project-management-capacity-assessment-tool.html. Accessed May 2021.

Canada TBoCS (2013b). *Guide to Using the Organizational Project Management Capacity Assessment Tool*. Ottawa: Treasury Board of Canada Secretariat. https://www.tbs-sct.gc.ca/pm-gp/doc/ompcag-ecogpg/ompcag-ecogpgpr-eng.asp. Accessed May 2021.

Canada TBoCS (2015). *Project Complexity and Risk Assessment Tool*. Ottawa: Treasury Board of Canada Secretariat. https://www.canada.ca/en/treasury-board-secretariat/services/information-technology-project-management/project-management/project-complexity-risk-assessment-tool.html. Accessed May 2021.

Canada TBoCS (2019). *Directive on the Management of Projects and Programmes*. Ottawa: Treasury Board of Canada Secretariat. https://www.tbs-sct.gc.ca/pol/doc-eng.aspx?id=32594&section=html. Accessed April 2021.

Canada TBoCS (2021a). *Treasury Board of Canada Secretariat*. Ottawa: Treasury Board of Canada Secretariat. https://www.canada.ca/en/treasury-board-secretariat.html. Accessed May 2021.

Canada TBoCS (2021b). *Office of the Comptroller General*. Ottawa: Treasury Board of Canada Secretariat. https://www.canada.ca/en/treasury-board-secretariat/corporate/organization.html#ocg. Accessed May 2021.

Caudill, S. (2014). *Public Projects in eProcurement Bureau at Commonwealth of Virginia, Richmond*. Virginia: Interview by S. Gasik on 2014.09.02.

Chile MINVU (2009). *Espacios Públicos: Recomendaciones para la Gestión de Proyectos*. Santiago de Chile: Ministerio de Vivienda y Urbanisto.

Clark, D. (2014). *Projects of Collaborative Project Management Office*. Toronto, Ontario: Interview by S. Gasik on 2014.09.17.

Collia, M. C. (2014). *System of Public Projects in Argentina*, Buenos Aires: Interview by S. Gasik on 2014.11.28.

Colorado GASC (2012). *House Bill 12-1288 Concerning the Administration of Information Technology Projects in State Government.* Denver: General Assembly of the State Colorado.

Colorado OoIT (2013). *Standard for Project Management Methodology.* Denver: Office of Information Technology. https://drive.google.com/file/d/0B_ZUv6gW8QZMUlNleVVMTEluSmc/view. Accessed May 2021.

Colorado OoSA (2013). *Policies and Procedures, Public Projects.* Denver: Office of the State Architect. https://www.colorado.gov/pacific/sites/default/files/PublicProjects.pdf. Accessed May 2021.

Cota, M. (2014). *Project Management in Central Bank of Brazil.* Brasilia: Interview by S. Gasik on 2014.11.17.

Cowra Council (2015). *Projects of Cowra Shire.* Cowra, NSW: Interview by S. Gasik on 2015.06.24.

Dominicana El Congreso Nacional (2006). *Ley de Planificación e Inversion Publica,* No. 498-06. Santo Domingo, Dominicana: El Congreso Nacional.

Durham R. M. (2013). *Public Sector Project Management Forum.* Durham: Regional Municipality of Durham. http://pspmf.ca/Sitemap.aspx. Accessed May 2021.

ECLAC (2021). *Network of National Public Investment Systems (SNIP).* Santiago de Chile: Economic Commission for Latin America and the Caribbean. https://www.cepal.org/en/infographics/network-national-public-investment-systems-snip. Accessed May 2021.

European Commission (2017). *A Guide to EU funding.* Luxembourg: Publications Office of the European Union.

European Commission (n.d.) *Find a Funding Opportunity.* https://ec.europa.eu/info/funding-tenders/how-eu-funding-works/how-get-funding/find-funding-opportunity_en. Accessed October 2021.

European Commission CoEPM (2018). *PM² Project Management Methodology Guide 3.0.* Brussels, Luxembourg: European Commission.

European Union (2020). *ImpleMentAll Advisory Body.* European Union. https://www.implementall.eu/10-advisory-board.html. Accessed November 2021.

Fernandes, J. (2014). *Projects of Toronto Panam 2015 Games.* Toronto, Ontario: Interview by S. Gasik on 2014.09.16.

Finland Prime Minister's Office (2021). *Management and Organization.* Helsinki: Government of Finland. https://vnk.fi/en/management-and-organisation. Accessed May 2021.

Florida DoMS (2013). *Real Estate Development & Management.* Tallahassee: Florida Department of Management Services. http://www.dms.myflorida.com/business_operations/real_estate_development_management/building_construction/project_information. Accessed September 2013.

Florida Legislature (2016). *The 2016 Florida Statutes.* Tallahassee: Florida Legislature. http://www.leg.state.fl.us/Statutes/index.cfm?App_mode=Display_Statute&Search_String=&URL=0200-0299/0282/Sections/0282.0051.html. Accessed May 2021.

Fuertes, A. I., & Herrera, J. C. (2014). *Project Management in the Argentinian Ministry of Labor, Employment and Social Security.* Buenos Aires: Interview by S. Gasik on 2014.11.24.

Gaino, M. (2014). *Projects in PMO SEFAZ, Government of Bahia, Brazil.* Salvador, Bahia: Interview by S. Gasik on 2014.11.18.

Gavazza, N. (2014). *Projects in PRODEB, Government of Bahia.* Salvador, Bahia: Interview by S. Gasik on 2014.11.19.

Georgia GTA (2014a). *Critical Projects Review Panel.* Atlanta: Georgia Technology Authority. http://gta.georgia.gov/critical-projects-review-panel. Accessed May 2021.

Georgia GTA (2014b). *Enterprise Portfolio Management Services.* Atlanta: Georgia Technology Authority. http://gta.georgia.gov/enterprise-portfolio-management-services. Accessed May 2021.

Georgia GTA (2014c). *State of Georgia Agency Project Assurance Self-Certification Guide.* Atlanta: Georgia Technology Authority. https://gta.georgia.gov/sites/gta.georgia.gov/files/related_files/site_page/Georgia%20Agency%20Project%20Assurance%20Self-Certification%20Guide.pdf. Accessed May 2021.

Georgia GTA (2014d). *Georgia Enterprise Management Suite (GEMS).* Atlanta: Georgia Technology Authority. https://gta.georgia.gov/georgia-enterprise-management-suite-gems. Accessed May 2021.

Gershon, P. (2004). *Releasing Resources to the Front Line.* Independent review of public sector efficiency, HM Treasury, London.

Grabowski, A. (2014). *Projects of Ministry of Health and Long-Term Care, Government of Ontario.* Toronto, Ontario: Interview by S. Gasik on 2014.09.15.

Greece GOfS (2009). *ELOT 1429 Standard: Managerial Capability of Organizations Implementing Projects of Public Interest–Requirements.* Athens, Greece: Greek Organization for Standardization (ELOT).

Hawaii ETS (2012). *PMO Development Plan, Office of Enterprise Technology Services.* Honolulu. http://ets.hawaii.gov/wp-content/uploads/2012/09/Governance_PMO-Development-Plan.pdf. Accessed May 2021.

Hawaii Ige, D. Y. (2015). *Administrative Directive on Program Governance Requirements for Act 119 and Enterprise Information Technology Project.* Honolulu: Governor of State of Hawaii.

Hodges, C. (2015). *Projects of Roads and Maritime Service (RMS), Parkes unit.* Parkes, NSW: Interview by S. Gasik on 2015.06.29.

Hong Kong ASD (2021). *Project Management Branch.* Hong Kong: Architectural Services Department. https://www.archsd.gov.hk/en/about-us/about-us-organization-structure.html. Accessed May 2021.

Hong Kong DB (2021). *Contractors.* Development Bureau. https://www.devb.gov.hk/en/construction_sector_matters/contractors/index.html. Accessed May 2021.

Iceland Althingi (1970). *Lög Um Skipan Opinberra Framkvæmda.* No. 63/1970. Reykjavik: Althingi.

Iceland Althingi (2001). *Lög Um Skipan Opinberra Framkvæmda.* No. 84/2001. Reykjavik: Althingi.

IMF (2021). *World Economic Outlook Database.* Washington, DC: International Monetary Fund. https://www.imf.org/en/Publications/WEO/weo-database/2021/October/download-entire-database. Accessed December 2021.

India CPWD (2016). *Central Public Works Department, Ministry of Urban Development;* Government of India; https://web.archive.org/web/20170119051059/http://cpwd.gov.in/cpwdnew/default.aspx. Accessed May 2021.

India CSID (2014). *Online Projects Management System User Manual (Version 2.0).* New Delhi: Cabinet Secretariat Informatics Division. https://www.esuvidha.gov.in/power/files/usermanual.pdf. Accessed June 2021.

India DMEO (2021). *Development Monitoring and Evaluation Office.* New Delhi: Development Monitoring and Evaluation Office. https://dmeo.gov.in/. Accessed May 2021.

India Lokh Sabha (2012). *Public Procurement Bill.* Bill No. 58 of 2012. India, New Delhi: Lokh Sabha,

India MoF (2003). *Office Memorandum: Guidelines for Formulation, Appraisal and Approval of Government Funded Plan Schemes/projects.* India, New Delhi: Ministry of Finance, Department of Expenditure Plan Finance-II Division.

India MoF (2014). *Enhancement in the Financial Powers of Ministries/Departments with regard to expenditures on Non-Plan Schemes/Projects. No. 1 (5). / 2016-E, II(A).* New Delhi: Ministry of Finance, Department of Expenditure Plan Finance-II Division.

India MoPI (1989). Project implementation manual, cit. after R. Dayal, P. Zachariach, & K. Rajpal (1996). *Project Management* (pp: 199 – 216). New Delhi, India: Mittal Publications.

India MoSPI (2005). *Project Implementation Status Report on Central Sector Projects (Costing Rs. 20 Crores & Above).* Ministry Of Statistics & Programme Implementation Infrastructure & Project Monitoring Division.

India MoSPI (2009). *Project Implementation Status Report of Central Sector Projects Costing Rs.20 Crore & Above (April-June, 2009).* India, New Delhi: Government of India, Ministry of Statistics and Programme Implementation; Infrastructure and Project Monitoring Division.

India MoSPI (2013). *Infrastructure and Project Monitoring Division.* http://mospi.nic.in/Mospi_New/site/inner.aspx?status=2&menu_id=108. Accessed May 2021.

India MoSPI (2014). *Guidelines on Members of Parliament Local Area Development Scheme (MPLADS).* Ministry of Statistics and Programme Implementation, Government of India.

India MoSPI (2015). *Members of Parliament Local Area Development Scheme (MPLADS) Annual Report 2013-14.* India, New Delhi: Government of India.

India MoSPI (2021). *Ministry of Statistics and Programme Implementation.* New Delhi: Ministry of Statistics and Programme Implementation, Government of India. http://www.mospi.nic.in/ Accessed June 2021.

India NHSRCL (n.d.) *National High Speed Rail Corporation Limited*. New Delhi: National High Speed Rail Corporation Limited. https://nhsrcl.in/index.php/en. Accessed June 2021.

India NITI (2015). *Guidelines for grant of in Principle Approval (IPA) and Appraisal of Schemes/ Programmes / Projects in NITI Aayog for 2015–16*. India, New Delhi: Government of India National Institution for Transforming India (NITI) Aayog (Project Appraisal and Management Division).

India NITI (2021). *Project Appraisal & Management Division (PAMD)*. https://niti.gov.in/verticals/project-appraisal-and-management-division, Accessed May 2021.

India Planning Commission (2010). *Reference Material Notes on the Functioning of Various Divisions*. India, New Delhi: Planning Commission, Government of India.

India Planning Commission (2011). *Faster, Sustainable and More Inclusive Growth. An Approach to the Twelfth Five Year Plan (2012–17)*. India, New Delhi: Government of India.

India Planning Commission (2013). *Twelfth Five-year Plan (2012/2017)*. Planning Commission, Government of India. SAGE Publications India Pvt Ltd. New Delhi, India.

India PMD (2012). *RFD Results-Framework Documents (2012-2013)*. Performance Management Division, Cabinet Secretariat, Government of India.

Infrastructure Australia (n.d.) *What We Do*. Canberra: Australian Government. https://www.infrastructureaustralia.gov.au/. Accessed April 2021.

Ireland DoT (2021). *Strategic Research and Analysis Division*. Dublin: Department of Transport. https://www.gov.ie/en/organisation-information/4d1c54-strategic-research-and-analysis-division/. Accessed June 2021.

Ireland NRA (2010). *Project Management Guidelines*. Dublin, Ireland: National Roads Authority.

Ireland PAI (2013). *Course Certificate in Public Sector Project Management*. Dublin: Public Affairs Ireland. https://www.pai.ie/leadership-management/project-management-skills/. Accessed May 2021.

Ireland Q. Q. (2013). *Dublin: Quality and Qualification Ireland*. https://www.qqi.ie/Articles/Pages/About-Us.aspx. Accessed May 2021.

Isakowitz, S. J. (2002). *NASA Cost Estimating Handbook*. Washington, DC: NASA HQ.

ISO (1996). *International Electrotechnical Commission (ISO/IEC) Guide 2*. Geneva: ISO Press.

ISO (2012). *ISO 21500:2012 Guidance on Project Management*. Geneva: International Organization for Standardization.

ISO (2015). *ISO 9000:2015. Quality Management Systems—Fundamentals and Vocabulary*. Geneva: International Organization for Standardization.

ISO (2021). *ISO 21502:2021, Project, Programme and Portfolio Management – Concepts and Context*. Geneva, Switzerland: International Standardization Organization.

Kansas KITO (2020). *Project Management Training Schedule*. Topeka: Kansas Information Technology Office. https://ebit.ks.gov/kito/resources/events/2020/10/19/default-calendar/kansas-project-management-methodology. Accessed May 2021.

Kansas OoITS (2008a). *Project Management Overview, Project Management Methodology*. Topeka: Enterprise Project Management Office, Office of Information Technology Services. http://oits.ks.gov/kito/itpmm.htm. Accessed October 2013.

Kansas OoITS (2008b). *Project Management Methodology, Appendix E, Information Technology Policy 2530 – Project Management, Enterprise Project Management Office*. Topeka: Office of Information Technology Services. http://oits.ks.gov/kito/Rel23/E_appendix.pdf. Accessed October 2013.

Kenya Government (2013). *Second Medium Term Plan, 2013 – 2017*. Kenya Vision 2030. Nairobi: Ministry of Devolution and Planning.

Kinnaird, M. (2003). *Report on the Defence Procurement Review*. Sydney, Australia: Department of the Prime Minister and Cabinet.

Koenig, N. (2015). *Managing Projects in Parkes County*. Parkes, NSW: Interview by S. Gasik on 2015.06.26.

KPMG (2011). *Portfolio, Programme and Project Management (P3M) Capabilities in Government – Increasing Success Rates and Reducing Costs*. Wellington, NZ: KPMG New Zealand.

Kuwait MPA (2020). *Mega Projects Agency*. Kuwait City: Ministry of Public Works. https://www.zawya.com/mena/en/company/Mega_Projects_Agency-1003740/. Accessed May 2021.

Lefman, J. (2014). *USA Department of Energy Projects*. Washington, DC: Interview by S. Gasik on 2014.11.19.

Liem, W. (2014). *Projects of Ministry of Health and Long-Term Care, Government of Ontario*. Toronto, Ontario: Interview by S. Gasik on 2014.09.15.

Lithuania Government (2021). *STRATA Government Strategic Analysis Center*. Vilnius: Government of Lithuania. https://strata.gov.lt/en/. Accessed May 2021.

Maine DoT (2013). *MaineDOT Projects*, Augusta: Department of Transportation. https://www.maine.gov/mdot/projects/. Accessed May 2021.

Maine MOIT (2013a). *Cross Functional Work Flow document, Project Management Office*. Augusta: Maine Office of Information Technology. http://www.maine.gov/oit/project_management/CrossFunctional WorkFlowforallOITWorkandProjectRequestsFinal_V1%200.htm. Accessed October 2013.

Maine MOIT (2013b). *Frequently Asked Questions, Project Management Office*. Augusta, Maine: Office of Information Technology. http://www.maine.gov/oit/project_management/faq/index.html. Accessed October 2013.

Maine OIT (2004). *Information Technology Portfolio Management Policy*. Augusta: Office of Information Technology. http://www.maine.gov/oit/policies/ITPortfolioManagement.doc. Accessed October 2013.

Maine OIT (2009). *Information Technology Project Management Policy*. Augusta: Office of Information Technology. http://www.maine.gov/oit/policies/ProjectManagementPolicy.doc. Accessed October 2013.

Malaysia Dahlan, D. Z. (2015). *National Development Planning Practices in Malaysia*. Kuala Lumpur: Implementation Coordination Unit of the Prime Minister's Department.

Malaysia JKRM (2013). *Jabatan Kerja Raya Malaysia*. Kuala Lumpur: Jabatan Kerja Raya Malaysia. https://www.jkr.gov.my/page/64. Accessed May 2021.

Maryland DoIT (2009). *Project Management Oversight*. Annapolis: Department of Information Technology. https://doit.maryland.gov/epmo/Documents/project_mgmt_oversight.pdf. Accessed May 2021.

Maryland DoIT (2013a). *Major IT Development Projects*. Annapolis: Department of Information Technology. https://doit.maryland.gov/epmo/Pages/MITDP/default.aspx. Accessed May 2021.

Maryland DoIT (2013b). *Project Planning and Project Implementation Process*. Annapolis: Department of Information Technology. http://msa.maryland.gov/megafile/msa/speccol/sc5300/sc5339/000113/013000/013991/unrestricted/20110815e-002.pdf. Accessed May 2021.

Maryland Government (2019). *Maryland Code, State Finance and Procurement 3A-301*. Annapolis: Maryland Government. https://www.lawserver.com/law/state/maryland/md-laws/maryland_laws_state_finance_and_procurement_3a-301. Accessed June 2021.

Maryland MDoIT (2021a). *Project Management Oversight*. Annapolis: Department of Information Technology. https://doit.maryland.gov/epmo/Pages/MITDP/oversight.aspx. Accessed May 2021.

Maryland MDoIT (2021b). *System Development Life-cycle (SDLC)*. Annapolis: Department of Information Technology. https://doit.maryland.gov/SDLC/Pages/agile-sdlc.aspx. Accessed May 2021.

Maryland OMAG (2021). *Open Meetings Act Manual*. 10th Edition. Annapolis: Office of the Maryland Attorney General. https://www.marylandattorneygeneral.gov/OpenGov%20Documents/omaManual Print.pdf. Accessed June 2021.

Mayer, C. (2014). *Public Projects in Canada*. Toronto, Ontario: Interview by S. Gasik on 2014.09.18.

Michigan DoITS (2000). *Project Management Tool Standard*. Lansing: Strategic Project Office. Department of Information Technology Solutions. Lansing. http://www.michigan.gov/documents/138001_36352_7.pdf. Accessed May 2021.

Michigan MDoIT (2003). *Dashboard Reporting User Guide*. Lansing: Michigan Department of Information Technology, Strategic Project Office. http://www.michigan.gov/documents/Dashboard_Users_Guide_60672_7.pdf. Accessed May 2021.

Michigan MDoIT (2004). *State of Michigan Project Management Methodology, Project Management Resource Center*. Lansing: Michigan Department of Information Technology.

Michigan MDTB (2004). *PM Certification Program*. Lansing: Michigan Department of Technology https://www.michigan.gov/documents/project_manager_training_93930_7.pdf. Accessed May 2021.

Michigan MDTMB (2013). *I Have a New Project*. Lansing: Michigan Department of Technology. http://www.michigan.gov/dtmb/0,5552,7-150-56355_56581_31294---,00.html. Accessed October 2013.

Missouri OoA (2006). *Missouri Project Management Best Practices Manual Change Management Process.* Jefferson City: Office of Administration. http://oa.mo.gov/itsd/cio/projectmgmt/PDF/bestpractices060224.pdf. Accessed September 2013.

Missouri OoA (2013a). *Project Management & Oversight, Information Technology Services Division.* Jefferson City: Office of Administration. https://oa.mo.gov/information-technology-itsd/it-governance/project-management-oversight. Accessed May 2021.

Missouri OoA (2013b). *Missouri Project Management Methodology, Information Technology Services Division.* Jefferson City: Office of Administration. http://oa.mo.gov/itsd/cio/projectmgmt/V4_1/MOBestPracticeManual_V4_1.pdf. Accessed September 2013.

Missouri OoA (2013c). *Information Technology Advisory Board (ITAB).* Jefferson City: Office of Administration. https://oa.mo.gov/itsd/it-governance/information-technology-advisory-board-itab. Accessed May 2021.

Missouri OoA (2020). *Project Management, Facilities Management, Design & Construction.* Jefferson City: Office of Administration. https://oa.mo.gov/facilities/project-management. Accessed May 2021.

Montana SIT PMO (2013a). *Project Scaling Worksheet.* Helena: State IT Project Management Office. http://pmo.mt.gov/documents.mcpx. Accessed October 2013.

Montana SIT PMO (2013b). *State Project Management Office.* Helena: State IT Project Management Office. http://pmo.mt.gov. Accessed October 2013.

Montana SIT PMO (2013c). *Project Life-cycle Framework.* Helena: State IT Project Management Office. http://pmo.mt.gov/Methodology/methodology.mcpx. Accessed October 2013.

Montana SITSD (2021a). *Project Management Advisory Workgroup (PMAW).* Helena: State Information Technology Services Division. http://sitsd.mt.gov/Governance/Boards-Councils/PMAW. Accessed May 2021.

Montana SITSD (2021b). *SITSD Project Status.* Helena: State Information Technology Services Division. https://sitsd.mt.gov/About-Us/Project-Status. Accessed May 2021.

Morrison, S. (2015). *Regional Development Projects in Australia.* Cowra, NSW: Interview by S. Gasik on 2015.06.25.

Naffah Ferreira, S. (2014). *Project Management in Bahia Government.* Salvador, Bahia, Brazil: Interview by S. Gasik on 2014.11.14.

Netherlands Elverding, P. (2008). *Sneller en Beter! – Advies Commissie VBIP.* Commissie Versnelling Besluitvorming Infrastructurele Projecten.

Netherlands MvIeW (2021). *Ministerie van Infrastructuur en Waterstaat.* Den Haag: Ministerie van Infrastructuur en Waterstaat -en-waterstaat. Accessed May 2021.

Nevada NDoT (2013). *Project Management Division.* Carson City: Nevada Department of Transportation. https://www.dot.nv.gov/doing-business/about-ndot/ndot-divisions/engineering/project-management. Accessed May 2021.

New York NYS Forum (2013). *Project Management Working Group.* Albany: The NYS Forum, Inc. http://www.nysforum.org/committees/projectmanagement/. Accessed May 2021.

New York OoIT (2013). *Enterprise Program Management Office.* Albany: Office of Information Technology Service. https://its.ny.gov/its-vocabulary/enterprise-program-management-office. Accessed September 2013.

New York SOT (2013). *New York State Project Management Methodology, Project Management Guidebook Release 2,* Albany: NY State Office of Technology. https://its.ny.gov/sites/default/files/documents/tocandpreface.pdf. Accessed May 2021.

New Zealand GCDO (2020). *System Assurance.* Wellington: Government Chief Digital Officer. https://www.digital.govt.nz/standards-and-guidance/governance/system-assurance/. Accessed May 2021.

New Zealand Government (2019). *Assuring Digital Government Outcomes. All-of-Government Portfolio, Programme and Project Assurance Framework.* Version 3.2 May 2019. Wellington: New Zealand Government.

New Zealand Infrastructure Commission (2019). *Major Infrastructure Project Governance Guidance.* Wellington: NZ Infrastructure Commission. https://infracom.govt.nz/assets/Uploads/4190498_ITU-Project-Governance-Guidance.pdf. Accessed May 2021.

New Zealand Infrastructure Commission (2021). *Major Projects*. Wellington: Infrastructure Commission Te Waihanga. https://infracom.govt.nz/major-projects/. Accessed May 2021.

New Zealand SSC (2011). *Guidance for Monitoring Major Projects and Programmes*. Wellington: State Services Commission. http://www.ssc.govt.nz/sites/all/files/monitoring-guidance_0.pdf. Accessed October 2013.

New Zealand SSC (2013). *Major Projects Assurance Group*. Wellington: State Services Commission. http://www.ssc.govt.nz/major-projects-assurance/. Accessed September 2013.

New Zealand Treasury (2016). *Investor Confidence Rating*. NZ Treasury. https://www.treasury.govt.nz/information-and-services/state-sector-leadership/investment-management/review-investment-reviews/investor-confidence-rating-icr. Accessed May 2021.

New Zealand Treasury (2017). *Gateway Reviews*. Wellington: The Treasury. https://www.treasury.govt.nz/information-and-services/state-sector-leadership/investment-management/review-investment-reviews/gateway-reviews. Accessed May 2021.

New Zealand Treasury (2019). *Investment Logic Mapping*. Wellington: Treasury. https://www.treasury.govt.nz/information-and-services/state-sector-leadership/investment-management/better-business-cases-bbc/bbc-methods-and-tools/investment-logic-mapping. Accessed May 2021.

New Zealand Treasury (2021a). *Investment Management System*. Wellington: The Treasury. https://www.treasury.govt.nz/information-and-services/state-sector-leadership/investment-management-system. Accessed May 2021.

New Zealand Treasury (2021b). *About the IMAP Team*. Wellington: The Treasury. https://www.treasury.govt.nz/information-and-services/state-sector-leadership/investment-management/about-imap-team. Accessed May 2021.

NI DoF (2014a). *Northern Ireland Gateway Review Process*. Belfast: Department of Finance. https://www.finance-ni.gov.uk/articles/northern-ireland-gateway-review-process. Accessed April 2021.

NI DoF (2014b). *Post Programme or Project Review*. Belfast: Department of Finance. https://www.finance-ni.gov.uk/articles/post-programme-or-project-review. Accessed May 2021.

NI DoF (2014c). *Programme Management*. Belfast: Department of Finance. https://www.finance-ni.gov.uk/articles/programme-management. Accessed May 2021.

NI DoF (2021). *Centre of Expertise for Programme and Project Management*. Belfast: Department of Finance. https://www.finance-ni.gov.uk/topics/programme-and-project-management-and-assurance/centre-expertise-programme-and-project-management. Accessed May 2021.

North Carolina Mays, G., & Bromead, K. (2012). *2011 Customer Satisfaction Survey*. Raleigh: Enterprise Project Management Office, Office of the State Chief Information Officer. http://www.epmo.scio.nc.gov/library/pdf/2011Survey.pdf. Accessed October 2013.

North Carolina OSCIO (2013a). *Enterprise Project Management Office*. Raleigh: Department of Information Technology. https://it.nc.gov/services/project-management. Accessed October 2013.

North Carolina OSCIO (2013b). *Project Managers Advisory Group, Enterprise Project Management Office*. Raleigh: Office of the State Chief Information Officer. http://www.epmo.scio.nc.gov/TaskGroups/PMWorkingGroup.aspx. Accessed September 2013.

North Dakota Dalrymple, J. (2011). *Executive Order. Additional Oversight of the Contracting and Implementation Process for Large Scale IT Projects*. Bismarck: Office of the Governor. (Archived). https://www.governor.nd.gov/executive-orders/executive-order-archive. Accessed May 2021.

Norway Berg, P. (2012). Updates on Project Governance in Norway. Paper presented at 5th Concept Symposium on Project Governance. Losby Gods.

Norway Christensen, T. (2008). Governance of Major Investment Projects. Paper presented at Concept Symposium, Trondheim, Norway, September 2008.

Norway NTNU (2013). *Quality Assurance Scheme, Concept Research Program*. Trondheim: Norwegian University of Science and Technology. https://www.ntnu.edu/concept/qa-scheme. Accessed May 2021.

NSW AO (2015). *Large Construction Projects: Independent Assurance*. Sydney, Australia: Audit Office of New South Wales.

NSW Buy.nsw (2021). *Supplier List*. Sydney, Australia: NSW Government. https://suppliers.buy.nsw.gov.au/supplier. Accessed June 2021.

NSW DFSI (2011). *Project Management Guideline*. Version 1.2. Sydney, Australia: NSW Government Finance, Services & Innovation.

NSW DFSI (2017). *ICT Assurance Framework*. Sydney, Australia: NSW Government Finance, Services & Innovation.

NSW Government (2017). *Unsolicited Proposals. Guide for Submission and Assessment*. Sydney, Australia: NSW Government.

NSW Government (2018). *Benefits Realisation Management Framework. Parts 1 – 5*. Sydney, Australia: NSW Government.

NSW Infrastructure (2020). *Oversight Framework for the NSW Infrastructure Program*. Sydney, Australia: Infrastructure New South Wales.

NSW Parliament (1993). *Local Government Act 1993 No 30*. Sydney: NSW Parliament.

NSW Parliament (2021). *Public Works and Procurement Act 1912 no 45*. Sydney, Australia: Parliament of New South Wales.

NSW Procure Point (2013). *Gateway Review System*. Sydney: Procure Point. https://arp.nsw.gov.au/tc10-13-gateway-review-system/. Accessed May 2021.

NSW Procurement (2020). *Accreditation Program for Goods & Services Procurement. Accreditation Program Requirements*. Sydney, Australia: NSW Procurement.

NSW PSC (2020a). *The NSW Public Sector Capability Framework*. Sydney, Australia: NSW Public Service Commission.

NSW PSC (2020b). *Annual Report 2019–2020*. Sydney, Australia: NSW Public Service Commission.

NSW PW (2013). *Project Services*. Sydney: Public Works. Accessed September 2013.

NSW PWA (2021). *Delivering a Lasting Legacy for Regional NSW*. Sydney: Public Works Advisory. https://www.publicworksadvisory.nsw.gov.au/. Accessed May 2021.

NSW Transport (2021). *Projects*. Sydney: Transport for NSW. https://www.transport.nsw.gov.au/projects. Accessed May 2021.

NSW Treasury (2017a). *NSW Gateway Policy*. TPP 17-01. Sydney, Australia: NSW Government Treasury.

NSW Treasury (2017b). *Guide to Cost-Benefit Analysis*. TPP 17-03. Sydney, Australia: NSW Government Treasury

NSW Treasury (2018a). *Circular TC18-03 Program Evaluation*. Sydney, Australia: Treasury, NSW Government.

NSW Treasury (2018b). *Major Projects Policy for Government Businesses*. TPP 18-05. Sydney, Australia: NSW Government Treasury.

NSW Treasury (2018c). *NSW Government Business Case Guidelines*. TPP 18-06. Sydney, Australia: NSW Government Treasury.

OECD (1999). *European Principles for Public Administration*. SIGMA Papers: No. 27. Paris: Organisation for Economic Co-operation and Development.

OECD (2005). *Modernising Government: The Way Forward*. Paris: Organisation for Economic Co-operation and Development. https://www.oecd.org/gov/modernisinggovernmentthewayforward.htm. Accessed June 2021.

Oliveira, M. (2014). *Projects of Department of Public Safety (World Cup 2014 and Carnival)*. Salvador, Bahia: Interview by S. Gasik on 2014.11.18.

Ontario Government (2021a). *Being Agile in the Ontario Public Service*. Toronto: Government of Ontario. https://www.ontario.ca/page/being-agile-ontario-public-service. Accessed September 2021.

Ontario Government (2021b). *Digital Service Standard, 2021*. Toronto: Government of Ontario. https://www.ontario.ca/page/digital-service-standard. Accessed September 2021.

Ontario OPS (2011). *OPS Unified I&IT Project Methodology (UPM) v2.6*. Toronto: I&IT Central Agencies Cluster, Ontario Public Services, Toronto.

Otoo, S., Agapitova, S., & Behrens, J. (2009). *The Capacity Development Results Framework*. Washington, DC: World Bank Institute.

Pakistan Planning Commission (2010). *Manual for Development Projects*. Islamabad: Planning Commission. https://www.pc.gov.pk/uploads/psdp/Manual_PDF.pdf. Accessed May 2021.

Pakistan Planning Commission (2013). *Project Wing*. Islamabad: Planning Commission. http://www.pc.gov.pk/sections.htm. Accessed September 2013.

Pakistan Shah, S. R. A., Khan, A. Z., & Khalil, M. S. (2011). Project management practices in e-government projects: a case study of Electronic Government Directorate (EGD) in Pakistan, *International Journal of Business and Social Science*, 2(7): 235–243.

Paulk M. C., Curtis B., Chrissis, M. B., & Weber, C. V. (1993). *Capability Maturity Model SM for Software*. Version 1.1. Technical Report CMU/SEI-93-TR-024 ESC-TR-93-177. Pittsburg, Pennsylvania, USA: Carnegie Mellon University.

Pennsylvania DoT (2013). *Bureau of Project Delivery*. Harrisburg: Pennsylvania Department of Transportation. http://www.dot.state.pa.us/Internet/Bureaus/pdDesign.nsf/DesignHomepage?OpenFrameSet. Accessed October 2013.

Pennsylvania PoA (2012). *Information Technology Bulletin ITB-EPM006*. Harrisburg: Pennsylvania Office of Administration. https://www.oa.pa.gov/Policies/Documents/itp_epm006.pdf. Accessed December 2018.

Pennsylvania PoA (2013). *Enterprise Project Management Methodology*. Harrisburg: Pennsylvania Office of Administration. http://www.portal.state.pa.us/portal/server.pt/document/1324879/project_management_process_guide_docx. Accessed October 2013.

Peru El Congreso de la Republica (2000). *Ley Del Sistema Nacional de Inversión Pública*. Ley No. 27293, Lima: El Congreso de la Republica del Peru.

Peru MEyF (2014). *Sistema Nacional De Inversión Pública*. Lima, Peru: Ministerio de Economia y Finanzas. https://www.mef.gob.pe/es/?option=com_content&language=es-ES&Itemid=100674&view=article&catid=180&id=306&lang=es-ES. Accessed May 2021.

Philippines DPWH (2020). *Department of Public Works and Highways*. Manila: Department of Public Works and Highways https://www.dpwh.gov.ph/dpwh/content/about-dpwh. Accessed May 2021.

PMI (2003). *Organizational Project Management Maturity Model (OPM3®)*. Newtown Square: Project Management Institute.

PMI (2004). *A Guide to the Project Management Body of Knowledge (PMBOK® Guide)*. 3rd Edition. Newtown Square: Project Management Institute.

PMI (2006a). *Practice Standard for Work Breakdown Structures*. Newtown Square: Project Management Institute.

PMI (2006b). *Government Extension to the PMBOK® Guide Third Edition*. Newtown Square: Project Management Institute.

PMI (2008a). *Organizational Project Management Maturity Model (OPM3^{TM})*. 2nd Edition. Newtown Square: Project Management Institute.

PMI (2008b). *A Guide to the Project Management Body of Knowledge (PMBOK® Guide)*. 3rd Edition. Newtown Square: Project Management Institute.

PMI (2013a). *A Guide to the Project Management Body of Knowledge (PMBOK® Guide)*. 5th Edition. Newtown Square: Project Management Institute.

PMI (2013b). *The Standard for Program Management*. 3rd Edition. Newtown Square: Project Management Institute.

PMI (2017a). *A Guide to the Project Management Body of Knowledge (PMBOK® Guide)*. 6th Edition. Newtown Square: Project Management Institute.

PMI (2017b). *The Standard for Program Management*. 4th Edition. Newtown Square: Project Management Institute.

PMI (2017c). *The Standard for Portfolio Management*. 4th Edition. Newtown Square: Project Management Institute.

PMI (2017d). *Project Management Competency Development Framework*. 3rd Edition. Newtown Square: Project Management Institute.

PMI (2018a). *The Standard for Organizational Project Management*. Newtown Square: Project Management Institute.

PMI (2018b). *Organizational Project Management Maturity Model (OPM3)*. 4th Edition. Newtown Square, PA: Project Management Institute.

PMI (2019). *The Standard for Earned Value Management*. Newtown Square: Project Management Institute.

Poland MS (2014). *Zarządzenie Ministra Sprawiedliwości z dnia 4 czerwca 2014 r. w sprawie ustanowienia i zadań Rady Portfela Projektów oraz zasad zarządzania Portfelem Projektów w Ministerstwie*

Sprawiedliwości, Dziennik Urzędowy Ministra Sprawiedliwości z dnia 6 czerwca 2014, poz. 110. Warszawa: Ministerstwo Sprawiedliwości.

Poland NFZ (2012). *Narodowy Fundusz Zdrowia.* Warszawa: Narodowy Fundusz Zdrowia. https://www. nfz.gov.pl/. Accessed June 2021.

Poland NIK (2013). *Wykonywanie Przez GDDKIA Obowiązków Inwestora Przy Realizacji Inwestycji Drogowych, nr Ewid.* 191/2012/P12076/KIN. Warszawa: Najwyższa Izba Kontroli.

Poland NIK (2014). *Informacja o Wynikach Kontroli: Realizacja Inwestycji Dotyczących Budowy Terminalu do Odbioru Skroplonego Gazu Ziemnego w Świnoujściu,* 187/2014/P/13/058/KGP. Warszawa: Najwyższa Izba Kontroli.

Poland Rada Ministrów (2017). *Strategia na Rzecz Odpowiedzialnego Rozwoju do Roku 2020 (z Perspektywą do 2030 r.).* Warszawa: Council of Ministers.

Poland Sejm (1990). *Ustawa z dnia 8 marca 1990 r., o Samorządzie Gminnym, Dz.U. 1990 Nr 16, poz. 95.* Warszawa: Sejm Rzeczpospolitej Polskiej.

Poland Sejm (2004). *Ustawa z dnia 29 Stycznia 2004 r., Prawo Zamówień publicznych.* Warszawa: Sejm Rzeczpospolitej Polskiej.

Poland Sejm (2007). *Ustawa o Przygotowaniu Finałowego Turnieju Mistrzostw Europy w Piłce Nożnej UEFA EURO 2012.* Warszawa: Sejm Rzeczpospolitej Polskiej.

Poland Sejm (2009). Ustawa z dnia 24 kwietnia 2009 r., o inwestycjach w zakresie terminalu regazyfikacyjnego skroplonego gazu ziemnego w Świnoujściu, Dz.U. z 2009 r., Nr 84, poz. 700.

Poland Sejm (2016). Ustawa z dnia 29 kwietnia 2016 r., o szczególnych zasadach wykonywania niektórych zadań z zakresu informatyzacji działalności organów administracji podatkowej, Służby Celnej i kontroli skarbowej, Dz.U. z 2016 r., poz. 781.

Practical Concepts Incorporated (1979). *The logical framework. A Manager's Guide to a Scientific Approach to Design & Evaluation.* Washington, DC: Practical Concepts Incorporated.

Prado, D. (2014). *Public Projects in Falconi Private Consulting Company.* Belo Horizonte, Minas Gerais: Interview by S. Gasik on 2014.11.19.

Prado, D. & Andrade, C. A. (2015). *Project Management Maturity.* Archibald & Prado Research. Report 2014. https://maturityresearch.com/wp-content/uploads/2020/05/RelatorioMaturidade2014-Governo-Parte-A-Indicadores-1.pdf. Accessed January 2022.

PwC (2012). *Insights and Trends: Current Portfolio, Programme, and Project Management Practices, the Third Global Survey on the Current State of Project Management.* PriceWaterhouseCooper. https://www.pwc. com.tr/en/publications/arastirmalar/pages/pwc-global-project-management-report-small.pdf. Accessed May 2021.

Queensland QTT (2013). *Projects Queensland.* Brisbane: Queensland Treasury and Trade. http://www. treasury.qld.gov.au/projects-queensland/about/index.shtml. Accessed September 2013.

Queensland Treasury (2015). *Project Assessment Framework. Policy Overview.* Brisbane: Department of Infrastructure and Planning. https://s3.treasury.qld.gov.au/files/paf-policy-overview.pdf. Accessed May 2021.

Queensland Treasury (2019). *Project Assessment Framework.* Brisbane: Queensland Treasury. https://www. treasury.qld.gov.au/programs-and-policies/project-assessment-framework/. Accessed May 2021.

Queensland Treasury (2020). *Gateway Reviews (assurance).* Brisbane: Queensland Treasury. https://www. treasury.qld.gov.au/programs-and-policies/project-assessment-framework/gateway-reviews/ Accessed April 2021.

Rand (2014). *Measuring the Capacity of Government to Deliver Policy.* Rand Corporation https://www.rand. org/pubs/corporate_pubs/CP755.html. Accessed June 2022.

Republic of China President (2011). *Taiwan Government Procurement Act.* Taipei, Taiwan: Government of Republic of China.

Ross, K. (2014). *IT Public Projects in Maryland State Board of Election.* Washington, DC: Interview by S. Gasik on 2014.09.01.

RSA CPCMP (2013). *Pretoria: South African Council for Project and Construction Management Professions.* http://www.sacpcmp.org.za/. Accessed May 2021.

RSA DoPW (2013). *Project Management Forms*. Pretoria: Department of Public Works. http://www. publicworks.gov.za/prm_forms.html. Accessed May 2021.

Saskatchewan MeO (2013). *Approved Major Capital Projects*. Regina: Government of Saskatchewan. https:// publications.saskatchewan.ca/#/products/76286. Accessed May 2021.

Scottish Government (2013a). *Portfolio Management*. Edinburgh: Scottish Government. http://www.gov. scot/Topics/Government/ProgrammeProjectDelivery/PortfolioManagement. Accessed May 2021.

Scottish Government (2013b). *Programme and Project Management Principles*. Edinburgh: Scottish Government. http://www.gov.scot/Topics/Government/ProgrammeProjectDelivery/Principles Accessed May 2021.

Scottish Government (2013c). *Portfolio Management*. Edinburgh: Scottish Government. http://www.gov. scot/Topics/Government/ProgrammeProjectDelivery/PortfolioManagement. Accessed May 2021.

SEI (2010). *CMMI® for Development, Version 1.3. CMU/SEI-2010-TR-033, Software Engineering Institute*. Pittsburg, PA: Carnegie Mellon University.

Siles, R. (2014). *Project Management in Argentinian Branch of Inter-American Development Bank*. Washington, DC: Interview by S. Gasik on 2014.11.23.

Singapore CPPM (2011). *Centre for Public Project Management*. Singapore: Government of Singapore. www.sgdi.gov.sg/ministries/mof/departments/cp2m/departments/cp2m. Accessed May 2021.

Singapore MoF (2013). *Centre for Public Project Management*. Singapore: Ministry of Finance. https://www. sgdi.gov.sg/ministries/mof/departments/cp2m. Accessed May 2021.

Smith, D. (2015). *Project Management in Australian Local Governments*. Cowra, NSW: Interview by S. Gasik on 2015.06.25.

Southwell, M. (2015). *Managing Projects in Department of Infrastructure and Regional Development*. Canberra: Interview by S. Gasik on 2015.07.04.

Switzerland FIT SU (2011). *HERMES die Schweizerische Projektführungsmethode*. Bern: Federal IT Steering Unit FITSU.

Tasmania DSS (2021a). *Communities of Practice*. Hobart: Digital Strategy and Services. http://www.dpac. tas.gov.au/divisions/digital_strategy_and_services/governance/communities_of_practice. Accessed May 2021.

Tasmania DSS (2021b). *Archive – Project Templates*. Hobart: Office of eGovernment. http://www.dpac.tas.gov. au/divisions/digital_strategy_and_services/projects/archive_-_project_templates. Accessed May 2021.

Tasmania OeG (2011). *Tasmanian Government Project Management Guidelines Version 7.0*. Hobart: Office of eGovernment. http://www.dpac.tas.gov.au/__data/assets/pdf_file/0019/147511/Tasmanian_ Government_Project_Management_Guidelines_V7_0_July_2011.pdf. Accessed May 2021.

Tasmania OeG (2013). *Project Management Advisory Committee*. Hobart: Office of eGovernment. http:// www.egovernment.tas.gov.au/project_management/project_management_advisory_committee_pmac. Accessed September 2013.

Tennessee TDoT (2021). *Program Development and Administration Division*. Nashville: Tennessee Department of Transportation. https://www.tn.gov/tdot/program-development-and-administration-home.html. Accessed May 2021.

Texas DIR (2013). *Statewide Project Delivery*. Austin, TX: Department of Information Resources. https:// dir.texas.gov/View-Resources/Pages/Content.aspx?id=16. Accessed May 2021.

Texas DIR (2021). *Project Management Essentials. Agile. Standard Operating Procedure*. Austin, TX: Department of Information Resources. https://dir.texas.gov/sites/default/files/2021-08/PM %20Essentials%20-%20Agile_ver1.0.docx. Accessed September 2021.

Texas SAO (2013). *Quality Assurance Team*. Austin: State Auditor's Office. https://qat.dir.texas.gov/. Accessed May 2021.

Thailand Puthamont, G. C. S., & Charoenngam, C. (2007). Strategic project selection in public sector: construction projects of the Ministry of Defence in Thailand, *International Journal of Project Management*, 25: 178–188.

Townsend, A. C. (2014). *Public Projects in the USA*. Washington, DC: Interview by S. Gasik on 2014.09.05.

Tziortsis, J. (2014). *Projects of Government of Ontario*. Toronto, Ontario: Interview by S. Gasik on 2014.09.15.

UK APMG (2021). *Certifications & Solutions*. High Wycombe, UK: APMG International. https://apmg-international.com/our-services/certifications. Accessed September 2021.

UK Axelos (2021). *Our Certifications*. London, UK: Axelos. https://www.axelos.com/certifications. Accessed September 2021.

UK Brett, R.B. (1904). *Esher Report*. London. http://en.wikipedia.org/wiki/Esher_Report. Accessed June 2021. Last edited day: 30 August 2022.

UK BSI (1968). *BS 4335. Glossary of Terms Used in Project Network Analysis*. London: BSI.

UK BSI (2000a). *Project Management. Part 2: Vocabulary*. London: BSI.

UK BSI (2000b). *Project Management. Part 3: Guide to the Management of Business Related Project Risk*. London: BSI.

UK BSI (2002). *Project Management. Part 1: Guide to Project Management*. London: BSI.

UK BSI (2006). *Project Management. Part 4: Guide to Project Management in the Construction Industry*. London: BSI.

UK Cabinet Office (2011). *Overview of the Major Projects Authority*. London: Cabinet Office.

UK Cabinet Office (2013). *Policy Managing Major Projects More Effectively*, London: Cabinet Office and Efficiency and Reform Group. https://www.gov.uk/government/policies/managing-major-projects-more-effectively. Accessed September 2013.

UK EHI (2002). *New Director General of NHS IT Appointed*. EHI eHealth Insider, 05.09.2002. https://web.archive.org/web/20110927170626/http://www.e-health-insider.com/news/item.cfm?ID=257. Accessed June 2022.

UK Gershon, P. (1999). *Review of Civil Procurement in Central Government*. London, UK: HM Treasury.

UK HM Government (2016a). *Agile and Government Services: An Introduction*. London, UK: HM Government. https://www.gov.uk/service-manual/agile-delivery/agile-government-services-introduction. Accessed September 2021.

UK HM Government (2016b). *Core Principles of Agile*. London, UK: HM Government. https://www.gov.uk/service-manual/agile-delivery/core-principles-agile#related-guides. Accessed September 2021.

UK HM Government (2018). *Government Functional Standard GovS 002: Project delivery. Portfolio, Programme and Project Management V1.2*. London, UK: HM Government.

UK HM Government (2021). *Government Launches New Projects Academy*. London: UK Government. https://www.gov.uk/government/news/government-launches-new-projects-academy. Accessed May 2021.

UK HM Treasury (2007). *Project Governance: A Guideline Note for Public Sector Projects*. London, UK: HM Treasury.

UK HM Treasury (2018). *Guide to Developing the Project Business Case. Better Business Cases for Better Outcome*. London, UK: HM Treasury.

UK HM Treasury, Cabinet Office (2011). *Major Project Approval and Assurance Guidance*. London, UK: HM Treasury.

UK IPA & Cabinet Office (2021). *Infrastructure and Projects Authority: Assurance Review Toolkit*. London, UK: UK Government. https://www.gov.uk/government/collections/infrastructure-and-projects-authority-assurance-review-toolkit. Accessed January 2022.

UK IPA (2016). *Guidance for Departments and Review Teams. Project Assessment Review (PAR) v1.0*. London, UK: Infrastructure and Project Authority.

UK IPA (2018). *Project Delivery Capability Framework for Project Delivery Professionals in Government Version 2*. London, UK: Infrastructure and Project Authority.

UK IPA (2020). *Annual Report on Major Projects 2019–20*. London, UK: Infrastructure and Project Authority.

UK IPA (2021). *Infrastructure and Projects Authority Mandate London*. UK: Infrastructure and Project Authority.

UK Latham, M. (1994). *Constructing the Team*. London, UK: HMSO.

UK Lord Browne of Madingley (2013). *Getting A Grip: How to Improve Major Project Execution and Control in Government*. London, UK: Cabinet Office.

UK NAO (2010). *Financial Management Maturity Model*. London, UK: National Audit Office.

UK NAO (2011). *Guide Initiating Successful Projects*. London, UK: National Audit Office.

UK NAO (2012). *Assurance for Major Project. Detailed Methodology.* London, UK: National Audit Office.

UK NAO (2016). *Delivering Major Projects in Government: A Briefing for the Committee of Public Accounts.* HC 713 Session 2015–16 6 January. London, UK: National Audit Office.

UK NAO (2020). *Lessons Learned from Major Programmes Cross-government. Report by the Comptroller and Auditor General.* Session 2019–2021 20 November 2020 HC 960. London, UK: National Audit Office.

UK OGC (2007). *The OGC Gateway™ Process.* A manager's checklist. London: TSO.

UK OGC (2010). *Portfolio, Programme and Project Management Maturity Model (P3M3®) Introduction and Guide to P3M3®.* Version 2.1. London: TSO.

UK OGC (2017). *Managing Successful Projects with Prince2 ©.* 6th Edition. London, UK: TSO.

UK OPSR (2003). *Improving Programme and Project Delivery.* London: Cabinet Office.

UK Oxford University (2021). *MSc in Major Programme Management.* Oxford: Oxford University. https://www.sbs.ox.ac.uk/programmes/degrees/msc-major-programme-management. Accessed May 2021.

UK POST (2003). *Government IT Projects.* London, UK: Parliamentary Office of Science and Technology.

UK University of Oxford (2012). *Oxford Teams up with Cabinet Office to Teach Leadership.* Oxford: University of Oxford. http://www.ox.ac.uk/media/news_stories/2012/120107.html. Accessed October 2013.

UNDP (1997). *Governance for Sustainable Human Development.* New York: United Nations Development Programme.

UNDP (2003). *Democratic Governance Group, Public administration practice note.* New York: United Nations Development Programme, Bureau for Development Policy.

UNDP (2009). *Frequently Asked Questions: The UNDP Approach to Supporting Capacity Development.* New York: Capacity Development Group. Bureau for Development Policy. United Nations Development Programme.

United Nations (2002). *Benchmarking E-government: A Global Perspective – Assessing the UN Member Status.* New York: Division for Public Economics and Public Administration and American Society for Public Administration.

United Nations (2006). *Definition of Basic Concepts and Terminologies in Governance and Public Administration, E/C.16/2006/4.* New York: Economic and Social Council, United Nations.

United Nations (2019). *World Economic Situation and Prospects 2019.* New York: United Nations.

USA ANSI/EIA (1998). *EIA Standard. Earned Value Management Systems.* ANSI/EIA-748. Arlington: Electronic Industries Alliance.

USA Congress (1928). *The Boulder Canyon Project Act.* Washington, DC: US Congress.

USA Congress (1966). *The Freedom of Information Act (FOIA).* Washington, DC: US Congress.

USA Congress (1972). *Federal Advisory Committee Act.* Washington, DC: US Congress.

USA Congress (1993). *Government Performance and Results Act.* Washington, DC: US Congress.

USA Congress (2010a). *GPRA Modification Act, GPRA MA.* Washington, DC: US Congress.

USA Congress (2010b). *Patient Protection and Affordable Care Act.* Washington, DC: US Congress.

USA Congress (2015). *Program Management Improvement and Accountability Act, PMIAA.* Washington, DC: US Congress.

USA Congress (2018). *Foundations for Evidence-Based Policy-Making Act of 2018.* Washington, DC: US Congress.

USA DoD (1967). *Performance Measurement for Selected Acquisitions. Department of Defense Instruction 7000.2.* Washington, DC: Department of Defense.

USA DoD (1996). *Mandatory Procedures for Major Defense Acquisition Programs and Major Automated Information System Acquisition Programs, Department of Defense Regulation (DODR) 5000.2.* Washington, DC: Department of Defense.

USA DoE (2012). *Electricity Subsector Cybersecurity Capability Maturity MODEL (ES-C2M2).* Version 1.0. Washington, DC: Department of Energy.

USA DoT (2021). *Policy and Strategy Analysis Team.* Washington, DC: Department of Transportation. https://www.fhwa.dot.gov/policy/otps/policyanalysis.cfm. Accessed May 2021.

USA EPA (n.d.) *Public Participation Guide. United States Environmental Protection Agency.* Washington, DC: United States Environmental Protection Agency. https://www.epa.gov/international-cooperation/public-participation-guide-introduction-guide. Accessed November 2021.

USA FAI (2015). *Project Manager's Guidebook*. Washington, DC: Federal Acquisition Institute.

USA FAI (2021). *Program and Project Managers (FAC-P/PM)*. Washington, DC: Federal Acquisition Institute. https://www.fai.gov/certification/program-and-project-managers-fac-ppm. Accessed June 2021.

USA GAO (2012). *Schedule Assessment Guide. Best Practices for Project Schedule. GAO-12-120G*. Washington, DC: United States Government Accountability Office.

USA GAO (2016). *Technology Readiness Assessment Guide. Best Practices for Evaluating the Readiness of Technology for Use in Acquisition Programs and Projects, GAO-16-410G*. Washington, DC: United States Government Accountability Office.

USA GAO (2019). *Improving Program Management. Key Actions Taken, but Further Efforts Needed to Strengthen Standards, Expand Reviews, and Address High-Risk Areas. GAO-20-44*. Washington, DC: Government Accountability Office.

USA GAO (2020a). *Cost Estimating and Assessment Guide. Best Practices for Developing and Managing Program Costs. GAO-20-195G*. Washington, DC: Government Accountability Office.

USA GAO (2020b). *Agile Assessment Guide. Best Practices for Agile Adoption and Implementation. GAO-20-590G*. Washington, DC: Government Accountability Office.

USA Government (2010). *Max.gov*. Washington, DC: US Government. https://max.gov. Accessed May 2021.

USA GSA (2019). *Project Management Tools*. Washington, DC: US General Services Administration. https://www.gsa.gov/node/81730. Accessed May 2021.

USA GSA, DoD, NASA USA (2019). *Federal Acquisition Regulations*. Washington, DC: General Services Administration.

USA NAVY / EC (1999). *Software Requirements Specification (SRS)*. http://everyspec.com/DATA-ITEM-DESC-DIDs/DI-IPSC/DI-IPSC-81433A_3709/ Accessed May 2020.

USA OMB (1976). *Circular A-109 Major System Acquisitions*. Washington, DC: Office of Management and Budget.

USA OMB (2005). *Guidance for Completing the Program Assessment Rating Tool (PART)*. Washington, DC: Office of Management and Budget.

USA OMB (2007). *PARTWeb User's Manual, Office of Management and Budget*. Washington, DC: Office of Management and Budget.

USA OMB (2021). *Office of Management and Budget. Management*. Washington, DC: Office of Management and Budget. https://www.whitehouse.gov/omb/management/. Accessed April 2021.

USA OPM (2019). *Interpretive Guidance for Project Manager Positions. Attracting, Hiring, and Retaining Project Managers*. Washington, DC: U.S. Office of Personnel Management.

USA OPM (2021). *Federal Workforce Data*. Washington, DC: Office for Personnel Management. https://www.fedscope.opm.gov/employment.asp. Accessed June 2021.

USAID (2019). *Promoting Accountability & Transparency*. Washington, DC: USAID. https://www.usaid.gov/what-we-do/democracy-human-rights-and-governance/promoting-accountability-transparency. Accessed January 2020.

Uzbekistan NAPM (2021). *National Agency of Project Management*. Tashkent: Government of Uzbekistan. https://www.napm.uz. Accessed June 2021.

Van Krieken, T. (2014). *Projects of City of Toronto*. Toronto, Ontario: Interview by S. Gasik on 2014.09.16.

Vermont AoA (2021). *Information Technology Retainer Contracts*. Montpelier: Agency of Administration. https://bgs.vermont.gov/sites/bgs/files/files/purchasing-contracting/Publications/Retainer%20Master%20List%20May%2017%2C%202021.xls. Accessed May 2021.

Vermont AoDS (2020). *Enterprise Project Management Office*. Statutes. Montpelier: Agency of Digital Services. https://epmo.vermont.gov/statutes. Accessed May 2021.

Vermont AoDS (2021). *Project Templates*. Montpelier: Agency of Digital Services. epmo.vermont.gov/project-process/project-templates. Accessed May 2021.

Vermont AoT (1996). *Project Development Process*. Montpelier, Vermont: Vermont Agency of Transportation. https://vtrans.vermont.gov/sites/aot/files/highway/documents/publications/PDManual.pdf. Accessed April 2021.

Vermont AoT (2020). *Project Delivery Bureau*. Montpelier, Vermont: Agency of Transportation. https://vtrans.vermont.gov/about. Accessed May 2020.

Vermont DoII (2010). *Enterprise Project Management Office Charter Revision 2.* Montpelier: Department of Information and Innovation. http://dii.vermont.gov/sites/dii/files/pdfs/EPMO-Charter.pdf. Accessed September 2013.

Vermont DoII (2013). *Project Management Procedure.* Montpelier: Department of Information and Innovation.

Victoria Development (2021). *Development Victoria* https://www.development.vic.gov.au/. Accessed May 2021.

Victoria DJPR (2020). *Significant Projects. Department of Jobs, Precincts and Regions.* Melbourne: Victoria State Government. https://djpr.vic.gov.au/significant-projects. Accessed May 2021.

Victoria DPAC (2016). *End-to-end project delivery framework, Department of Premier and Cabinet.* Melbourne: Victoria State Government. https://www.vic.gov.au/sites/default/files/2019-08/PM-GUIDE-03-End-to-End-Project-Delivery-Framework.docx. Accessed May 2021.

Victoria DPAC (2018). *ICT Project Quality Assurance Framework.* Melbourne: Victoria State Government. https://www.vic.gov.au/sites/default/files/2019-08/IT-Project-quality-assurance-framework.pdf. Accessed May 2021.

Victoria DSDBI (2013). *Annual Report 2012–13.* Melbourne: Department of State Development Business and Innovation. https://www.parliament.vic.gov.au/file_uploads/DSDBI_annual_report_book_web_complete_lr_WznTTPQn.pdf. Accessed May 2021.

Victoria DTF (2012). *Project Governance.* Melbourne: Department of Treasury and Finance. https://www.dtf.vic.gov.au/sites/default/files/2018-03/Project%20governance%20-%20Technical%20guide.doc. Accessed May 2021.

Victoria DTF (2019). *Gateway Review Process.* Melbourne: Department of Treasury and Finance. https://www.dtf.vic.gov.au/infrastructure-investment/gateway-review-process. Accessed April 2021.

Victoria DTF (2021). *High Value High Risk Framework.* Melbourne: Treasury and Finance. https://www.dtf.vic.gov.au/infrastructure-investment/high-value-high-risk-framework. Accessed May 2021.

Victoria MPV (2013). *Major Projects Victoria.* Melbourne, Victoria, http://www.majorprojects.vic.gov.au/. Accessed September 2013.

Victoria OPV (2016). *Office of Projects Victoria.* Melbourne, Victoria: Government of Victoria. http://www.opv.vic.gov.au/Home. Accessed September 2021.

Victoria OPV (2021a). *Australian Major Projects Leadership Academy. Office of Projects Victoria.* Melbourne, Victoria: Government of Victoria. http://www.opv.vic.gov.au/System-wide-improvements/Australian-Major-Projects-Leadership-Academy. Accessed May 2021.

Victoria OPV (2021b). *Advisory Board. Office of Projects Victoria.* Melbourne, Victoria: Government of Victoria http://www.opv.vic.gov.au/About-us/Advisory-Board. Accessed May 2021.

Victoria Parliament (1994). *Project Development and Construction Management Act 1994 No. 101 of 1994.* Melbourne: Parliament of Victoria.

Virginia VDoT (2021). *Project Management Office.* Richmond: Virginia Department of Transportation. http://www.virginiadot.org/business/locdes/project_management_office.asp. Accessed May 2021.

Virginia VITA (2013). *CTP – Oracle Primavera Portfolio Management (OPPM).* Richmond: Virginia Information Technologies Agency. http://www.vita.virginia.gov/oversight/projects/default.aspx?id=505. Accessed October 2013.

Virginia VITA (2016). *Best Practices.* Richmond: Virginia Information Technologies Agency. https://www.vita.virginia.gov/current-copy/vitavirginiagov/integrated-services/psc-9-1-1-services/best-practices/. Accessed June 2021.

Virginia VITA (2017). *Information Technology Investment Management (ITIM).* Richmond: Virginia Information Technologies Agency. https://www.vita.virginia.gov/policy--governance/it-investment-management/information-technology-investment-management-itim/. Accessed June 2021.

Virginia VITA (2018). *Project Manager Selection and Training Standard.* Richmond: Virginia Information Technologies Agency. https://www.vita.virginia.gov/media/vitavirginiagov/it-governance/psgs/pdf/Project_Manager_Selection_and_Training_Standard_CPM_111.pdf. Accessed May 2021.

Virginia VITA (2019). *Commonwealth IT Project Manager Orientation.* Richmond: Virginia Information Technologies Agency. https://www.vita.virginia.gov/media/vitavirginiagov/it-governance/pdf/Commonwealth_IT_Project_Manager_Orientation_Workbook.pdf. Accessed May 2021.

Wade, J. (2014). *US Federal Public Projects*. Washington, DC: Interview by S. Gasik on 2014.09.10.

Washington OCIO (2011). *Policy No. 131: Managing Information Technology Projects*. Olympia, Washington: Office of the Chief Information Officer. https://ocio.wa.gov/sites/default/files/public/policies/131.docx?n7bd. Accessed May 2021.

Washington WSDoT (2013a). *Project Management Online Guide*. Olympia: Washington State Department of Transportation. https://wsdot.wa.gov/construction-planning/project-management/online-guide/home. Accessed April 2021.

Washington WSDoT (2013b). *Strategic Analysis and Estimating Office*. Olympia: Washington State Department of Transportation. http://www.wsdot.wa.gov/Design/SAEO/. Accessed May 2021.

Washington WSDoT (2013c). *Project Management – Delivering the Capital Construction Programs at the Project Level*. Olympia: Washington State Department of Transportation. http://www.wsdot.wa.gov/Projects/ProjectMgmt/. Accessed October 2013.

Washington WSDoT (2013d). *Design – Value Engineering*. Olympia: Washington State Department of Transportation Management Principles https://wsdot.wa.gov/Design/ValueEngineering/default.htm. Accessed May 2021.

Washington WSHR (2013). *Project Management Certificate Program*. Olympia: Washington State Human Resources. http://des.wa.gov/training/category/49/ProjectManagementTraining. Accessed May 2021.

Western Australia DoF (2020a). *Buildings and Contracts. Contact List Agency Contacts for Procurement, Project Delivery and Maintenance*. Perth: Department of Finance. https://www.wa.gov.au/sites/default/files/2020-12/Finance%20agency%20services%20contracts%20and%20building%20contact%20list%20Nov%202020.pdf. Accessed May 2021.

Western Australia DoF (2020b). *Strategic Projects*. Perth, WA: Department of Finance. https://www.wa.gov.au/government/multi-step-guides/building-projects/strategic-projects. Accessed May 2021.

Western Australia DoSD (2013a). *Resources and Industry Development*. Perth: Department of State Development. http://www.dsd.wa.gov.au/what-we-do. Accessed January 2017.

Western Australia DoSD (2013b). *Project Approvals Framework*. Australia: Department of State Development, Perth, Western Australia. http://www.dsd.wa.gov.au/6737.aspx. Accessed September 2013.

Western Australia DPaC (2002). *Consulting Citizens: A Resource Guide*. Perth: Citizen and Civics Unit, Dept. of the Premier and Cabinet.

Western Australia DPaC (2009). *Lead Agency Framework. A Guidance Note for Implementation*. Perth: Department of the Premier and Cabinet.

Western Australia ExpoTrade (2020). 11th Annual WA Major Projects Conference 2020, Perth. http://www.waconference.com.au/. Accessed May 2021.

World Bank (1989). *Sub-Sahara Africa. From Crisis to Sustainable Growth: A Long-Term Perspective Study*. Washington, DC: World Bank.

World Bank (1993). *Governance*. Washington, DC: World Bank.

World Bank (2002). *World Development Report 2002: Building Institutions for Markets*. Washington, DC: The World Bank.

World Bank (2007). *Little Data Book, International Bank for Reconstruction and Development*. Washington, DC: Development Data Group.

World Bank (2012). *Guide to Evaluating Capacity Development Results*. Washington, DC: The World Bank.

Wyoming WWP (2013). *Project/Program Management*. Cheyenne: Wyoming Workforce Planning. http://wyomingworkforceplanning.state.wy.us/wyoming_competencies/performance_management_cluster/project_program_management.htm. Accessed May 2021.

References: Literature

Abba, W. F. (1997). *Earned Value Management—Reconciling Government and Commercial Practices.* PM: Special Issue January-February 58–63.

Abbott, K. W., & Snidal, D. (2001). International 'standards' and international governance, *Journal of European Public Policy*, 8 (3): 345–370.

Abrahamsson, A., & Agevall, L. (2010). Immigrants caught in the crossfire of projectification of the Swedish public sector: short-term solutions to long-term problems, *Diversity in Health and Care*, 7: 201–209.

Abuya, I. O. (2016). Development projects as mechanisms for delivery of public services in Kenya. Development projects as mechanisms for delivery, *PM World Journal*, 5 (5): 1–10.

Acemoglu, D., & Robinson, J. (2008) The Role of Institutions in Growth and Development. *Australian Government, AusAID.* Canberra, Australia: Commission on Growth and Development.

Acemoglu, D., & Robinson, J. (2008). *The Role of Institutions in Growth and Development.* Washington, DC: World Bank.

Achterstraat, P. (2013). *Why Large Public Sector Projects Sometimes Fail.* Sydney, NSW: Audit Office of NSW.

Adelman, I. (1999). The Role of Government in Economic Development, Department of Agricultural and Resource Economics and Policy Division of Agricultural and Natural Resources University of California at Berkeley Working Paper No. 890.

Adler, T. R., Pitz, T. G., & Meredith, J. (2016). An analysis of risk sharing in strategic R&D and new product development projects, *International Journal of Project Management*, 34 (6): 94–922.

Agnafors, M. (2013). Quality of government: toward a more complex definition, *American Political Science Review*, 107 (3): 433–445.

Ahsan, K., & Gunawan, I. (2008). *Performance Analysis of International Development Projects.* Auckland, New Zealand: Auckland University of Technology. https://www.anzam.org/wp-content/uploads/pdf-manager/1191_AHSAN_KAMRUL-161.PDF

Al Mamun, M., Sohag, K., & Hassan, M. K. (2017). Governance, resources and growth, *Economic Modelling*, 63: 238–261. Doi: 10.1016/j.econmod.2017.02.015

Alari, C., & Thomas, P. (2016). *Improving Government Effectiveness across the World Can Lessons from the UK's Reform Experience Help?* London, UK: Institute for Government.

Albrecht, J. C., & Spang, K. (2016). Disassembling and reassembling project management maturity, *Project Management Journal*, 47 (5): 18–35.

Al-Haddad, S., & Kotnour, T. (2015). Integrating the organizational change literature: a model for successful change, *Journal of Organizational Change Management*, 28 (2): 234–262. Doi: 10.1108/JOCM-11-2013-0215

Allas, T. (2018). How the public sector fits in the productivity puzzle. In R. Dobbs, D. Fine, S. Hieronimus, & N. Singh (eds.), *McKinsey on Government* (pp. 13–15). New York: McKinsey&Company.

Allison, G. (1971). *Essence of Decision: Explaining the Cuban Missile Crisis.* 1st Edition. Little Brown. ISBN 0-673-39412-3.

Almarabeh, T., & AbuAli, A. (2010). A general framework for E-government: definition maturity challenges, opportunities, and success, *European Journal of Scientific Research*, 39 (1): 29–42.

Althaus, C., Bridgman, P., & Davis G. (2013). *Australian Policy Handbook.* 5th Edition. Sydney: Allen & Unwin.

Alzahrani, J. I., & Emsley, M. W. (2013) The impact of contractors' attributes on construction project success: a post construction evaluation, *International Journal of Project Management,* 31 (2): 313–322. Doi: 10.1016/j.ijproman.2012.06.006

Ambler, S. (2021) Five Ways Agencies Can Become More Agile. Government Executive. July 15, 2021. https://www.govexec.com/management/2021/07/five-ways-agencies-can-become-more-agile/183799/. Accessed September 2021.

Andersen, E. S., Grude, K. V., & Haug, T. (1987). *The Goal Directed Project Management.* London, UK: Kogan Page.

Andersen, J. A. (2010). Public versus private managers: how public and private managers differ in leadership behavior, *Public Administration Review,* 70 (1): 131–141.

Anderson, J. E. (1975). *Public Policy-making.* New York, Washington, DC: Praeger.

Anderson, J. E. (2003). *Public Policy-making: An Introduction.* Boston: Houghton Mifflin Company.

Anderson, J. E. (2015). *Public Policy-making: An Introduction.* 8th Edition. Stamford, CT: Cengage Learning.

Andrews, M. (2010). Good government means different things in different countries, *Governance,* 23: 7–35. Doi: 10.1111/j.1468-0491.2009.01465.x

Andrews, M., Pritchett, L., & Woolcock, M. (2017). *Building State Capability. Evidence, Analysis, Action.* Oxford: Oxford University Press.

Andrews, R., Beynon, M. J., & McDermott, A. M. (2016). Organizational capability in the public sector: a configurational approach, *Journal of Public Administration Research and Theory,* 26 (2): 239–258. Doi: 10.1093/jopart/muv005

Anthopoulos, L., Reddick, C. G., Giannakidou, I., & Mavridis, N. (2016). Why e-government projects fail? An analysis of the Healthcare.gov website, *Government Information Quarterly,* 33: 161–173. Doi: 10.1016/j.giq.2015.07.003

Archibald, R. D. (2008). Project Management in Support of Public Administration: Reflecting the State of the Art in International Project Management. Paper presented at ISIPM Seminario Il PM come leva di cambiamento nella PA, November 13, 2008. Roma: Luiss Business School.

Arrain, F. M. (2005). Strategic management of variation orders for institutional buildings: leveraging on information technology, *Project Management Journal,* 36 (4): 66–77.

Asquer, A., & Mele, V. (2018). Policy-making and Public Management. In E. Ongaro & S. van Thiel (eds.), *The Palgrave Handbook of Public Administration and Management in Europe* (pp. 517–533). Doi: 10.1057/978-1-137-55269-3_27

Aubry, M., Richer, M-C., Lavoie-Tremblay, M., & Cyr, G. (2011). Pluralism in PMO performance: the case of a PMO dedicated to a major organisational transformation, *Project Management Journal,* 42 (6): 60–77.

Baccarini, D. (1996). The concept of project complexity – a review, *International Journal of Project Management,* 14 (4): 201–204.

Bach, T. (2014). The autonomy of government agencies in Germany and Norway: Explaining variation in management autonomy across countries and agencies, *International Review of Administrative Sciences,* 80 (2): 341–361. Doi: 10.1177/0020852313514527

Bach, T. (2016). Administrative autonomy of public organizations. In A. Farazmand (ed.), *Global Encyclopedia of Public Administration, Public Policy, and governance.* Springer. Doi: 10.1007/978-3-319-31816-5_143-1

Bachmann, R. (2001). Trust, power and control in trans-organizational relations, *Organization Studies,* 22 (2): 337–365.

Bachtler, J., & Wren, C. (2006). Evaluation of European Union cohesion policy. Research questions and policy challenges, *Regional Studies,* 40 (2): 143–153. Doi: 10.1080/00343400600600454

Bäck, H., & Hadenius, A. (2008). Democracy and state capacity: exploring a J-shaped relationship, *Governance: An International Journal of Policy Administration, and Institutions,* 21 (1): 1–24.

Bageis, A. S., & Fortune, C. (2009). Factors affecting the bid/no bid decision in the Saudi Arabian construction contractors, *Construction Management and Economics,* 27: 53–71.

Baldry, D. (1998). The evaluation of risk management in public sector capital project, *International Journal of Project Management*, 16 (1): 35–41.

Baldwin, J. N. (1990). Perceptions of public versus private sector personnel and informal red tape: their impact on motivation, *American Review of Public Administration*, 20 (1): 7–28. Doi: 10.1177/0275 07409002000102

Baqir, M. N., & Iyer, L. (2010). E-government maturity over 10 years: a comparative analysis of E-government maturity in select countries around the world. In C. G. Reddick (ed.), *Comparative E-government* (pp. 3–22). Integrated Series in Information Systems, 25.

Baranskaya, A. (2007). Project Management in Public Administration of Transitional Countries. Paper presented at the 15th NISPAcee Annual Conference, Kiev, May 2007.

Barkley B. T. (2011). *Government Program Management*. New York: McGraw Hill.

Barzelay, M. (2003). Introduction: the process dynamics of public management policy-making, *International Public Management Journal*, 6: 251–281.

Baser, H., & Morgan, P. (2008). *Capacity, Change and Performance: Study Report*. Maastricht: ECDPM.

Batool, A., & Abbas, F. (2017). Reasons for delay in selected hydro-power projects in Khyber Pakhtunkhwa (KPK), Pakistan, *Renewable and Sustainable Energy Reviews*, 73: 196–204. Doi: 10.1016/j.rser. 2017.01.040

Bauer, M. W. (2018). Public Administration and Political Science. In E. Ongaro & S. van Thiel (eds.), *The Palgrave Handbook of Public Administration and Management in Europe* (pp. 1049–1065). Doi: 10. 1057/978-1-137-55269-3_53

Bauhr, M., & Nasiritousi, N. (2012). How do international organizations promote quality of government? Contestation, Integration, and the limits of IO power, *International Studies Review*, 14: 541–566. Doi: 10.1111/misr.12009

Baum, W. C., & Tolbert, S. M. (1985). *Investing in Development. Lessons of World Bank Experience*. New York: Oxford University Press.

Baumgartner, F. R., & Jones, B. J. (1993). *Agendas and Instability in American Politics*. Chicago: University of Chicago Press.

Bay, A. F., & Skitmore, M. (2006). Project management maturity: some results from Indonesia, *Journal of Building and Construction Management*, 10: 1–5.

BearingPoint (2012). *Bearing Point's Contract Management Maturity Model*. Amsterdam: Bearing Point. Retrieved from www.bearingpoint.com/en-other/download/Contract_Management_Self_Assessment.pdf

Bayiley, Y. T., & Teklu, G. K. (2016). Success factors and criteria in the management of international development projects: evidence from projects funded by the European Union in Ethiopia, *International Journal of Managing Projects in Business*, 9 (3) 562–582. Doi: 10.1108/IJMPB-06-2015-0046

Beck, K., Beedle, M., van Bennekum, A., Cockburn, A., Cunningham, W., Fowler, M., Grenning, J., Highsmith, J., Hunt, A., Jeffries, R., Kern, J., Marick, B., Martin, R. C., Mellor, S. Schwaber, K. Sutherland, J., & Thomas, D. (2001). Manifesto for Agile Software Development. https:// agilemanifesto.org/

Becker, J., Knackstedt, R., & Pöppelbuß, J. (2009). Developing maturity models for IT management – a procedure model and its application, *Business & Information Systems Engineering*, 3: 213–222.

Beer, M., & Nohria, N. (2000). Cracking the code of change, *Harvard Business Review*, 78 (3) 133–141.

Bekker, M. C. (2014). Project governance: "schools of thought", *South African Journal of Economic and Management Sciences*, 17 (Special Issue): 22–32.

Bell, D. E., Raiffa, H., & Tversky, A. (eds.). (1988). *Decision Making: Descriptive, Normative, and Prescriptive Interactions*. Cambridge, UK: Cambridge University Press.

Berman, P. (1978). *Designing Implementation to Match Policy Situation: A Contingency Analysis of Programmed and Adaptive Implementation*. Santa Monica, CA: P-6211, RAND Corporation.

Besner, C., & Hobbs, B. (2008). Discriminating contexts and project management best practices on innovative and noninnovative projects, *Project Management Journal*, 39: S1–S123.

Bester, A. (2015). Capacity development. A report prepared for the United Nations Department of Economic and Social Affairs for the 2016 Quadrennial Comprehensive Policy Review. QPCR2016 – Study on Capacity Development.

Bhuiyan, N., & Thomson, V. (1999). The use of continuous approval methods in defence acquisition project, *International Journal of Project Management*, 17 (2): 121–129.

Biesenthal, C., & Wilden, R. (2014). Multi-level project governance: trends and opportunities, *International Journal of Project Management*, 32: 1291–1308. Doi: 10.1016/j.ijproman.2014.06.005

Biggs, S., & Smith, S. (2003). A paradox of learning in project cycle management and the role of organizational culture, *World Development*, 31 (10): 1743–1757. Doi:10.1016/S0305-750X(03)00143-8

Bititci, U. S., Garengo, P., Ates, A., & Nudurupati, S. S. (2015). Value of maturity models in performance measurement, *International Journal of Production Research*, 53 (10): 3062–3085.

Blau, P. M., & Meyer, M. W. (1987). *Bureaucracy in Modern Society*. New York, NY: Random House.

Blomkamp, E. (2018). The promise of co-design for public policy, *Australian Journal of Public Administration*, 77 (4): 729–743. Doi: 10.1111/1467-8500.12310

Boersma, K., Kingma, S. F., & Veenswijk, M. (2007). Paradoxes of control: the (electronic) monitoring and reporting system of the Dutch High Speed Alliance (HSA), *International Journal of Project Management*, 38 (2): 75–83.

Boettke, P. (2018). Presidential address. Economics and Public Administration, *Southern Economic Journal*, 84 (4): 938–959. Doi: 10.1002/soej.12265

Bogucki, D. (2015). *Studium Wykonalności (in Polish)*. Warszawa: Presscom.

Boh, W. F. (2007). Mechanisms for sharing knowledge in project-based organizations, *Information and Organization*, 17 (1): 27–58.

Boräng, F., Nistotskaya, M., & Xezonakis, G. (2017). The quality of government determinants of support for democracy, *Journal of Public Affairs*, 17 (1–2): e1643. Doi: 10.1002/pa.1643

Borell, K., & Westermark, Å. (2016). Siting of human services facilities and the not in my back yard phenomenon: a critical research review, *Community Development Journal*, 53 (2): 246–262. Doi: 10.1093/cdj/bsw039

Bosch-Rekveldt, M., Jongkind, Y., Mooi, H., Bakker, H., & Verbraeck, A. (2011). Grasping project complexity in large engineering projects: the TOE (Technical, Organizational and Environmental) Framework, *International Journal of Project Management*, 29: 728–739.

Botlhale, E. (2017). Enhancing public project implementation in Botswana during the NDP 11 period, *Africa's Public Service Delivery and Performance Review*, 5 (1): 1–9.

Bouckaert, G., & Halligan, J. (2008). *Managing Performance: International Comparisons*. London: Routledge.

Bouckaert, G., Nakrošis, V., & Nemec, J. (2011). Public administration and management reforms in CEE: main trajectories and results, *The NISPA CEE Journal of Public Administration and Policy*, IV (1): 9–29. Doi: 10.2478/v10110-011-0001-9

Bourne, L. (2008). SRMM: Stakeholder Relationship Maturity Model. Paper presented at PMI EMEA Global Congress, Malta.

Bourne, L., & Walker, D. H. T. (2008). Project relationship management and the stakeholder circle, *International Journal of Managing Projects in Business*, 1 (1): 125–130.

Bovaird, T., & Löffler, E. (2009). Understanding public management and governance. In T. Bovaird & E. Löffler (eds.), *Public Management and Governance*. 2nd Edition (pp. 3–14). New York: Routledge.

Bowden, P. (1986). Problems of implementation, *Public Administration and Development*, 6 (1): 61–71.

Bowen, G. (2008). Naturalistic inquiry and the saturation concept: a research note, *Qualitative Research*, 8 (1): 137–142.

Boyne, G. A. (1998). *Public Choice Theory and Local Government: A Comparative Analysis of the UK and USA*. London: Macmillan.

Boyne, G. A. (2002). Public and private management: what's the difference, *Journal of Management Studies*, 39 (1): 97–122.

Boyne, G. A. (2003a). What is public service improvement? *Public Administration*, 81 (2): 211–227.

Boyne, G. A. (2003b). Sources of public service improvement: a critical review and research agenda, *Journal of Public Administration Research and Theory*, 13 (3): 367–394. Doi: 10.1093/jopart/mug027

Bozeman, B., & Bretschneider S. (1994). The "Publicness Puzzle" in organization theory: a test of alternative explanations of differences between public and private organizations, *Journal of Public Administration Research and Theory*, 4 (2): 197–223. Doi: 10.1093/oxfordjournals.jpart.a037204

Bozeman, B., & Kingsley G. (1998). Risk culture in public and private organizations, *Public Administration Review*, 58 (2): 109–118.

Bozeman, B., Reed P. N., & Scott P. (1992). Red tape and task delays in public and private organizations, *Administration and Society*, 24 (3): 290–322. Doi: 10.1177/009539979202400302

Brady, T., & Davies, A. (2014). Managing structural and dynamic complexity: a tale of two projects, *Project Management Journal*, 45 (4): 21–38.

Bredgaard, T., Dalsgaard, L., & Larsen, F. (2003). *An Alternative Approach for Studying Public Policy – The Case of Municipal Implementation of Active Labour Market Policy in Denmark*. Working Paper from Department of Economics, Politics and Public Administration Aalborg University.

Bredillet, C. N. (2003). Genesis and role of standards: theoretical foundations and socio-economical model for the construction and use of standards, *International Journal of Project Management*, 21 (6): 463–470.

Bretschneider, S. (1990). Management information systems in public and private organizations: an empirical test, *Public Administration Review*, 50 (5): 536–545. Doi: 10.2307/976784

Brock, D. M. (2003). Autonomy of individuals and organizations: towards a strategy research agenda, *International Journal of Business and Economics*, 2 (1): 57–73.

Brown, M. (2012). Enhancing and measuring organizational capacity: assessing the results of the U.S. Department of Justice rural pilot program evaluation, *Public Administration Review*, 72 (4): 506–515.

Brown, R. S., Ali, M., & Joseph, R. C. (2015). *Organizational Capabilities as a Solution to the Problem of Government Inertia: The Case of T-government and Big Data*. Middletown, PA: School of Business Administration, Pennsylvania State University.

Brunet, M., & Aubry, M. (2016). The three dimensions of a governance framework for major public projects, *International Journal of Project Management*, 34 (8): 1596–1607. Doi: 10.1016/j.ijproman. 2016.09.004

Brunetto, Y., & Farr-Wharton, R. (2003). The impact of government practice on the ability of project managers to manage, *International Journal of Project Management*, 21(2): 125–133.

Buchanan, B. (1974). Government managers, business executives, and organizational commitment, *Public Administration Review*, 34 (4): 339–347.

Buchanan, B. (1975). Red-tape and the service ethic: some unexpected differences between public and private managers, *Administration and Society*, 6 (4): 423–428.

Buelens, M., & Van den Broeck, H. (2007). An analysis of differences in work motivation between public and private sector organizations, *Public Administration Review*, 67 (1): 65–74.

Burnstein, P. (1991). Policy domains: organization, culture, and policy outcomes, *Annual Review of Sociology*, 17: 327–350.

Cats-Baril, W., & Thompson, R. (1995). Managing information technology projects in the public sector, *Public Administration Review*, 55 (6): 559–566. Doi: 10.2307/3110347

Cepiku, D. (2017a). Collaborative Governance. In Public Administration. In T. R. Klassen, D. Cepiku, & T. J. Lah (eds.), *The Routledge Handbook of Global Public Policy and Administration* (pp. 141–156). New York, NY: Routledge.

Cepiku, D. (2017b). Performance Management. In Public Administration. In T. R. Klassen, D. Cepiku, T. J. Lah (eds.), *The Routledge Handbook of Global Public Policy and Administration* (pp. 293–308). New York, NY: Routledge.

Chang, A., Chih, Y-Y., Chew, E., & Pisarski, A. (2013). Reconceptualising mega project success in Australian Defence: recognising the importance of value co-creation, *International Journal of Project Management*, 31 (8): 1139–1153.

Chapman, J. R. (2016). A framework for examining the dimensions and characteristics of complexity inherent with rail megaprojects, *International Journal of Project Management*, 34 (6): 937–956.

Charron, N., & Lapuente, L. (2010). Does democracy produce quality of government? *European Journal of Political Research*, 49 (4): 443–470. Doi: 10.1111/j.1475-6765.2009.01906.x

Charron, N., Dijkstra, L., & Lapuente, V. (2014). Regional governance matters: quality of government within European Union Member States, *Regional Studies*, 48 (1): 68–90. Doi: 10.1080/00343404. 2013.770141

Chauvet, L., Collier, P., & Duponchel, M., (2010). What explains aid project success in post-conflict situations? *The World Bank Policy Research Working Paper*, 5418.

Chetankumar, P., & Ramachandran, M. (2009). Agile maturity model (AMM): a software process improvement framework for agile software development practices, *International Journal of Software Engineering*, 2 (1): 3–28.

Chih, Y-Y, & Zwikael, O. (2015). Project benefit management: a conceptual framework of target benefit formulation, *International Journal of Project Management*, 33 (2): 352–362.

Childe, V. G. (1964). *How Labour Governs: A Study of Workers' Representation in Australia*. 2nd Edition. Parkville: Melbourne University Press.

Chou, J-S., & Yang J-G. (2012). Project management knowledge and effects on construction project outcomes: an empirical study, *Project Management Journal*, 43 (5): 47–67.

Chuaire, M. F., Scartascini, C., & Tommasi, M. (2014). *State Capacity and the Quality of Policies. Revisiting the Relationship between Openness and Government Size*. New York, USA: Inter-American Development Bank.

CIA (2020). *The CIA World Factbook 2020–2021*. Langley, McLean, USA: Central Intelligence Agency.

CIoG (2002). *What is Governance?* Ottawa: Canada's Institute of Governance. https://iog.ca/what-is-governance/. Accessed June 2021.

Clegg, D., & Barker, R. (1994). *Case Method Fast-track. A RAD Approach*. Reading, USA: Addison Wesley.

Cochran, C. E., Mayer, L. C., Carr, T. R., & Cayer, N. J. (2010). *American Public Policy: An Introduction*. 10th Edition. Boston, MA: Cengage Wadsworth.

Cochran, C. L., & Malone, E. F. (1995). *Public Policy: Perspectives and Choices*. New York: McGraw-Hill.

Cochran, C. L., & Malone, E. F. (2005). *Public Policy: Perspectives and Choices*. 3rd Edition. Boulder, CO: Lynne Rienner Publishers.

Cochran, C. L., & Malone, E. F. (2010). *Public Policy: Perspectives and Choices*. 4th Edition. Boulder, CO: Lynne Rienner Publishers.

Cochran, C. L., Mayer, L. C., Carr, T. R., & Joseph, N. (2010). *American Public Policy: An Introduction*. 10th Edition. Boston, MA: Cengage Wadsworth.

Cohen, M. D., March, J. G., & Olsen, J. P. (1972). A garbage can model of organizational choice, *Administrative Science Quarterly*, 17 (1): 1–25. Doi: 10.2307/2392088

Colebatch H. (1998). *Policy*. Buckingham: Open University Press.

Colebatch, H. (2006). What work makes policy? Paper for the panel 'Working for policy: who does what, where and how?' organized by the Research Committee on Public Policy and Administration World Congress International Political Science Association Fukuoka, Japan, July 2006.

Conyers, D., & Kaul, M. (1990). Strategic issues in development management: learning from successful experience, *Part I. Public Administration and Development*, 10: 127–140. Doi: 10.1002/pad.4230100202

Cooper, R. G. (1990). Stage-gate systems: a new tool for managing new products, *Business Horizons*, 33 (3): 44–54.

Coster, C. J., & Van Wijk, S. (2015). Lean project management: an exploratory research into lean project management in the Swedish public and private sector. Retrieved from http://www.diva-portal.org/smash/record.jsf?pid=diva2%3A850350&dswid=9324

Cottam, I., Kawalek, P., & Shaw, D. (2004). A local government CRM maturity model: a component in the transformational change of UK councils, Proceedings of the Americas Conference on Information Systems, New York, NY.

Couillard, J., Garon, S., & Riznic, J. (2009). The logical framework approach–millennium, *Project Management Journal*, 40 (4): 31–44.

Crawford, J. K. (2006). The project management maturity model, *Information Systems Management*, 23 (4): 50–58.

Crawford, J. K. (2014). *Project Management Maturity Model*. 3rd Edition. Boca Raton: CRC.

Crawford, L., & Helm, J. (2009). Government and governance: the value of project management in the public sector, *Project Management Journal*, 40 (1): 73–87.

Crawford, L., Polack, J., & England, D. (2007). How standard are standards: an examination of language emphasis in project management standards, *Project Management Journal*, 38 (3): 6–21.

Crosby, P. (1979). *Quality Is Free. The Art of Making Quality Certain*. New York, NY: McGraw-Hill.

Dabla-Norris, E., Brumby, J., Kyobe, A., Mills, Z., & Papageorgiou, C. (2010). *Investing in Public Investment: An Index of Public Investment Efficiency*. IMF Working Paper. Washington, DC: International Monetary Fund.

Daft, R. L. (2010). *Management*. 9th Edition. Mason, OH: South-Western Cengage Learning.

Dahl, R. (1947). The science of Public Administration: three problems, *Public Administration Review*, 7 (1): 1–11.

Damm, D., & Schindler, M. (2002). Security issues of a knowledge medium for distributed project work, *International Journal of Project Management*, 20 (1): 30–47.

Damoah, I. S., & Akwei, C. (2016). Government project failure in Ghana: a multidimensional approach, *International Journal of Managing Projects in Business*, 10 (1): 32–59. Doi: 10.1108/IJMPB-02-2016-0017

Damoah, I. S., & Kumi, D. K. (2017). Causes of government construction projects failure in an emerging economy. Evidence from Ghana, *International Journal of Managing Projects in Business*, 11 (3): 558–582. Doi: 10.1108/IJMPB-04-2017-0042

Damoah, I. S., Akwei, C. A., Amoako, I. O., & Botchie, D. (2018). Corruption as a source of government project failure in developing countries: evidence from Ghana, *Project Management Journal*, 49 (3): 17–33. Doi: 10.1177/8756972818770587

De Bruijn, J. A., & ten Heuvelhof, E. F. (1991). *Sturingsinstrumenten voor de overheid*. Leiden/Antwerpen: Stenfert Kroese.

De Bruin, T., Freeze, R., Kulkarni, U., & Rosemann, M. (2005). Understanding the main phases of developing a Maturity Assessment Model. In B. Campbell, J. Underwood, & D. Bunker (eds.), Australasian Conference on Information Systems (ACIS), November 30–December 2, 2005. Sydney, Australia.

de Souza Silva, A. J., & Gomes Feitosa, M. G. (2012). Maturity in project management: a study of existing practices in government bodies, *Pernambuco, Revista de Gestão e Projetos*, 3 (2): 7–234.

deLeon, P., & deLeon, L. (2002). What ever happened to policy implementation? An alternative approach, *Journal of Public Administration Research and Theory*, 12 (4): 467–492.

deLeon, P., & Steelman, T. A. (2001). Making public policy programs effective and relevant: the role of the political sciences, *Journal of Policy Analysis and Management*, 20 (1): 163–171.

Denizer, C., Kaufmann, D., & Kraay, A. (2013). Good countries or good projects? Macro and micro correlates of World Bank project performance, *Journal of Development Economics*, 105: 288–302. Doi: 10.1016/j.jdeveco.2013.06.003

Dettbarn, Jr. J. L., Ibbs, C. W., & Murphree, Jr. E. L. (2005). Capital project portfolio management for federal real property, *Journal of Management in Engineering*, 21 (1): 44–53.

Dey, P. K. (2000). Managing projects in fast track – a case of public sector organization in India, *International Journal of Public Sector Management*, 13 (7): 588–609.

Di Pierro, G., & Piga, G. (2017). The road ahead for public procurement in Europe. In T. R. Klassen, D. Cepiku, & T. J. Lah (eds.), *The Routledge Handbook of Global Public Policy and Administration* (pp. 373–386). New York, NY: Routledge.

Diallo, A., & Thuillier, D. (2004). The success dimensions of international development projects: the perceptions of African project coordinators, *International Journal of Project Management*, 22 (1): 19–31. Doi: 10.1016/S0263-7863(03)00008-5

Dilts, D. M., & Pence K. R. (2006). Impact of role in the decision to fail: an exploratory study of terminated projects, *Journal of Operations Management*, 24 (4): 378–396.

Dimitrova, A., Mazepus, H., Toshkov, D., Chulitskaya, T., Rabava, N., & Ramasheuskaya, I. (2020). The dual role of state capacity in opening socio-political orders: assessment of different elements of state capacity in Belarus and Ukraine, *East European Politics*. Doi: 10.1080/21599165.2020.1756783

do Céu Alves, M., & Amaral Matos, S. I. (2013). ERP adoption by public and private organizations – a comparative analysis of successful implementations, *Journal of Business Economics and Management*, 14 (3): 500–519.

Dowding, K., Hindmoor, A., & Martin, A. (2016). The Comparative Policy Agendas Project: theory, measurement and findings, *Journal of Public Policy*, 36 (1): 3–25. Doi: 10.1017/S0143814X15000124

Downs, A. (1957). *An Economic Theory of Democracy*. New York: Harper & Row.

Drechsler, W. (2005). The rise and demise of the New Public Management, *Post-Autistic Economics Review*, 33 (14). http://www.paecon.net/PAEReview/issue33/Drechsler33.htm. Accessed September 2020.

Drew, D., & Skitmore, M. (1997). The effect of contract type and size on competitiveness in bidding, *Construction Management and Economics*, 15 (5): 469–489.

Du, Y., & Yin, Y. (2010). Governance-Management-Performance (GMP) framework: a fundamental thinking for improving the management performance of public projects, *iBusiness*, 2 (03): 282–294.

Duffield, S. M., & Whitty, S. J. (2016). Application of the systemic lessons learned knowledge model for organisational learning through projects, *International Journal of Project Management*, 34 (6): 1280–1293.

Dunn, W. N. (1981). *Public Policy Analysis: An Introduction*. Englewood Cliffs: Prentice Hall.

Dunn, W. N. (2004). *Public Policy Analysis*. 5th Edition. Boston, MA: Pearson.

Dvir, D., Raz, T., & Shenhar, A. J. (2003). An empirical analysis of the relationship between project planning and project success, *International Journal of Project Management*, 21 (2): 89–95.

Dye, T. R. (2001). *Top-Down Policy-making*. New York: Chatham House.

Dye, T. R. (2013). *Understanding Public Policy*. 14th Edition. Boston, MA: Pearsons.

Easton, D. (1965). *A Systems Analysis of Political Life*. New York: John Wiley.

Edelenbos, J., & Klijn, E-H. (2009). Project versus process management in public-private partnership: relation between management style and outcomes, *International Public Management Journal*, 12 (3): 310–331. Doi: 10.1080/10967490903094350

Eden, C., Ackerman, F., & Williams, T. (2005). The amoebic growth of project costs, *Project Management Journal*, 36 (2): 15–27.

Edmunds, S. W. (1984). Strengthening administration of the development process, *International Journal of Public Administration*, 6 (2): 217–243.

Ekstedt, E., Lundin, R. A., Soderholm, A., & Wirdenius, H. (1999). *Neo-industrial Organising*. London: Routledge.

Elazar, D. J. (1984). *American Federalism: A View from the States*. 3rd Edition. New York: Harper & Row.

El-Gohary, N. M., Osman, H., & El-Diraby, T. E. (2006). Stakeholder management for public private partnerships, *International Journal of Project Management*, 24 (7): 595–604.

Enteman, W. F. (1993). *Managerialism: The Emergence of a New Ideology*. Madison, Wisconsin: University of Wisconsin Press.

EPRS (2017). Understanding capacity-building/capacity development. A core concept of development policy. *European Parliamentary Research Service PE 599.411*. https://www.europarl.europa.eu/RegData/etudes/BRIE/2017/599411/EPRS_BRI(2017)599411_EN.pdf

Evans, P., & Rauch, J. E. (1999). Bureaucracy and growth: a cross-national analysis of the effects of "Weberian" state structures on economic growth, *American Sociological Review*, 64 (5): 748–765.

Farazmand, A. (2009). Building administrative capacity for the age of rapid globalization: a modest prescription for the twenty-first century, *Public Administration Review* 69 (6): 1007–1020.

Faridian, P. H. (2015). Innovation in public management: is public E-procurement a wave of the future? A theoretical and exploratory analysis, *International Journal of Public Administration*, 38 (9): 654–662.

Fath-Allah, A., Cheikhi, L., Al-Qutaish, R. E., & Idri, A. (2014). E-government maturity models: a comparative study, *International Journal of Software Engineering & Applications*, 5 (3): 71–91.

Ferlie, E. (1996). *The New Public Management in Action*. Oxford: Oxford University Press.

Fischer, M. (2015). Institutions and coalitions in policy processes: a cross-sectoral comparison, *Journal of Public Policy*, 35 (2): 245–268. Doi: 10.1017/S0143814X14000166

Fitzpatrick, J., Goggin, M., Heikkila, T., Klingner, D. Machado, J., & Martell, C. (2011). A new look at comparative Public Administration: trends in research and an agenda for the future, *Public Administration Review*, 71 (6): 821–830.

Fleming, Q. W., & Koppelman, J. M. (2016). *Earned Value Project Management*. 4th Edition. Newtown Square: Project Management Institute.

Flinders, M. (2006). Public/Private: The boundaries of the state. In C. Hay, M. Lister, & D. Marsh (eds.), *The State: Theories and Issues* (pp. 223–247). New York: Palgrave Macmillan.

Flores, C. C., & de Boer, C. (2015). Symbolic implementation: governance assessment of the water treatment plant policy in the Puebla's Alto Atoyac sub-basin, Mexico, *International Journal of Water Governance*, 3 (4): 1–24. Doi: 10.7564/14-IJWG79

Flyvbjerg, B. (2014). What you should know about megaprojects and why: an overview, *Project Management Journal*, 45 (2): 6–19.

Flyvbjerg, B. (2018). Oral evidence: The Government's Management of Major Projects, HC 1631. *UK House of Commons Public Administration and Constitutional Affairs Committee*. London, UK: House of Commons.

Flyvbjerg, B., Garbuio, M., & Lovallo, D. (2009). Delusion and deception in large infrastructure projects: two models for explaining and preventing executive disaster, *California Management Review*, 51 (2): 170–193.

Flyvbjerg, B., Holm, M., & Buhl, S. (2002). Underestimating costs in public works projects. Error or lie? *Journal of the American Planning Association*, 68 (3): 279–295.

Fong, P. S. W. (2003). Knowledge creation in multidisciplinary project teams: an empirical study of the processes and their dynamic interrelationships, *International Journal of Project Management*, 21 (7): 479–486.

Fong, P. S. W. (2005). Co-creation of knowledge by multidisciplinary project teams. W: Love, Peter E. D., Patrick S. W. Fong, and Zahir Irani (eds.) (2005) Management of Knowledge in Project Environments. Butterworth-Heinemann.

Fontaine, G. (2015). *El análisis de políticas públicas: conceptos, teorías y métodos*. Barcelona, España/Quito, Ecuador: Anthropos Editorial/FLACSO.

Fottler, M. D. (1981). Is management really generic? *Academy of Management Review*, 6 (1): 1–12.

Fred, M. (2015). Projectification in Swedish municipalities. A case of porous organizations. *Scandinavian Journal of Public Administration*, 19 (2): 49–68.

Fred, M., & Hall, P. (2017). A projectified Public Administration how projects in Swedish Local governments become instruments for political and managerial concerns, *Statsvetenskaplig tidskrift*, 119: 185–205.

Frederickson, H. G., Smith, K. B., Larimer, C. W., & Licari, M. J. (2012). *The Public Administration Theory Primer*. 2nd Edition. Boulder: Westview Press.

Fry, B. (1989). *Mastering Public Administration*. Chatham, NJ: Chatham House.

Fu, H-P., & Ou, J-R. (2013). Combining PCA with DEA to improve the evaluation of project performance data: a Taiwanese Bureau of Energy Case Study, *Project Management Journal*, 44 (1): 94–106.

Gale (2008). *West's encyclopedia of American Law*. 2nd Edition. The Gale Group, Inc. Accessible at https://legal-dictionary.thefreedictionary.com/Sunset+laws. Accessed August 2020.

Gallivan, M. J., & Keil, M. (2003). The user–developer communication process: a critical case study, *Information Systems Journal*, 13 (1): 37–68.

Gantt, H. L. (1919). *Organizing for work*. New York: Harcourt, Brace and Howe, Inc.

Garland, R. (2009). *Project Governance: A Practical Guide to Effective Project Decision Making*. London, UK: Kogan Page.

Garzás, J., Pino, F. J., Piattini, M., & Fernández, C. M. (2013). A maturity model for the Spanish software industry based on ISO standards, *Computer Standards and Interfaces*, 35 (6): 616–628.

Gasik, S. (2008a). *Project Management Maturity Meta-Model*. Warszawa: Sybena Consulting. http://www.sybena.pl/dokumenty/Gasik-PM-maturity-meta-model.pdf. Accessed May 2021.

Gasik, S. (2008b) Project Families and the Unified Project Evaluation Model. Paper presented at 22nd IPMA World Congress "Project Management to Run" Roma, Italy: International Project Management Association.

Gasik, S. (2011a). A model of project knowledge management, *Project Management Journal*, 42 (3) 23–44.

Gasik, S. (2011b). A proposal of PMBOK® Chapter 13 Project Knowledge Management. http://www.sybena.pl/dokumenty/PMBoK-project-knowledge-management-area-Gasik.pdf. Accessed May 2021.

Gasik, S. (2014). P-government – a framework for public projects management, *PM World Journal*, 3 (7): 1–26. (Portuguese translation: p-Governo – Um Sistema para Gestão de Projetos Públicos; Gabinete do Governador do Estado da Bahia, Brazil; http://www.sybena.pl/dokumenty/Projekty-publiczne/PMWJ-Gasik-p-governo-um-sistema-para-gestao-de-projetos-publicos-POR.pdf. Accessed June 2021).

Gasik, S. (2016a). National Public Projects Implementation System for India. Project Management Institute. Project Management National Conference, Mumbai, India, 17–19 November 2016.

Gasik, S. (2016b). A conceptual model of national public projects implementation systems. In Ż. Ilmete (ed.), *Project Management Development – Practice and Perspectives*. Proceedings of the 5th International Scientific Conference on Project Management in the Baltic Countries, University of Latvia, Riga, Latvia.

Gasik, S. (2016c) Are public projects different than projects in other sectors? Preliminary results of empirical research. Proceedings of International Conference on Project MANagement, Porto, Portugal, October 5–7, 2016, Procedia Computer Science 100 (2016) 399–406.

Gasik, S. (2017). Jak zarządza się projektami publicznymi w Polsce i na świecie (in Polish) Eng. *How are Public Projects Managed in Poland and in the World*. Paper presented at PMI Poland Chapter Warsaw Branch seminar, Warsaw, Poland, June 2017.

Gasik, S. (2018). A framework for analyzing differences between public and other sector projects, *Public Governance*, 3 (45): 73–88. Doi: 10.15678/ZP.2018.45.3.05

Gasik, S. (2019). A proposal of governmental project management maturity model, *PM World Journal*, 8 (9): 1–25.

Gasik, S. Andreasik, T., Guzik, M., Madura, S., & Szubiela, T. (2014). Raport Projekty Publiczne w Polsce. Stan aktualny, analiza, propozycje. Warszawa, Sierpień 2014. http://www.sybena.pl/dokumenty/Projekty-Publiczne/Gasik-Andreasik-Guzik-Madura-Szubiela-Projekty-Publiczne-w-Polsce.pdf. Accessed May 2021.

George, J. M., & Jones, G. R. (2012). *Understanding and Managing Organizational Behavior*. 6th Edition. Boston, USA: Prentice Hall.

Geraldi, J., Maylor, H., & Williams, T. (2011). Now, let's make it really complex (complicated) a systematic review of the complexities of project, *International Journal of Operations & Production Management*, 31 (9): 966–990.

Giacchino, S., & Kakabadse, A. (2003). Successful policy implementation: the route to building self-confident government, *International Review of Administrative Sciences*, 69 (2): 139–160.

Glaser, B., & Strauss, A. L. (1967). *Discovery of Grounded Theory: Strategies for Qualitative Research*. Chicago, IL: Aldine.

Gleeson, D., Legge, D., O'Neill, D., & Pfeffer, M. (2011). Negotiating tensions in developing organizational policy capacity: comparative lessons to be drawn, *Journal of Comparative Policy Analysis*, 13 (3): 237–263.

Godenhjelm, S., Lundin, R. A., & Sjöblom, S. (2015). Projectification in the public sector – the case of the European Union, *International Journal of Managing Projects in Business*, 8 (2): 324–348. Doi: 10.1108/IJMPB-05-2014-0049

Goldfinch, S. (2007). Pessimism, computer failure, and information systems development in the public sector, *Public Administration Review*, 67 (5): 917–929.

Gomes, C. F., Yasin, M. M., & Small, M. H. (2012). Discerning interrelationships among the knowledge, competencies, and roles of project managers in the planning and implementation of public sector projects, *International Journal of Public Administration*, 35 (5): 315–328. Doi: 10.1080/01900692.2012.655461

Goodnow, F. J. (1900). *Politics and Administration: A Study in Government*. New York: Macmillan.

Gosling F. G. (2010). The Manhattan Project. Making the atomic bomb. *National Security History Series, DOE/MA-0002 Revised*. Washington, DC: US Department of Energy.

Gottschalk, P. (2009). Maturity levels for interoperability in digital government, *Government Information Quarterly*, 26 (1): 75–81.

Goulet, L. R., & Frank M. L. (2002). Organizational commitment across three sectors: public, non-profit, and for-profit, *Public Personnel Management*, 31 (2): 201–210.

Grandia, J. (2018). Public Procurement in Europe. In E. Ongaro & S. van Thiel (eds.), *The Palgrave Handbook of Public Administration and Management in Europe* (pp. 363–380). Doi: 10.1057/978-1-137-55269-3_19.

Green, J., & Thorogood, N. (2004). *Qualitative Methods for Health Research*. London: Sage.

Griffith-Cooper, B., & King, K. (2007). The partnership between project management and organizational change: integrating change management with change leadership, *Performance Improvement*, 46 (1): 14–20.

Grundy, T. (1998). Strategy implementation and project management, *International Journal of Project Management*, 16 (1): 43–50.

Grzeszczyk, T. A. (2012). *Modelowanie ewaluacji projektów europejskich*. Warszawa: Wydawnictwo Placet (in Polish).

Gulick, L. (1937). Notes on the Theory of Organization. In L. Gulick & L. Urwick (eds.), *Papers on the Science of Administration*. New York: Institute of Public Administration.

Hall, M., Holt, R., & Purchase, D. (2003). Project sponsors under New Public Management: lessons from the frontline, *International Journal of Project Management*, 21 (7): 495–502.

Halleröd, B., Ekbrand, H., & Gordon, D. (2014). Good Governance – what we think it is and what we really measure. Paper presented at The Quality of Government and the Performance of Democracies seminar, Gothenburg, May 20–22, 2014.

Hansen, M. T., Nohria, N., & Tierney, T. (1999). What's your strategy for managing knowledge? *The Knowledge Management Year Book 2000–2001*, 77 (2): 106–116.

Hansmann, T. (2017). Empirical Development and Evaluation of a Maturity Model for Big Data Applications. Ph.D. thesis. Lüneburg: Universität Lüneburg.

Happe, R. (2009). The Community Maturity Model. *The Community Roundtable*. Retrieved from https://communityroundtable.com/what-we-do/research/community-maturity-model

Hasty, B. K., Schechtman G. M., & Killaly M. (2012). Cloud computing: differences in public and private sector concerns, *International Journal of the Academic Business World*, 6 (1): 51–62.

Heady, F. (2001). *Public Administration: A Comparative Perspective*. 6th Edition. Englewood Cliffs, NJ: Prentice Hall.

Heiskanen, A., Newman, M., & Eklin, M. (2008). Control, trust, power, and the dynamics of information system outsourcing relationships: a process study of contractual software development, *The Journal of Strategic Information Systems*, 17(4): 268–286.

Helliwell, J. F., Layard, R., Sachs, J., & De Neve, J-E. (eds.) (2020). *World Happiness Report 2020*. New York: Sustainable Development Solutions Network.

Helm, J., & Remington, K. (2005). Effective project sponsorship. An evaluation of the role of the executive sponsor in complex infrastructure projects by senior project managers, *Project Management Journal*, 36 (2): 51–58.

Hill, M., & Hupe, P. (2014). *Implementing public policy. An introduction to the study of operational governance*. 3rd Edition. London, UK: Sage.

Hill, M., & Varone, F. (2017). *The Public Policy Process*. New York: Routledge.

Hillson, D. (1997). Towards a risk maturity model, *The International Journal of Project & Business Risk Management*, 1 (1): 35–45.

Hitch, C. J. (1965). *Decision-Making for Defense*. Berkeley and Los Angeles, CA: University of California Press.

Hobbs B., & Aubry M. (2008). An empirically grounded search for a typology of project management office, *Project Management Journal*, 39 (Supplement): S69–S82. Doi: 10.1002/pmj.20061

Hobday, M. (2000). The project-based organisation, an ideal form for managing complex products and systems? *Research Policy*, 29 (7–8): 871–893.

Hodgetts, R., & Luthans, F. (1997). *Managing Organizational Culture and Diversity*. New York, NY: McGraw-Hill.

Hodgson, D., Fred, M., Bailey, S., & Hall, P. (2019). Introduction. In D. Hodgson, M. Fred, S. Bailey, & P. Hall (eds.), *The projectification of the public sector* (pp. 14–42). New York: Taylor and Francis.

Hogwood, B. W., & Gunn, L. (1984). *Policy Analysis for the Real World*. Oxford: Oxford University Press.

Holmberg, S., Rothstein, B., & Nasiritousi, N. (2008). *Quality of Government: What You Get*. QoG Working Paper Series 2008:21. Gothenburg: University of Gothenburg.

Holt, G. D., Olomolaiye, P. O., & Harris, F. C. (1995). A review of contractor selection practice in the U.K. construction industry, *Building and Environment*, 30 (4): 553–561.

Hood, C. (1991). A public management for all seasons? *Public Administration*, 69: 3–19.

Hood, C. (1995). The "New Public Management" in the 1980s: variations on a theme, *Accounting, Organizations and Society*, 20 (3): 93–109.

Hossain, L., & Kuti, M. (2008). CordNet: toward a distributed behavior model for emergency response coordination, *Project Management Journal*, 39 (4): 68–94.

Howlett, M. (2005). What Is a policy instrument? Tools, mixes, and implementation styles. In Eliadis P., Hill M. M., & Howlett M. (eds.), *Designing Government. From Instruments to Governance*. Montreal & Kingston: McGill-Queen's University Press.

Howlett, M. (2009). Governance modes, policy regimes and operational plans: a multi-level nested model of policy instrument choice and policy design, *Political Sciences*, 42: 73–89. Doi 10.1007/s11077-009-9079-1

Howlett, M. (2014). From the 'old' to the 'new' policy design: design thinking beyond markets and collaborative governance, *Political Sciences*, 47: 187–2014. Doi: 10.1007/s11077-014-9199-0

Howlett, M. (2019). *Designing Public Policies Principles and Instruments*. 2nd Edition. New York: Routledge.

Howlett, M. (2020). Re-thinking the political sciences after the 'governance turn' – identifying and creating a (more) capable state. In K. Crowley, J. Stewart, A. Kay, & B. W. Head (eds.), *Reconsidering Policy* (pp. xi–xiv). Complexity, Governance and the State. Bristol, UK: Policy Press, University of Bristol.

Howlett, M., & Rayner, J. (2007). Design principles for policy mixes: cohesion and coherence in new governance arrangements, *Policy and Society*, 26 (4): 1–18. Doi: 10.1016/S1449-4035(07)70118-2

Howlett, M., & Rayner, J. (2013). Patching vs packaging in policy formulation: assessing policy portfolio design, *Politics and Governance*, 1 (2): 170–182.

Humble, J., & Russell, R. (2009). *The Agile Maturity Model Applied to Building and Releasing Software*. ThoughtWorks, Inc. https://info.thoughtworks.com/rs/thoughtworks2/images/agile_maturity_model.pdf. Accessed May 2021.

Hung, Ch-L., & Chou, J., (2013). Resource commitment, organizational diversity, and research performance: a case of the National Telecommunication Program in Taiwan, *Project Management Journal*, 44 (3): 32–47.

Hvidman, U., & Andersen S. C. (2014). Impact of performance management in public and private organizations, *Journal of Public Administration Research and Theory*, 24 (1): 35–58.

Hwang, B. G., Zhao X., & Ng S. Y. (2013). Identifying the critical factors affecting schedule performance of public housing project, *Habitat International*, 38: 214–221.

Ibrahim, L. (2000). Using an Integrated Capability Maturity Model – The FAA Experience. Proceedings of the Tenth Annual International Symposium of the International Council on Systems Engineering (INCOSE), 643–648. Minneapolis, MN.

IIARF (2009). *Internal Audit Capability Model (IA-CM) for the Public Sector*. Altamont Springs, FL: The Institute of Internal Auditors Research Foundation (IIARF).

Ika, L. A., & Donelly, J. (2017). Success conditions for international development capacity building projects, *International Journal of Project Management*, 35 (1): 44–63. Doi: 10.1016/j.ijproman.2016.10.005

Ika, L. A., & Saint-Macary, J. (2014). Special issue: why do projects fail in Africa? *Journal of African Business*, 15 (3): 151–155. Doi: 10.1080/15228916.2014.956635

Ika, L. A., Diallo, A., & Thuillier, D. (2012). Critical success factors for World Bank projects: an empirical investigation, *International Journal of Project Management*, 30 (1): 105–116.

Ikejemba, E. C. X., Mpuan, P. B., Schuur, P. C., & Van Hillegersberg, J. (2017). The empirical reality & sustainable management failures of renewable energy projects in Sub-Saharan Africa, *Renewable Energy*, 102 (2017): 234–240. Doi: 10.1016/j.renene.2016.10.037

IMF (1997). *Good Governance: The IMF's role*. Washington, DC: International Monetary Fund. https://www.imf.org/external/pubs/ft/exrp/govern/govindex.htm

Ingraham, P. W., Joyce, P. G., & Donahue, A. K. (2003). *Government Performance: Why Management Matters*. Baltimore: Johns Hopkins University Press.

Ishikawa, K. (1976). *Guide to Quality Control*. Tokyo, Japan: Asian Productivity Organization.

Jackson, B. J. (2004). *Construction Management Jump Start*. San Francisco: Sybex.

Jackson, C. J. (2007). *Analysis of the 314th Contracting Squadrons Contract Management Capability Using the Contract Management Maturity Model (CMMM)*. Naval Postgraduate School, Monterey.

Jacobson, C., & Choi, S. O. (2008). Success factors: public works and public-private partnerships, *International Journal of Public Sector Management*, 21 (6): 637–657.

Jałocha, B. (2019). The European Union's multi-level impact on member state projectification in light of neoinstitutional theory, *International Journal of Managing Projects in Business*, 12 (3): 578–601. Doi: 10.1108/IJMPB-09-2018-0198

Jałocha, B., & Prawelska-Skrzypek, G. (2017). Public Policies and Projectification Processes. In B. Jałocha, Lenart-Gansiniec, E. Bogacz-Wojtanowska, & G. Prawelska-Skrzypek (eds.), *Public Management: Aims, Attitudes, Approaches* (pp. 135–147). Monographs and Studies of the Jagiellonian University. Kraków: Institute of Public Affairs.

Jałocha, B., Krane, H. P., Ekambaram, A., & Prawelska-Skrzypek, G. (2014). Key competences of public sector project managers, *Procedia – Social and Behavioral Sciences*, 119: 247–256.

Jänicke, M., Jörgens, H., Jörgensen, K., & Nordbeck, R. (2001). *Governance for Sustainable Development in Germany: Institutions and Policy Making*. Paris: OECD Publishing.

Jann, W., & Wegrich, K. (2007). Theories of the policy cycle. In F. Fischer, G. J. Miller, & M. S. Sidney (eds.), *Handbook of Public Policy Analysis Theory, Politics, and Methods* (pp. 43–62). Boca Raton, FL: CRC Press, Taylor & Francis Group.

Jenkins, W. I. (1978). *Policy Analysis: A Political and Organisational Perspective*. Martin Robertson & Co Ltd.

Jensen, C., Johansson, S., & Lofstrom, M. (2017). Policy implementation in the era of accelerating projectification: synthesizing Matland's conflict–ambiguity model and research on temporary organizations, *Public Policy and Administration*, 33 (4): 447–465. Doi: 10.1177/0952076717702957

Jia, G., Chen, Y., Xue, X., Chen, J., Cao, J., & Tang, K. (2011). Program management organization maturity integrated model for mega construction programs in China, *International Journal of Project Management*, 29 (7): 834–845.

Jiang, J. J., Klein, G., Hwang, H., Huang, J., & Hung, S. (2004). An exploration of the relationship between software development process maturity and project performance, *Information Management*, 41 (3): 279–288.

Jin, D., Chai, K-H., & Tan, K-C. H. (2014). New service development maturity model, *Managing Service Quality*, 24 (1): 86–116.

Johansson, J.-E. (2009). Strategy formation in public agencies, *Public Administration*, 87 (4): 872–891. Doi: 10.1111/j.1467-9299.2009.01767.x

Jordan, A. (2017). Portfolio management in government. ProjectManagement.com. Accessible at https://www.projectmanagement.com/articles/417817/Portfolio-Management-in-Government. Accessed May 2021.

Jørgensen, T. B., & Bozeman, B. (2007). Public values: an inventory, *Administration & Society*, 39 (3): 354–381. Doi: 10.1177/0095399707300703

Joyce, P. (2017). Strategic management and public governance in the public sector. In T. R. Klassen, D. Cepiku, & T. J. Lah (eds.), *The Routledge Handbook of Global Public Policy and Administration* (pp. 280–292). New York, NY: Routledge.

Jreisat, J. E. (2010). Comparative Public Administration and Africa, *International Review of Administrative Sciences*, 76 (4): 612. Doi: 10.1177/0020852310381205

Jreisat, J. E. (2011). Commentary – comparative Public Administration: a global perspective, *Public Administration Review*, 71 (6): 834–838.

Jreisat, J. E. (2012). *Globalization and Comparative Public Administration*. New York: CRC Press.

Juchniewicz, M. (2012). Analiza czynników kształtujących poziom i strukturę dojrzałości projektowej organizacji w Polsce, [w:] P. Wyrozębski, M. Juchniewicz, W. Metelski, Wiedza, dojrzałość, ryzyko w zarządzaniu projektami, Oficyna Wydawnicza SGH, Warszawa.

Jurkiewicz, C. L., Massey, T. K. Jr., & Brown, R. G. (1998). Motivation in public and private organizations: a comparative study, *Public Productivity & Management Review*, 21 (3): 230–250.

Judah, T. D. (1857). *A Practical Plan for Building the Pacific Railroad*. San Francisco: Civil Engineer.

Juerges, N., & Hansjürgens, B. (2018). Soil governance in the transition towards a sustainable bioeconomy – a review, *Journal of Cleaner Production*, 170 (2018): 1628e1639. Doi: 10.1016/j.jclepro.2016.10.143

Karaburun, N. (2009) Urban transformation projects in Ankara: challenge for a holistic urban planning system. M.Sc. Thesis. Ankara: Middle East Technical University.

Kassel, D. S. (2010). *Managing Public Sector Projects: A Strategic Framework for Success in an Era of Downsized Government*. Boca Raton: CRC Press.

Kaufmann, D., Kraay, A., & Mastruzzi, M. (2007). Governance matters VI: aggregate and individual governance indicators 1996–2006. *World Bank Policy Research Working Paper*, No. 4280. Washington, DC: World Bank.

Kaufmann, D., Kraay, A., & Mastruzzi, M. (2008). Governance matters VII: aggregate and individual governance indicators 1996-2007, *Policy Research Working Paper 4654*, Washington, DC: World Bank.

Kaufmann, D., Kraay, A., & Mastruzzi, M. (2009). Governance matters VIII. aggregate and individual governance indicators 1996–2008. *World Bank Policy Research Working Paper* No. 4978. Washington, DC: World Bank.

Kaufmann, D., Kraay, A., & Mastruzzi, M. (2010). The worldwide governance indicators: methodology and analytical issues. *Draft Policy Research Working Paper.* Washington DC: World Bank.

Kelley, J. E., & Walker, M. R. (1959). Critical-path planning and scheduling. IRE-AIEE-ACM '59 (Eastern): Papers presented at the December 1-3, 1959, eastern joint IRE-AIEE-ACM computer conference. 160–173. Doi: 10.1145/1460299.1460318

Kennedy, J. F. (1961). *Special Message to Congress on Urgent National Needs.* Boston, MA: John F. Kennedy Presidential Library and Museum. https://www.jfklibrary.org/learn/about-jfk/historic-speeches/address-to-joint-session-of-congress-may-25-1961. Accessed June, 2021.

Kerzner, H. (2005). Using the Project Management Maturity Model. *Strategic planning for project management.* Hoboken, NJ: John Wiley and Sons.

Kettl, D. F. (2010). Governance, contract management and public management. In S. P. Osborne (ed.), *The New Public Governance? Emerging perspectives on the theory and practice of public governance* (pp. 239–254). London, UK: Routledge.

Khal, A. (2011). Introduction to Comparative Public Administration. http://docshare02.docshare.tips/files/7048/70483927.pdf. Accessed June 2021.

Kiel, G., Nicholson, G., Tunny, J. A., & Beck, J. (2012). *Directors at work: A practical guide for boards* (pp. 375–392). Sydney: Thomson Reuters. https://www.effectivegovernance.com.au/do-you-need-a-policy-on-policies/

Kilpatrick, D. G. (2000). *Definitions of Public Policy and the Law.* https://mainweb-v.musc.edu/vawprevention/policy/definition.shtml

Kim, S. E., & Lee, J. W. (2009). The impact of management capacity on government innovation in Korea: an empirical study, *International Public Management Journal*, 12 (3): 345–369.

Kingdon, J. W. (2003). *Agendas, Alternatives, and Public Policies.* 2nd Edition. New York, NY: Pearson.

Klakegg, O. J. (2009). Pursuing relevance and sustainability Improvement strategies for major public project, *International Journal of Managing Projects in Business*, 2 (4): 499–518.

Klakegg, O. J. (2010). Governance of major public investment projects. In pursuit of relevance and sustainability. PhD Thesis. Trondheim, Norway: Norwegian University of Science and Technology, Faculty of Engineering Science and Technology, Department of Civil and Transport Engineering.

Klakegg, O. J., Williams, T., Magnussen, O. M., & Glasspool, H. (2008). Governance framework for public project development, *Project Management Journal*, 39 (S1): 27–42.

Klakegg, O. J., Samset, K., & Magnussen, O. M. (n.d.). Improving success in public investment projects. *Lessons from a Government Initiative in Norway to Improve Quality at Entry*, http://www.researchgate.net/publication/237639077. Accessed May 2021.

Koenig, N. (2015). *Project Management Maturity in Local Government.* Toowoomba: University of Southern Queensland, Faculty of Health, Engineering & Sciences.

Kohsaka, A. (2006). Overview: Infrastructure Development in the Pacific Region. In A. Kohsaka (ed.), *Infrastructure Development in the Pacific Region* (pp. 1–20). London: Taylor and Francis.

Koning, E. A. (2016). The three institutionalisms and institutional dynamics: understanding endogenous and exogenous change, *Journal of Public Policy*, 36 (4): 639–664.

Kooiman, J. (2010). Governance and governability. In S. P. Osborne, *The New Public Governance? Emerging perspectives on the theory and practice of public governance* (pp. 72–86). London, UK: Routledge.

Kooiman, J. (ed.) (1993). *Modern Governance: New Government–Society Interactions.* London: Sage.

Koops, L., Bosch-Rekveldt, M., Coman, L., Hertogh, M., & Bakker, H. (2016). Identifying perspectives of public project managers on project success: Comparing viewpoints of managers from five countries in

North-West Europe, *International Journal of Project Management*, 34 (5): 874–889. Doi: 10.1016/j.ijproman.2016.03.007

Kotnour, T., & Landaeta, R. (2002). Developing a Theory of Knowledge Management Across Projects. IIE Annual Conference Proceedings.

Kouzmin, A. (1979). Building the New Parliament House: An Opera House Revisited? [in:] G. Hawker et al., Working Papers on Parliament, *Canberra College of Advanced Education, Canberra Series in Administrative Studies, No. 5*: 115–171.

Kozak-Holland, M., & Procter, C. (2014). Florence Duomo Project (1420-1436): Learning best project management practice from history, *International Journal of Project Management*, 32 (2): 242–255.

Krukowski, K. (2015). Różnice we wdrażaniu business process reengineeringu w organizacjach publicznych i biznesowych, *Acta Universitatis Nicolai Copernici*, 42 (2): 165–176.

Kuipers, B. S., Higgs, M., Kickert, W., Tummers, L., Grandia, J., & van der Voet, J. (2014). The management of change in public organizations: a literature review, *Public Administration*, 92 (1): 1–20. Doi: 10.1111/padm.12040

Kuokkanen, K. (2014). *Projectified urban policy and citizen participation: An analysis of EU-funded urban projects in Finland*. Helsinki: Swedish School of Social Science, University of Helsinki.

Kuula, M., Putkiranta, A., & Tulokas, P. (2013). Parameters in a successful process outsourcing project: a case from the Ministry of Foreign Affairs, Finland, *International Journal of Public Administration*, 36 (12): 857–864.

Kwak, Y. H., Liu, M., Patanakul, P., Zwikael, O., & Allison, G. T. (2014). *Challenges & Best Practices of Managing Government Projects & Programs*. Newtown Square, PA: Project Management Institute.

Kwak, Y. H., & Anbari, F. T. (2011). History, practices, and future of earned value management: perspectives from NASA, *Project Management Journal*, 43 (1): 77–90. Doi: 10.1002/pmj.20272

Kwak, Y. H., & Smith B. M. (2009). Managing risks in mega defense acquisition projects: performance, policy, and opportunities, *International Journal of Project Management*, 27 (8): 812–820.

Kwak, Y. H. (2003). Brief history of project management. In E. G. Carayannis, Y. H. Kwak, F. Anbari (eds.), *The Story of Managing Projects*. Westport: Quorum Books.

La Porta, R., Lopez-de-Salanes, F., Shleifer, A., & Vishny, R. (1999). The quality of government, *Journal of Law, Economics, and Organization*, 15 (1): 222–279. Doi: 10.1093/jleo/15.1.222

Lane, J. E. (1994). Will public management drive out Public Administration? *Asian Journal of Public Administration*, 16 (2): 139–151.

Lane, J. E. (2000). *The Public Sector. Concepts, Models and Approaches*. 3rd Edition. London, UK: Sage.

Larman, C., & Basili, R. V. (2003). Iterative and incremental development: a brief history, *Computer*, 36 (6): 47–56.

Larsen, A. S. A., Volden, G. H., & Andersen, B. (2021). Project governance in state-owned enterprises: the case of major public projects' governance arrangements and quality assurance schemes, *Administrative Sciences*, 11 (3): 66. Doi: 10.3390/admsci11030066

Lascoumes, P., & Le Gales, P. (2007). Introduction: understanding public policy through its instruments—from the nature of instruments to the sociology of public policy instrumentation, *Governance: An International Journal of Policy, Administration, and Institutions*, 20 (1): 1–21.

Lasswell, H. D. (1951). The policy orientation. In D. Lerner & H. D. Lasswell (eds.), *The Political Sciences*. Stanford, USA: Stanford University Press.

Lau A. W., Pavett, C. M., & Newman, A. R. (1980). The nature of managerial work: a comparison of public and private sector jobs, *Academy of Management Proceedings*, 80 (1): 339–343.

Layne, K., & Lee, J. (2001). Developing fully functional e-government: a four stage model, *Government Information Quarterly*, 18 (2): 122–136.

Le Gales, P. (2011). Policy instruments and governance. In M. Bevir (ed.), *The SAGE Handbook of Governance*. London: Sage.

Lenfle, S., & Loch, C. (2010). Lost roots. How project management came to emphasize control over flexibility and novelty, *California Management Review*, 53 (1): 32–55.

Levy, B. (2004). Governance and Economic Development in Africa: meeting the challenge of capacity building. In B. Levy, & S. Kpundeh (eds.), *Building State Capacity in Africa. New Approaches, Emerging Lessons*. Washington, DC: World Bank Institute.

Lewis-Beck, M. S. (1995). *Data Analysis: An Introduction.* Thousand Oaks: Sage Publications.

Liddle J. (2018). Public value management and new public governance: key traits, issues and developments. In E. Ongaro, & S. Van Thiel (eds.), *The Palgrave Handbook of Public Administration and Management in Europe.* London: Palgrave Macmillan.

Linder, S. H., & Peters, B. G. (1989). Instruments of government: perceptions and contexts, *Journal of Public Policy,* 9 (1): 35–58.

Lindquist, E. (2006). Organizing for policy implementation: the emergence and role of implementation units in policy design and oversight, *Journal of Comparative Policy Analysis,* 8 (4): 311–324, Doi: 10.1080/13876980600970864

Ling, F. Y. Y., Ning, Y., Ke, Y., & Kumaraswamy, M. M. (2013). Modeling relational transaction and relationship quality among team members in public projects in Hong Kong, *Automation in Construction,* 36: 16–24.

Lipsky, M. (1971). Street level bureaucracy and the analysis of urban reform, *Urban Affairs Quarterly,* 6 (4): 391–409.

Locatelli, G., Mariani, G., Sainati, T., & Greco, M. (2017). Corruption in public projects and mega-projects: there is an elephant in the room! *International Journal of Project Management,* 35 (3): 252–268. Doi: 10.1016/j.ijproman.2016.09.010

Lockamy, A., & McCormack, K. (2004). The development of a supply chain management process maturity model using the concepts of business process orientation, *Supply Chain Management International Journal,* 9: 272–278.

Locke, R. R., & Spender, J.-C. (2011). *Confronting Managerialism: How the Business Elite and their Schools Threw Our Lives Out of Balance.* London, UK: Zed Books.

Löffler, E. (2009). Public governance in a network society. In T. Bovaird & E. Löffler, *Public Management and Governance.* 2nd Edition (pp. 215–232). New York, USA: Routledge.

Lopes, C., & Theisohn, T. (2003). *Ownership, Leadership and Transformation: Can We Do Better for Capacity Development?* London: Earthscan/UNDP.

Loveman, B. (1976). The comparative administration group, development administration and anti-development, *Public Administration Review,* 36 (6): 616–621. Doi: 10.2307/975052

Lundin, R. A., & Soderholm, A. (1995). A theory of the temporary organization, *Scandinavian Journal of Management,* 11 (4): 437–455.

Lynn, L. E., & Robichau, R. W. (2013). Governance and organisational effectiveness: towards a theory of government performance, *Journal of Public Policy,* 33 (2): 201–228.

Mackhaphonh, N., & Jia, G. (2017). Megaprojects in developing countries and their challenges, *International Journal of Business Economics and Management,* 4 (11): 6–12.

Magnussen, O. M., & Olsson, N. O. E. (2006). Comparative analysis of cost estimates of major public investment projects, *International Journal of Project Management,* 24 (4): 281–288.

Mahmood, M. N., Dhakal, S., Wiewiora, A., Keast, R. L., Robyn L., & Brown, K. A. (2015). Towards an Integrated Maturity Model of asset management capabilities. In W. B. Lee & J. Mathew (eds.), *Proceedings of the 7th World Congress on Engineering Asset Management* (pp. 1–11). London: Springer.

Mainwaring, S., & Scully, T. R. (2008). Latin America: eight lessons for governance, *Journal of Democracy,* 19 (3): 113–127. Doi: 10.1353/jod.0.0001

Maldonado, N. (2010). The World Bank's evolving concept of good governance and its impact on human rights. Doctoral workshop on development and international organizations, May 29–30, 2017 Stockholm, Sweden.

Male, S., Kelly, J., Gronqvist, M., & Graham, D. (2007). Managing value as a management style for project, *International Journal of Project Management,* 25 (2): 107–114.

Mannion, M., & Keepence, B. (1995). SMART requirements, *ACM SIGSOFT Software Engineering Notes,* 20 (2): 42–47.

Maranny, E. A. (2011). Stage Maturity Model of m-Government (SMM m-Gov). *Improving e-Government Performance by Utilizing m-Government Features.* Enschede, the Netherlands: University of Twente, School of Management and Governance.

Marks, G., Hooghe, L., & Blank, K. (1996). European integration from the 1980s: State-centric v. multi-level governance, *Journal of Common Market Studies,* 34 (3): 341–378.

Markus, M. L., & Mao, J. (2004). Participation in development and implementation — updating an old, tired concept for today's IS contexts, *Journal of the Association for Information Systems*, 5 (11–12): 514–544.

Marsh, E. R. (1976). The harmonogram: an overlooked method of scheduling work, *Project Management Quarterly*, 7 (1): 21–25. https://www.pmi.org/learning/library/harmonogram-overlooked-method-scheduling-work-5666

Marshall, S. (2005). Determination of New Zealand Tertiary Institution E-learning capability: an application of an e-learning maturity model, *Journal of Distance Learning*, 9 (1): 58–63.

Martinsuo, M., Hensman, N., Artto, K., Kujala, J., & Jaafari, A. (2006). Project-based management as an organizational innovation: drivers, changes, and benefits of adopting project-based management, *Project Management Journal*, 37 (3): 87–97.

Marx, F., Wortmann, F., & Mayer, J. H. (2012). A maturity model for management control systems five evolutionary steps to guide development, *Business and Information Systems Engineering*, 4 (4): 193–207.

Masci, P. (2007). Implementing U.S. Airline Industry deregulation: lessons for emerging countries, *Journal of the Washington Institute of China Studies*, 2 (2): 1–17. https://www.bpastudies.org/bpastudies/article/view/34

Maslow, A. H. (1962). *Toward a Psychology of Being*. 2nd Edition. New York: Van Nostrand Reinhold Company.

Matland, R. E. (1995). Synthesizing the implementation literature: the ambiguity–conflict model of policy implementation, *Journal of Public Administration Research and Theory*, 5 (2): 145–174.

Maximilian, J. R., & Jens, P. (2012). Maturity models in business process management, *Business Process Management Journal*, 18 (2): 328–346.

May, P. J., Sapotichne, J., & Workman, S. (2006). Policy coherence and policy domains, *Policy Studies Journal*, 34 (3): 381–403.

McConnell, A. (2010). Policy success, policy failure and grey areas in-between, *Journal of Public Policy*, 30 (3): 345–362. Doi: 10.1017/S0143814X10000152

McDonnell, K., & Elmore, R. (1987). Alternative policy instruments, *Educational Evaluation and Policy Analysis*, 9 (2): 133–152.

McEvoy, P., Brady, M., & Munck, R. (2016). Capacity development through international projects: a complex adaptive systems perspective, *International Journal of Managing Projects in Business*, 9 (3): 528–545. Doi: 10.1108/IJMPB-08-2015-0072

McKinsey (2018). *Delivering for Citizens How to Triple the Success Rate of Government Transformations*. McKinsey Center for Governments.

McPhee, I. (2008). Project Management in the Public Sector, *International Journal of Government Auditing*, October 35 (4): 10–12.

Medeiros, B. C., Danjour, M. F., & Sousa Neto, M. V. (2017). Gerenciamento de projetos: contribuições para a governança de TI no setor público brasileiro (in Portuguese), *Revista Gestão & Tecnologia, Pedro Leopoldo*, 17 (1): 54–78.

Mees, H. L. P., Dijk, J. J., van Soes, D. P., Driessen, P. P. J., van Rijswick, H. F. M. W., & Runhaar, H. (2014). A method for the deliberate and deliberative selection of policy instrument mixes for climate change adaptation, *Ecology and Society*, 19 (2): 58–72. Doi: 10.5751/ES-06639-190258

Meier, K. J., & O'Toole, Jr. L. J. (2011). Comparing public and private management: Theoretical expectations, *Journal of Public Administration Research and Theory*, 21 (Supplement 3): i283–i299. Doi: 10.1093/jopart/mur027

Meinert, L., & Whyte, S. R. (2014). Epidemic projectification. AIDS responses in Uganda as event and process, *Cambridge Anthropology*, 32 (1): 77–94. Doi: 10.3167/ca.2014.320107

Mengel, T., Cowan-Sahadath, K., & Follert, F. (2009). The value of project management to organizations in Canada and Germany, or do values add value? Five case studies, *Project Management Journal*, 40 (1): 28–41.

Merton, R. K. (1963). Bureaucratic Structure and Personality. In N. J. Smelser & W. T. Smelser (eds.), *Personality and Social Systems* (pp. 255–264). John Wiley & Sons Inc. 10.1037/11302-024

Midler, C. (1995). Projectification of the firm: the Renault case, *Scandinavian Journal of Management*, 11 (4): 363–376.

Mihăescu, C., & Ţapardel, A-C. (2013). A Public Administration based on project management, *Administration and Public Management*, 20: 97–107.

Misra, D. C., & Dhingra, A. (2002). E-governance maturity model, *Electronics and Information Planning*, 29 (6–7): 269–275.

Moe, T. L., & Pathranaraku, P. (2006). An integrated approach to natural disaster management. Public project management and its critical success factors, *Disaster Prevention and Management*, 15 (3): 396–413.

Moore, M. (1995). *Creating Public Value: Strategic Management in Government*. Massachusetts: Harvard University Press.

Morris, P. W. G. (1994). *The Management of Projects*. London: Thomas Telford.

Mouly, S. V., & Sankaran, J. K. (2007). Public- versus private-sector research and development. a comparative analysis of two Indian R&D Project Groups, *International Studies of Management and Organization*, 37 (1): 80–102.

Moutinho, J. D. A., & Rabechini Junior, R. (2020). Gestão de projetos no contexto público: mapeamento do campo de investigação (in Portuguese), *Revista de Administração Pública*, 54 (5): 1260–1285.

Müller, R. (2009). *Project Governance. Fundamentals of Project Management*. Farnham, Surrey, UK: Gover Publishing. Doi: 10.4324/9781315245928

Munck af Rosenschöld, J. (2017). Projectified environmental governance and challenges of institutional change toward sustainability. Doctoral thesis. Helsinki, Finland: Faculty of Social Sciences, University of Helsinki.

Munshi, S. (2004). Concern for good governance in comparative perspective. In S. Munshi & P. A. Biju (eds.), *Good Governance, Democratic Societies and Globalization*. New Delhi, India: Sage Publications.

Murray, M. (1975). Comparing public and private management: an exploratory essay, *Public Administration Review*, 35: 364–371.

Musgrave, R. (1994). Teoría múltiple de la hacienda pública. In J. F. Corona (ed.), *Lecturas de Hacienda pública*. Madrid, España: Minerva Ediciones.

Nagadevara, V. (2012). Project success factors and inter-sectoral differences, *Review of Business Research*, 12 (1): 115–120.

NARA (2014). *Federal RIM Program Maturity Model User's Guide*. College Park, MD: Joint Working Group of the Federal Records Council and National Archives and Records Administration.

Negandhi, A. R., & Prasad, S. B. (1971). *Comparative Management*. New York: Appleton Century-Crofts.

Ng, T. S. T., Li, T. H. Y., & Wong, J. M. W. (2012). Rethinking public participation in infrastructure project, *Proceedings of the Institution of Civil Engineers. Municipal Engineer*, 165 (2): 101–113.

Ning, Y. (2014). Quantitative effects of drivers and barriers on networking strategies in public construction projects, *International Journal of Project Management*, 32 (2): 286–297.

Nolan, R. L. (1973). Managing the computer resource: a stage hypothesis, *Communications of the ACM*, 7 (16): 399–405.

Nolan, R. L. (1979). Managing the crisis in data processing, *Harvard Business Review*, 2 (57): 115–126.

Nonaka, I., & Takeuchi, H. (1995). *The Knowledge-Creating Company: How Japanese Companies Create the Dynamics of Innovation*. Oxford University Press.

Nutt P. C. (2006). Comparing public and private sector decision-making practices, *International Journal of Public Administration Research and Theory*, 16 (2): 289–318.

O'Connor, T. R. F. (2014). Introduction to comparative Public Administration. http://www.drtomoconnor.com/4090/4090lect01.htm. Accessed June 2014].

O'Dowd, A., & Cross, M. (2007). Richard Granger resigns as chief executive of connecting for health, *BMJ*, 334 (7607): 1290–1291.

O'Leary, T., & Williams, T., (2008). Making a difference? Evaluating an innovative approach to the project management Centre of Excellence in a UK Government department, *International Journal of Project Management*, 26 (5): 556–565.

O'Toole, L. J. Jr. (2004). The theory–practice issue in policy implementation research, *Public Administration*, 82 (2): 309–329.

Obradović, V. (2019). Project management competences as a driver of growing economies, *Advances in Economics, Business and Management Research*, 108: 1–4. Doi: 10.2991/senet-19.2019.1

Ofori, D., & Deffor, E. W. (2013). Assessing project management maturity in Africa: a Ghanaian perspective, *International Journal of Business Administration*, 4 (6): 41–61.

Öjehag-Pettersson, A. (2017). Working for change: projectified politics and gender equality, *NORA—Nordic Journal of Feminist and Gender Research*, 25 (3): 163–178. Doi: 10.1080/08038740. 2017.1370011

Oliphant, S., & Howlett, M. (2010). Assessing policy analytical capacity: insights from a study of the Canadian Environmental Policy Advice System, *Journal of Public Administration Research and Theory*, 12 (2): 217–240.

Olsson, N., & Magnussen, O. M. (2007). Flexibility at different stages in the life-cycle of projects: An empirical illustration of the "Freedom to Maneuver", *Project Management Journal*, 38 (4): 25–32.

Orchard, L. (2014). Understanding Public Policy. https://flo.flinders.edu.au/pluginfile.php/1243010/mod_resource/content/4/1%20%20Understanding%20Public%20Policy.pdf. Accessed March 2020.

Osborne, S. P. (2006). Editorial. The new public governance? *Public Management Review*, 8 (3): 377–387.

Ostrom, E. (2008). Institutions and the environment, *Economic Affairs*, 28 (3): 24–31.

Ostrom, E. (2010). Beyond markets and states: polycentric governance of complex economic systems, *American Economic Review*, 100 (3): 641–672. Doi: 10.1257/aer.100.3.641

Othman, A. (2014). A conceptual model for overcoming the challenges of mega construction projects in developing countries, *African Journal of Engineering Research*, 2 (4): 73–84.

Ott, R., & Longnecker, M. (2010). *An Introduction to Statistical Methods and Data Analysis*. Pacific Grove: Duxbury.

Oudot, J. (2010). Performance and risks in the defense procurement sector, *Journal of Public Policy*, 30 (2): 201–218.

Padhi, S. S., & Mohapatra, P. K. J. (2010). Centralized bid evaluation for awarding of construction projects – a case of India government, *International Journal of Project Management*, 28 (3): 275–284.

Pal, L. (2014). *Beyond Policy Analysis: Public Issue Management in Turbulent Times*. Toronto, Canada: Nelson Higher Education.

Pandey, S. K., & Garnett, J. L. (2006). Exploring public sector communication performance: testing a model and drawing implications, *Public Administration Review*, 66 (1): 37–51. Doi: 10.1111/j.1540-6210.2006.00554.x

Parsons, W. (1995). *Public Policy: An Introduction to the Theory and Practice of Policy Analysis*. Edward Elgar.

Patanakul, P. (2014). Managing large-scale IS/IT projects in the public-sector: problems and causes leading to poor performance, *Journal of High Technology Management Research*, 25 (1): 21–35.

Patapas, A., & Smalskys, V. (2014). New public governance: the tracks of changes, *International Journal of Business and Social Research*, 4 (5): 25–32. Doi: 10.18533/ijbsr.v4i5.478

Patton, C. V., & Sawicki, D. S. (1993). *Basic Methods of Policy Analysis and Planning*. Englewood Cliffs: Prentice-Hall.

Peled, A. (2000). Creating winning information technology project teams in the public sector, *Team Performance Management*, 6 (1/2): 6–14.

Peled, M., & Dvir, D. (2012). Towards a contingent approach of customer involvement in defence projects: an exploratory study, *International Journal of Project Management*, 30 (3): 317–328.

Pells, D. (2018). Project management is a national competence. *Project Management Review*. Accessible at http://www.pmreview.com.cn/english/Home/article/detail/id/78.html

Perry, J. L., & Rainey, H. G. (1988). The public–private distinction in organization theory: a critique and research strategy, *Academy of Management Review*, 13 (2): 182–201. Doi: 10.5465/amr.1988.4306858

Peters, B. G. (2000). Policy instruments and public management: bridging the gaps, *Journal of Public Administration Research and Theory*, 10 (1): 35–47.

Peters, B. G. (2010). Meta-governance and public management. In S. P. Osborne (ed.), *The New Public Governance? Emerging Perspectives on the Theory and Practice of Public Governance* (pp. 36–51). London, UK: Routledge.

Pierre, J. (2000). *Debating Governance: Authority, Steering, and Democracy*. Oxford, UK: Oxford University Press.

Pilkaitė, A., & Chmieliauskas, A. (2015). Changes in public sector management: establishment of project management offices – a comparative case study of Lithuania and Denmark, *Public Policy and Administration*, 14 (2): 291–306.

PMI (2014). *Projects Fit for Purpose: Delivering More with Less in the Public Sector*. Newtown Square: Project Management Institute.

PMI (2020). *PMI Today, September/October 2020*. Newtown Square, PA: Project Management Institute.

PMI, & NAPA (2020). Building an Agile Federal Government: a call to action. *Newtown Square: Project Management Institute*. Washington, DC: National Academy of Public Administration.

Polanyi, M. (1958). *Personal Knowledge*. Chicago: University of Chicago Press.

Polanyi, M. (1966). *The Tacit Dimension*. Garden City, NY: Doubleday.

Pollitt, C. (1990). *Managerialism and the Public Services*. Oxford: Basil Blackwell.

Pollitt, C. (2007). The New Public Management: its current status, *Administratie Si Management Public*, 8: 110–115.

Pollitt, C., & Bouckaert, G. (2004). *Public Management Reform: A Comparative Analysis*. 2nd Edition. New York: Oxford University. Press.

Pöppelbuß, J., Niehaves, B., Simons, A., & Becker, J. (2011). Maturity models in information systems research: literature search and analysis, *Communications of AIS*, 29: 505–532.

Prado, D. (2016). MMGP Maturity by project Category Model, http://www.maturityresearch.com/novosite/en/index.html. Accessed December 2016.

Prado, D., & Andrade, C. E. (2015). Maturidade Em Gerenciamento De Projetos. Relatório (2014). Retrieved from http://www.maturityresearch.com/novosite/en/index.html

Prado, D., Archibald, R., & Dias, A., (2009). Maturity in Project Management, The Brazilian Experience. Paper presented at 23rd International Project Management Association World Congress, Helsinki.

Prasad, S., Tata, J., Herlache, L., & McCarthy, E. (2013). Developmental project management in emerging countries, *Operations Management Research*, 6 (1): 53–73.

Premfors, R. (1999). Det historiska arvet – Sverige. In P. Lægreid & O. K. Pedersen (eds.), *Fra opbygning til ombygning i staten. Organisationsforandringer i tre nordiske lande*. Copenhagen: DJØF Publishing.

Pressman, J., & Wildavsky, A. (1973). Implementation: how great expectations in Washington are dashed in Oakland: or, why it's amazing that Federal Programs work at all, this being a saga of the economic development administration as told by two sympathetic observers who seek to build morals on a foundation of ruined hopes. University of California Press, London.

Procca, A. E. (2008). Development of a project management model for a government research and development organization, *Project Management Journal*, 39 (4): 6–27.

Pullen, W. (2007). A public sector HPT maturity model, *Performance Improvement*, 46 (4): 9–15.

Pūlmanis, E. (2015a). Micro-economical aspects of public projects: impact factors for project efficiency and sustainability, *PM World Journal*, 4 (6): 1–12.

Pūlmanis, E. (2015b). Public funding and ensuring public project efficiency: micro-economical perspective of the EU funds management frame in the Republic of Latvia, *PM World Journal*, 4 (2): 1–16.

Pülzl, H., & Treib, O. (2007). Implementing Public Policy. In F. Fischer, G. J. Miller, & M. S. Sidney (eds.), *Handbook of Public Policy Analysis. Theory, Politics, and Methods* (pp. 89–107). Boca Raton, FL: CRC Press.

PWN (2006). *Słownik języka polskiego*. Warszawa: Wydawnictwo Naukowe PWN (in Polish).

Radu, I., Şendroiu, C., & Ioniţă, F. (2008). The re-engineering of managerial process in Public Administration, *Annals of the University of Oradea, Economic Science Series*, 17 (4): 535–541.

Rahman, M. M., & Kumaraswamy, M. M. (2004). Contracting relationship trends and transitions, *Journal of Management in Engineering*, 20 (4): 147–161.

Rainey, H. G. (2014). *Understanding and Managing Public Organizations*. 5th Edition. San Francisco: Jossey-Bass.

Rainey, H. G., & Bozeman B. (2000). Comparing public and private organizations: empirical research and the power of the a priori, *Journal of Public Administration Research and Theory*, 10 (2): 447–469.

Rainey, H. G., Traut, C., & Blunt, B. (1986). Reward expectancies and other work-related attitudes in public and private organizations: a review and extension, *Review of Public Personnel Administration*, 6 (3): 50–72.

Ramesh, M., & Howlett, M. (2017). The role of policy capacity in policy success and failure. In T. R. Klassen, D. Cepiku & T. J. Lah (eds.), *The Routledge Handbook of Global Public Policy and Administration* (pp. 319–334). New York: Routledge.

Ramos, P., Mota, C., & Correa, L. (2016). Exploring the management style of Brazilian project managers, *International Journal of Project Management*, 34 (6): 902–913.

Recker, J., Indulska, M., Roseman, M., & Green, P. (2009). Business process modeling - a comparative analysis, *Journal of the Association for Information Systems*, 10 (4): 333–363.

Reich, B. H. (2007). Managing knowledge and learning in IT projects: a conceptual framework and guidelines for practice, *International Journal of Project Management*, 38 (2): 5–17.

Reichard, C. (2003). Local public management reforms in Germany, *Public Administration*, 81 (2): 345–363.

Rendon, R. G. (2006). Measuring contract management process maturity: a tool for enhancing the value chain. Paper presented at 91st Annual International Supply Management Conference.

Repenning, N. P., & Sterman, J. D. (2002). *Capability Traps and Self-confirming Attribution Errors in the Dynamics of Process Improvement*. MIT Sloan School of Management Working Paper 4372-02. http://ssrn.com/abstract_id=320380

Richardson, G. L. (2010) *Project Management Theory and Practice*. Boston: Auerbach Pub./CRC Press.

Riggs, F. W. (1954). Notes on literature available for the study of comparative Public Administration, *American Political Science Review*, 48 (2): 515–537.

Riggs, F. W. (1970). The context of development administration. In F. W. Riggs (ed.), *Frontiers of Development Administration* (pp. 74–75). Durham, USA: Duke University Press.

Rogers, M. Z., & Weller, N. (2014). Income taxation and the validity of state capacity indicators, *Journal of Public Policy*, 34 (2): 183–206.

Rolland, V. W., & Roness, P. G. (2009). Mapping organizational units in the state – challenges and classifications, Paper presented at COST-CRIPO Meeting 20-21 April 2009, Brussels, Belgium.

Rosacker, K. M., & Rosacker, R. E. (2010). Information technology project management within the public-sector organizations, *Journal of Enterprise Information Management*, 23 (5): 587–594.

Rose, K. H. (2006). Cover to Cover. Government extension to a guide to the project management body of knowledge (PMBOK® Guide) – Third Edition, *Project Management Journal*, 37 (4): 72.

Rose, R. (1976). On the priorities of government: a developmental analysis, *European Journal of Political Research*, 4 (3): 247–289.

Rose, T. M., & Manley, K. (2010). Financial incentives and advanced construction procurement systems, *Project Management Journal*, 41 (1): 40–50.

Rosemann, M., & de Bruin, T. (2005). Towards a Business Process Management Maturity Model, Proceedings of the 13th European Conference on Information Systems (ECIS 2005). Regensburg, Germany.

Rosling, H., Rosling Rönnlund, A., & Rosling, O. (2018). Factfulness: ten reasons we're wrong about the world--and why things are better than you think. Factfulness AB.

Rothstein, B., & Tannenberg, M. (2015). The quality of government and development policy. *A Report to the Swedish Government's Expert Group for Aid Studies*. Gothenburg, Sweden: University of Gothenburg.

Rothstein, B., & Teorell, J. (2008). What is quality of government? A theory of impartial government institutions governance, *An International Journal of Policy Administration, and Institutions*, 21 (2): 165–190.

Royce, W. W. (1970). Managing the Development of Large Software Systems. Proceedings of 1970 WESCON Conference, August 1970, Los Angeles.

Rwelamila, P. D., & Purushottam, N. (2012). Project management trilogy challenge in Africa – where to from here? *Project Management Journal*, 43 (4): 5–13.

Sabatier, P. A. (1986). Top-down and bottom-up approaches to implementation research: a critical analysis and suggested synthesis, *Journal of Public Policy*, 6 (1): 21–48.

Sabatier, P. A. (1988). An advocacy coalition framework of policy change and the role of policy-oriented learning therein, *Political Sciences*, 1 (2): 129–168.

Sætren, H., & Hupe, P. L. (2018). Policy Implementation in an Age of Governance. In E. Ongaro & S. van Thiel (eds.), *The Palgrave Handbook of Public Administration and Management in Europe*, Doi: 10.1057/978-1-137-55269-3_29

Salamon, L. M. (2002). *The Tools of Government. A Guide to the New Governance.* New York: Oxford University Press.

Samset, K. F., Volden, G. H., Olsson, N., & Kvalheim, E. V. (2016). Governance schemes for major public investment projects. *A comparative study of principles and practices in six countries.* Concept report No 47. Ex Ante Academic Publisher. Trondheim, Norway: The Concept Research Program.

Samset, K., & Volden, G. H. (2013) *Investing for Impact. Concept Report No. 36.* Trondheim, Norway: Ex Ante Academic Publisher. Norwegian University of Science and Technology.

Sargeant, R. (2010). *Creating Value in Project Management Using PRINCE2.* Brisbane: Queensland University of Technology.

Sauser, B. J., Reilly, R. B., & Shenhar, A. J. (2009). Why projects fail? How contingency theory can provide new insights – a comparative analysis of NASA's Mars Climate Orbiter loss, *International Journal of Project Management,* 27 (7): 665–679. Doi: 10.1016/j.ijproman.2009.01.004

Savelsberg, C. M. J. H., Havermans, L., & Storm, P. (2016). Development paths of project managers: what and how do project managers learn from their experiences, *International Journal of Project Management,* 34 (4): 559–569.

Sayre, W. S. (1958). Premises of Public Administration, *Public Administration Review,* 18 (2): 102–105.

Schein, E. H. (1992). *Organizational Culture and Leadership.* San Francisco: Jossey Bass.

Schlichter, J., Friedrich, R., & Haeck, B., 2003. The History of OPM3. Paper presented at PMI Global Congress Europe, Den Haag, The Netherlands. PMI: Newtown Square.

Schmid, B., & Adams, J. (2008). Motivation in project management: the project manager's perspective, *Project Management Journal,* 39 (2): 60–71.

Schoper, Y-G., Wald, A., Ingason, H. T., & Fridgeirsson, T. V. (2018). Projectification in Western economies: a comparative study of Germany, Norway and Iceland, *International Journal of Project Management,* 36 (1): 71–82. Doi: 10.1016/j.ijproman.2017.07.008

Schuster, A. (2015). *Exploring Projectification in the Public Sector: The Case of the Next Stage Review Implementation Programme in the Department of Health.* DBA thesis at the School of Management, Cranfield University, UK.

Schwaber, K., & Sutherland, J. (2017). *The Scrum Guide™ The Definitive Guide to Scrum: The Rules of the Game.* https://www.scrumguides.org/docs/scrumguide/v2017/2017-Scrum-Guide-US.pdf

Scott, P. G., & Falcone, S. (1998). Comparing public and private organizations. An exploratory analysis of three frameworks, *American Review of Public Administration,* 28 (2): 126–145. Doi: 10.1177/027507409802800202

Seelhofer, D., Graf, C. O. (2018) National project management maturity: a conceptual framework, *Central European Business Review,* 7 (2): 1:20

Shapira, Z. (2011). "I've got a theory paper – do You?": conceptual, empirical, and theoretical contributions to knowledge in the organization sciences, *Organization Science,* 22 (5): 1312–1321.

Shen, L. Y., Li, Q. M., Drew, D., & Shen, Q. P. (2004). Awarding construction contracts on multicriteria basis in China, *Journal of Construction Engineering and Management,* 130: 385–393.

Shi, Q. (2011). Rethinking the implementation of project management: a value adding path map approach, *International Journal of Project Management,* 29 (3): 295–302.

Shiferaw, A. T. (2013a) The Norwegian Project Governance System: Weaknesses and improvements. In O. J. Klakegg, K. H. Kjølle, C. G. Mehaug, N. O. E. Ollsson, A. T. Shiferawt & R. Woods, (eds.), Proceedings from 7th Nordic Conference on Construction Economics and Organisation 2013. Trondheim, June 2013.

Shiferaw, A. T., & Klakegg, O. J. (2012). Linking policies to projects: the key to identifying the right public investment projects, *Project Management Journal,* 43 (4): 14–26.

Shiferaw, A. T. (2013). Dutch Project Governance System: Weaknesses and Improvements, Proceedings from 7th Nordic Conference on Construction Economics and Organisation, L. Postmyr (ed.). Akademika Publishing, Trondheim, 203–214.

SIDA (2002). SIDA 2002. *Good Governance.* Stockholm: SIDA. https://www.sida.se/contentassets/51f962f6acf0489f90fab393d807bd6d/good-governance_762.pdf

SIGMA (2014). *The Principles of Public Administration.* Paris: OECD and EU.

Simangunsong, E., & Da Silva, E. N. (2013). Analyzing project management maturity level in Indonesia, *The South East Asian Journal of Management*, 7 (1): 72–84.

Simon, H. (1947). *Administrative Behavior*. New York: Macmillan.

Sjöblom, S. (2006). Introduction: towards a projectified public sector: project proliferation as a phenomenon. In S. Sjöblom (ed.), *Project Proliferation and Governance: The Case of Finland* (pp. 9–31). Helsinki: SSKH Meddelanden 69.

Sjöblom, S., Löfgren, K., & Godenhjelm, S. (2013) Projectified politics – temporary organisations in a public context, *Scandinavian Journal of Public Administration*, 17(2): 3–12.

Skulmoski, G. J., & Hartman, F. T. (2010). Information systems project manager soft competencies: a project-phase investigation, *Project Management Journal*, 41 (1): 61–80.

Smith, M. L., & Erwin, J. (n.d.). Role & Responsibility Charting (RACI). https://pmicie.org/files/22/PM-Toolkit/85/racirweb31.pdf. Accessed March 2021.

Soderholm, A. (2008). Project management of unexpected events, *International Journal of Project Management*, 26: 80–86. Doi: 10.1016/j.ijproman.2007.08.016

Songer, A. D., & Molenaar, K. R. (1997). Project characteristics for successful public-sector design-build, *Journal of Construction Engineering and Management*, 123 (1): 34–40.

Spałek, S. (2014). Does Investment in project management pay off? *Industrial Management & Data Systems*, 114 (5): 832–856.

Spicker, P. (2009). The nature of a public service, *International Journal of Public Administration*, 32 (11): 970–991.

Spittler, J. R., & McCracken, C. J. (1996). Effective project management in bureaucracies, *Transactions of AACE International*, 6 (1): PM 6-1–PM 6-10.

Stahl, B. C., Obach, M., Yaghmaei, E., Ikonen, V., Chatfield, K., & Brem, A. (2017). The responsible research and innovation (RRI) maturity model: linking theory and practice, *Sustainability*, 9 (6): 1036–1054.

Stentoft, J., Freytag, P. V., & Thoms, L. (2015). Portfolio management of development projects in Danish municipalities, *International Journal of Public Sector Management*, 28 (1): 11–28.

Stewart, J. Jr., Hedge, D. M., & Lester, J. P. (2007). *Public Policy: An Evolutionary Approach*. 3rd Edition. Boston, MA: Wadsworth Cengage Learning.

Storm, I., Harting, J., Stronks, K., & Schuit, A. J. (2014). Measuring stages of Health In All Policies on a Local level: the applicability of a maturity model, *Health Policy*, 114 (2–3): 183–191.

Strauss, A., & Corbin, J. (1998). *Basics of Qualitative Research: Techniques and Procedures for Developing Grounded Theory*. Thousand Oaks, CA: Sage Publications.

Streeck, W., & Thelen, K. (2005). Introduction: Institutional change in advanced political economies. In W. Streeck & K. Thelen (eds.), *Beyond Continuity: Institutional Change in Advanced Political Economies* (pp. 1–39). Oxford: Oxford University Press.

Struyk, R. J. (2007). Factors in successful program implementation in Russia during the transition: pilot programs as a guide, *Public Administration and Development*, 27 (1): 63–83. Doi: 10.1002/pad.430

Stummer, M., & Zuchi, D. (2010). Developing roles in change processes – a case study from a public sector organisation, *International Journal of Project Management*, 28 (4): 384–394.

Subramanian, S., & Kruthika, J. (2012). Comparison between public and private sector executives on key psychological aspects, *Journal of Organisation and Human Behaviour*, 1 (1): 27–35.

Sun, M., Vidalakis, C., & Oza, T. (2009). A change management maturity model for construction projects. In A. Dainty (ed.), Proceedings of 25th Annual ARCOM Conference, 7–9 September 2009, 803–812. Nottingham, UK: Association of Researchers in Construction Management.

Sutterfield, J. S., Friday-Strout, S. S., & Shivers-Blackwell, S. L. (2007). A case study of project and stakeholder management failures: lessons learned, *Project Management Journal*, 37 (5): 26–35.

Szymczak, M. (1989). *Słownik Języka Polskiego, Polskie Wydawnictwo Naukowe, Warszawa*.

Tabish, S. Z. S., & Jha, K. N. (2011). Identification and evaluation of success factors for public construction project, *Construction Management and Economics*, 29 (8): 809–823.

Taylor, F. W. (1911 / 2010). *The Principles of Scientific Management*. Lawrence, KS: DigiRead.

Teah, H. Y., Pee, L. G., & Kankanhalli, A. (2006). Development and application of a general knowledge management maturity model. *PACIS 2006 Proceedings*. Retrieved from http://aisel.aisnet.org/pacis2006/12

Tembo, E., & Rwelamila, P. M. D. (2008). Project Management Maturity in Public Sector Organisations: the Case of Botswana, Paper presented at Joint W055/W065/W089 International Conference on Construction in a Changing World, "Transformation Through Construction", Dubai.

Tennant, G. (2001). *SIX SIGMA: SPC and TQM in Manufacturing and Services.* London: Gower Publishing.

Teorell, J., Dahlberg, S., Holmberg, S., Rothstein, B., Khomenko, A., & Svensson, R. (2017). The Quality of Government Standard Dataset, version Jan 17. University of Gothenburg: The Quality of Government Institute. http://www.qog.pol.gu.se. Doi: 10.18157/QoGStdJan17

Thynne, I. (1998). Government companies as instruments of state action, *Public Administration and Development*, 18 (3): 217–228.

Tiernan, A. (2012). Advising Australian Federal Governments: assessing the evolving capacity and role of the Australian public service, *The Australian Journal of Public Administration*, 70 (4): 335–346. Doi: 10.1111/j.1467-8500.2011.00742.x

Tinbergen, J. (1952). *On the Theory of Economic Policy.* Amsterdam, NL: North-Holland.

Tinbergen, J. (1998) Problems of planning economic policy, *International Social Science Journal*, 11 (3): 335–342.

Too, E. G., & Weaver, P. (2014). The management of project management: a conceptual framework for project governance, *International Journal of Project Management* 32 (8): 1382–1394.

Toor, S., & Ogunlana, S. O. (2010). Beyond the 'iron triangle': stakeholder perception of key performance indicators (KPIs) for large-scale public sector development project, *International Journal of Project Management*, 28: 228–236.

Torfing, J., & Triantafillou, P. (2013). What's in a name? Grasping new public governance as a political-administrative system, *International Review of Public Administration*, 18 (2): 9–25. Doi: 10.1080/12294659.2013.10805250

Torres, L., & Pina, V. (2004). Reshaping Public Administration: the Spanish experience compared to the UK, *Public Administration* 82 (2): 445–464.

Treisman, D. (2002). Decentralization and the Quality of Government. Paper published by Department of Political Science University of California, Los Angeles.

Turner, R., Huemann, M., Anbari, F., & Bredillet, C. (2010). Perspectives on Projects. London: Routledge.

Van de Walle, S. G. J., & Scott, Z. (2009). The role of public services in state- and nation building: Exploring lessons from European history for fragile states. Governance and Social Development Resource Centre, University of Birmingham. http://hdl.handle.net/1765/17084

Van Der Merwe, A. P. (2002). Project management and business development: integrating strategy, structure, processes and projects, *International Journal of Project Management*, 20 (5): 401–411.

Van der Voet, J. (2014). The effectiveness and specificity of change management in a public organization: transformational leadership and a bureaucratic organizational structure, *European Management Journal*, 32 (3): 373–382. Doi: 10.1016/j.emj.2013.10.001

Van der Waldt, G. (2011). The uniqueness of public sector project management: a contextual perspective, *Politeia. South African Journal for Political Science and Public Administration*, 30 (2): 67–88.

Van der Waldt, G. (2016). From policy to projects: a public service value-chain network model, *Journal of Social Sciences: Interdisciplinary Reflection of Contemporary Society*, 49 (1,2): 145–157.

Van Kersbergen, K., & Van Waarden, F. (2004). 'Governance' as a bridge between disciplines: cross-disciplinary inspiration regarding shifts in governance and problems of governability, accountability and legitimacy, *European Journal of Political Research*, 43 (2): 143–171. Doi: 10.1111/j.1475-6765.2004

Van Nispen, F. K. M. (2011). The Study of Police Instruments. Robert Schuman Centre for Advanced Studies, European University Institute: Fiesole, Italy.

Vargas, M. H. F., & Restrepo, D. R. (2019). The instruments of public policy. A transdisciplinary look, *Cuadernos de Administración, Journal of Management*, 35 (63): 101–113. Doi: 10.25100/cdea.v35i63.6893

Vit, G. B. (2011). Competing logics: project failure in Gaspesia, *European Management Journal*, 29 (3): 234–244. Doi: 10.1016/j.emj.2010.10.003

Volden, G. H. (2018). Public project success as seen in a broad perspective: lessons from a meta-evaluation of 20 infrastructure projects in Norway, *Evaluation and Program Planning*, 69: 109–117. Doi: 10.1016/j.evalprogplan.2018.04.008

Volden, G. H., & Samset, K. B. (2017). Governance of major public investment projects: principles and practices in six countries, *Project Management Journal*, 48 (3): 90–108.

Wang, Y-D., Yang, C., & Wang, K-Y. (2012). Comparing public and private employees' job satisfaction and turnover, *Public Personnel Management*, 41 (3): 557–572.

Weber, M. (1947). *The Theory of Social and Economic Organization*. Oxford: Oxford University Press.

Weber, M. (1952). The essentials of bureaucratic organization: an ideal-type construction. In R. K. Merton, A. P. Gray, B. Hockey, & H. C. Selvin (eds.), *Reader in Bureaucracy*. Glencoe, IL: The Free Press.

Weber, M. (1994). *Political Writings*. In P. Lassman & R. Speirs (eds.). Cambridge, UK: Cambridge University Press.

Wendler, R. (2012). The maturity of maturity model research: a systematic mapping study, *Information and Software Technology*, 54 (2012, December): 1317–1339.

Weske, M., van der Alst, W. M. P., & Verbeek, H. M. W. (2004). Advances in business process management, *Data & Knowledge Engineering*, 50 (1): 1–8. Doi: 10.1016/j.datak.2004.01.001

Wettenhall, R. (2003). Exploring types of public sector organizations: past exercises and current issues, *Public Organization Review*, 3 (3): 219–245. Doi: 10.1023/A:1025333414971

Whitford, A., & Lee, S. (2012). Disorder, dictatorship and government effectiveness: cross-national evidence, *Journal of Public Policy*, 32 (1): 5–31.

Whitty, S. J., & Maylor, H. (2009). And then came complex project management (revised), *International Journal of Project Management*, 27 (3): 304–310.

Wikipedia. (2022). Berlin Brandenburg Airport. https://en.wikipedia.org/wiki/Berlin_Brandenburg_Airport. Accessed July 2022. Last edited day: 20 July 2022.

Wild, N., & Lichter, H. (2017). A Meta-Model for Maturity Model classification. In H. Lichter, K. Fögen, M. Hacks, M. F. Harun, A. Nicolaescu, C. Plewnia & A. Steffens (supervisors) Seminar: CSE/FsSE 2018, pp. 61–66. Aachen: Aachen University.

Williams, T. (2019). The reasons governments struggle to run projects successfully. Accessible at https://www.linkedin.com/pulse/reasons-governments-struggle-run-projects-todd-williams/?trackingId=IyS3HYEBApmEspVEGQ15tQ%3D%3D

Williams, T. Klakegg, O. J. Magnussen, O. M., & Glasspool, H. (2010). An investigation of governance frameworks for public Project in Norway and the UK, *International Journal of Project Management*, 28 (1): 40–50.

Williams, T., Klakegg, O. J., Walker, D. H. T., Andersen, B., & Magnussen, O. M. (2012). Identifying and acting on early warning signs in complex projects, *Project Management Journal*, 43 (2): 37–53.

Wilson, D. (2011). Comparative analysis in public management reflections on the experience of a major research Programme, *Public Management Review*, 13 (2): 293–308. Doi: 10.1080/14719037.2010.532967

Wilson, J. Q. (1989). *Bureaucracy: What Government Agencies Do and Why They Do It*. Basic Books.

Wilson, W. (1887). The study of administration. *Political Science Quarterly*, 2 (2): 197–222.

Winter, S. C. (2003). Understanding dynamic capabilities, *Strategic Management Journal*, 24 (10): 991–995.

Winter, S. C. (2012). Implementation. In B. G. Peters, & J. Pierre (eds.), *The SAGE Handbook of Public Administration* (pp. 255–263). London: Sage.

Wirick, D. (2009). *Public-Sector Project Management: Meeting the Challenges and Achieving Results*. New Jersey: John Wiley & Sons.

Yalegama, S., Chileshe, N., & Ma, T. (2016). Critical success factors for community-driven development Project: a Sri Lankan community perspective, *International Journal of Project Management*, 34 (4): 643–659.

Yanwen, W. (2012). The study on complex project management in developing countries, *Physics Procedia*, 25: 1547–1552. Doi: 10.1016/j.phpro.2012.03.274

Yazici, H. J. (2009). The role of project management maturity and organizational culture in perceived performance, *Project Management Journal*, 40 (3): 14–33.

Young, M., Young, R., & Zapata, J. R. (2014). Project, programme and portfolio maturity: a case study of Australian Federal Government, *International Journal of Managing Projects in Business*, 7 (2): 215–230.

Zahariadis, N., & Exadaktylos, T. (2016). Policies that succeed and programs that fail? Ambiguity, conflict, and crisis in greek higher education, *Policy Studies Journal*, 44 (1): 59–82. Doi: 10.1111/psj.12129

Zeitoun, A. A. (2002). Who is the international development project manager? Paper presented at Project Management Institute Annual Seminars & Symposium, San Antonio, USA, Project Management Institute, Newtown Square, USA.

Zhang, Y. (2015). The formation of public-private partnerships in China: an institutional perspective, *Journal of Public Policy*, 35 (2): 329–354.

Zhu, Y-Q., & Kindarto, A. (2016). A garbage can model of government IT project failures in developing countries: the effects of leadership, decision structure and team competence, *Government Information Quarterly*, 33 (4): 629–637. Doi: 10.1016/j.giq.2016.08.002

Zwicker, J., Fettke, P., & Loos, P. (2010). Business process maturity in public administrations. In J. vom Brocke & M. Rosemann (eds.), *Handbook on Business Process Management 2, International Handbooks on Information Systems* (pp. 369–396). Berlin: Springer-Verlag.

Zwikael, O. (2009). The relative importance of the PMBOK® guide's nine knowledge areas during project planning, *Project Management Journal*, 40 (4): 94–103.

Zwikael, O. (2020). When doesn't formal planning enhance the performance of government projects? *Public Administration Quarterly*, 44 (3): 331–362.

Index

Note: **Bold** page numbers refer to tables and *italic* page numbers refer to figures.

Printed in the United States
by Baker & Taylor Publisher Services